Clean Energy Production Technologies

Series Editors

Neha Srivastava, Department of Chemical Engineering and Technology, IIT (BHU) Varanasi, Varanasi, Uttar Pradesh, India

P. K. Mishra, Department of Chemical Engineering and Technology, IIT (BHU) Varanasi, Varanasi, Uttar Pradesh, India

The consumption of fossil fuels has been continuously increasing around the globe and simultaneously becoming the primary cause of global warming as well as environmental pollution. Due to limited life span of fossil fuels and limited alternate energy options, energy crises is important concern faced by the world. Amidst these complex environmental and economic scenarios, renewable energy alternates such as biodiesel, hydrogen, wind, solar and bioenergy sources, which can produce energy with zero carbon residue are emerging as excellent clean energy source. For maximizing the efficiency and productivity of clean fuels via green & renewable methods, it's crucial to understand the configuration, sustainability and techno-economic feasibility of these promising energy alternates. The book series presents a comprehensive coverage combining the domains of exploring clean sources of energy and ensuring its production in an economical as well as ecologically feasible fashion. Series involves renowned experts and academicians as volume-editors and authors, from all the regions of the world. Series brings forth latest research, approaches and perspectives on clean energy production from both developed and developing parts of world under one umbrella. It is curated and developed by authoritative institutions and experts to serves global readership on this theme.

Dan Bahadur Pal • Amit Kumar Tiwari
Editors

Sustainable Valorization of Agriculture & Food Waste Biomass

Application in Bioenergy & Useful Chemicals

Springer

Editors
Dan Bahadur Pal
Department of Chemical Engineering
Harcourt Butler Technical University
Kanpur, India

Amit Kumar Tiwari
Department of Chemical Engineering
Birla Institute of Technology
Ranchi, Jharkhand, India

ISSN 2662-6861 ISSN 2662-687X (electronic)
Clean Energy Production Technologies
ISBN 978-981-99-0525-6 ISBN 978-981-99-0526-3 (eBook)
https://doi.org/10.1007/978-981-99-0526-3

This Springer imprint is published by the registered company Springer Nature Singapore Pte Ltd.
The registered company address is: 152 Beach Road, #21-01/04 Gateway East, Singapore 189721, Singapore

Preface

The increasing rate of the world population is currently a big and serious problem that is directly related to global issues such as global warming, air and water pollution, soil pollution, and food security. Irresponsible behavior of humans is the major reason for all these problems. Improper use and disposal of agricultural and food wastes contribute significantly to environment pollution. These wastes can be converted into value-added products using various biological, chemical, and biochemical methods and approaches. This conversion process not only helps reduce the aforementioned problems but also generates economic growth and employment opportunities. In the last few decades, scientist and researchers from different countries have discovered the potential of this type of waste material as a feedstock for manufacturing various valuable products. In recent years, there has been greater focus on bio-ethanol production from different types of agricultural biomass. Advanced technologies for producing clean energy from water bodies, air bodies, and solid biomass waste have also been discussed.

Chapter 1 discusses waste biomass valorization and its application in the environment, Chap. 2 discusses biomass valorization as energy production using waste biomass, Chap. 3 discusses volatile organic compounds impacts on the environment: bio-filtration as an effective control method, Chap. 4 discusses the utilization of waste biomass for producing useful chemicals, Chap. 5 discusses forestry biomass as a carbon neutral source for the production of biofuels and aromatics, Chap. 6 discusses biomass (algae) valorization from an energy perspective: review of process options and utilization, Chap. 7 discusses bio-hydrogen production using agricultural biowaste materials, Chap. 8 discusses the conversion of food waste into valuable products, Chap. 9 discusses food waste materials for bioenergy production, Chap. 10 discusses biochar for sustainable crop production, Chap. 11 discusses the production of alternative fuel from lignocellulosic kitchen waste through pyrolysis, and Chap. 12 discusses the generation of bioenergy from industrial waste materials.

Kanpur, India — Dan Bahadur Pal
Ranchi, Jharkhand, India — Amit Kumar Tiwari

Acknowledgments

We, the editors, are thankful to the host institution for providing all the necessary facilities. We also thankfully acknowledge our parents and family members for their support during these activities.

Contents

Editors and Contributors

About the Editors

Dan Bahadur Pal, B. Tech, M. Tech, Ph.D., is currently working as an Assistant Professor in the Department of Chemical Engineering at Harcourt Butler Technical University, Kanpur-208002, Uttar Pradesh, India. He has received his M. Tech and Ph.D. in the field of chemical engineering from Indian Institute of Technology (BHU), Varanasi, Uttar Pradesh, India. Before that, he completed his B. Tech in Chemical Engineering from UPTU, Lucknow. Dr. Pal's research interests are nano-technology, catalysis, energy and environment, and waste management with a special focus on developing process and materials using waste as raw materials. He also prefers to work on bio-waste processing and value addition. Dr. Pal has published more than 77 articles in reputed journals and books, as well as 25 book chapters, and has edited 6 books.

Amit Kumar Tiwari, B.Sc., M.Sc., Ph.D., is currently working as an Assistant Professor in the Department of Chemical Engineering at Birla Institute of Technology, Mesra, Ranchi, Jharkhand. He has received his Ph.D. in the field of food engineering from Birla Institute of Technology, Mesra. Before that, he completed his M. Sc. in Fruit and Vegetable Technology and B.Sc. in Agriculture from Kanpur University (Chhatrapati Shahu Ji Maharaj University), Kanpur. Dr. Tiwari's research interests lie in the development of new and novel food products with a special focus on developing process techniques for fruits, vegetables, food grains, etc. with enhanced nutritional and edible qualities. Dr. Tiwari published more than nine publications in reputed journals and ten book chapters. Dr. Tiwari is also working on agricultural waste utilization through value addition, energy and environment, and food waste management with a special focus on developing process and value-added products using agricultural and food waste as raw materials. Dr. Tiwari also prefers to work on biomass application.

Contributors

MulualemAdmas Alemu Department of Chemical Engineering, Faculty of Technology, Marwadi University, Rajkot, India

Neeru Anand University School of Chemical Technology, Guru Gobind Singh Indraprastha University, New Delhi, India

Shraddha Awasthi Department of Environmental Science, Ramanujan College, Delhi University, New Delhi, Delhi, India

Nilambar Bariha Assam Energy Institute, Sivasagar, Centre of Rajiv Gandhi Institute of Petroleum Technology, Sivasagar, Assam, India

Renu Bhardwaj Department of Botanical and Environmental Sciences, Guru Nanak Dev University, Amritsar, Punjab, India

Tamanna Bhardwaj Department of Botanical and Environmental Sciences, Guru Nanak Dev University, Amritsar, Punjab, India

Neelanjan Bhattacharjee Department of Chemical Engineering, University of Calcutta, Kolkata, India

Amandeep Bhatti S.R. Government College for Women, Amritsar, Punjab, India

Asit Baran Biswas Department of Chemical Engineering, University of Calcutta, Kolkata, India

Piyush R. Chauhan Department of Chemical Engineering, Birla Institute of Technology, Ranchi, Jharkhand, India

Bimal Das Department of Chemical Engineering, National Institute of Technology Durgapur, Durgapur, West Bengal, India

Kamini Devi Department of Botanical and Environmental Sciences, Guru Nanak Dev University, Amritsar, Punjab, India

Shalini Dhiman Department of Botanical and Environmental Sciences, Guru Nanak Dev University, Amritsar, Punjab, India

Rashmi Dhurandhar Department of Chemical Engineering, National Institute of Technology Durgapur, Durgapur, West Bengal, India

Deepak Garg University School of Chemical Technology, Guru Gobind Singh Indraprastha University, New Delhi, India

F. S. Idroos Department of Zoology, Centre for Water Quality and Algae Research, University of Sri Jayewardenepura, Gangodawila, Nugegoda, Sri Lanka
Faculty of Applied Sciences, University of Sri Jayewardenepura, Gangodawila, Nugegoda, Sri Lanka

Fiston Iradukunda Department of Chemical Engineering, Faculty of Technology, Marwadi University, Rajkot, India

Kakeeto Ismail Department of Chemical Engineering, Faculty of Technology, Marwadi University, Rajkot, India

Arinjay Kumar Jain University School of Chemical Technology, Guru Gobind Singh Indraprastha University, New Delhi, India

Sumit Kumar Jana Department of Chemical Engineering, Birla Institute of Technology, Ranchi, Jharkhand, India

Michael C. Kabesha Department of Chemical Engineering, Faculty of Technology, Marwadi University, Rajkot, India

Dhriti Kapoor School of Bioengineering and Biosciences, Lovely Professional University, Jalandhar, Punjab, India

Nitika Kapoor PG Department of Botany, Hans Raj Mahila Maha Vidyalaya, Jalandhar, Punjab, India

Makusalani Ole Kawanara Department of Chemical Engineering, Faculty of Technology, Marwadi University, Rajkot, India

Jaspreet Kour Department of Botanical and Environmental Sciences, Guru Nanak Dev University, Amritsar, Punjab, India

Aman Kumar Department of Chemical Engineering, Birla Institute of Technology, Ranchi, Jharkhand, India

Sanjay Kumar Department of Chemical Engineering, Faculty of Technology, Marwadi University, Rajkot, India

Shravan Kumar Department of Biochemical Engineering, Harcourt Butler Technical University, Kanpur, India

Steward Laishi Department of Chemical Engineering, Faculty of Technology, Marwadi University, Rajkot, India

Akanksha Majumder Department of Chemical Engineering, University of Calcutta, Kolkata, India

Pathmalal M. Manage Department of Zoology, Centre for Water Quality and Algae Research, University of Sri Jayewardenepura, Gangodawila, Nugegoda, Sri Lanka
Faculty of Applied Sciences, University of Sri Jayewardenepura, Gangodawila, Nugegoda, Sri Lanka

Ambneesh Mishra Department of Environmental Science, Ramanujan College, Delhi University, New Delhi, Delhi, India

Dan Bahadur Pal Department of Chemical Engineering, Harcourt Butler Technical University, Kanpur, India

Swapnajeet Pandey Department of Chemical Engineering, Birla Institute of Technology, Ranchi, India

Pankaj Parmar Department of Chemical Engineering, Indian Institute of Technology Kharagpur, Kharagpur, West Bengal, India

Nirupama Prasad Department of Chemical Engineering, Birsa Institute of Technology, Dhanbad, India

Rahul Department of Paint Technology, Government Polytechnic Bindki, Fatehpur, UP, India

Ila Jogesh Ramala Sarkar Department of Chemical Engineering, Faculty of Technology, Marwadi University, Rajkot, India

Reena Saxena School of Applied Science, Suresh Gyan Vihar University, Jaipur, Rajasthan, India

Neerja Sharma Department of Botanical and Environmental Sciences, Guru Nanak Dev University, Amritsar, Punjab, India

Amrit Pal Singh Department of Pharmaceutical Sciences, Guru Nanak Dev University, Amritsar, Punjab, India

Harsh Singh Department of Chemical Engineering, Birla Institute of Technology, Ranchi, India

Rajeev Singh Department of Chemistry, Atma Ram Sanatan Dharma College, Delhi University, New Delhi, Delhi, India

Chandrakant Thakur Department of Chemical Engineering, National Institute of Technology Raipur, Raipur, Chhattisgarh, India

Amit Kumar Tiwari Department of Chemical Engineering, Birla Institute of Technology, Ranchi, Jharkhand, India

Uplabdhi Tyagi University School of Chemical Technology, Guru Gobind Singh Indraprastha University, New Delhi, India

K. P. P. Udayagee Faculty of Technology, Department of Bio-systems Technology, University of Sri Jayewardenepura, Gangodawila, Nugegoda, Sri Lanka

P. A. K. C. Wijerathna Department of Zoology, Centre for Water Quality and Algae Research, University of Sri Jayewardenepura, Gangodawila, Nugegoda, Sri Lanka
Faculty of Applied Sciences, University of Sri Jayewardenepura, Gangodawila, Nugegoda, Sri Lanka
Faculty of Graduate Studies, University of Sri Jayewardenepura, Gangodawila, Nugegoda, Sri Lanka

Tefera Kassahun Zerfu Department of Chemical Engineering, Faculty of Technology, Marwadi University, Rajkot, India

Chapter 1
Waste Biomass Valorization and Its Application in the Environment

P. A. K. C. Wijerathna, K. P. P. Udayagee, F. S. Idroos, and Pathmalal M. Manage

Abstract There has been a significant increase in global waste generation owing to rapid urbanization and industrialization. Anthropogenic activities associated with exploiting natural resources pose severe threats to the long-term resilience of ecosystems. The buildup of waste biomass in ecosystems causes various adverse environmental conditions, such as greenhouse gas emissions, global warming, bioaccumulation and biomagnification of hazardous chemicals, surface and groundwater pollution, and acid rains suppress and lessen biological diversity. According to the World Bank predictions, 3.4 billion tons of municipal solid waste will have been generated by 2050. Thus, effective waste biomass management through valorization is critical in circular bio-economy and meeting environmental feasibility. Due to its abundance and renewability, lignocellulosic waste biomass can be a beneficial substrate to produce many high-value goods such as biofuels, biofertilizers, composts, biochar, pharmaceuticals, bioplastics, and food additives. This chapter summarizes the potential of hydrothermal conversion processes, including hydrothermal carbonization, hydrothermal liquefaction, and hydrothermal gasification, in

P. A. K. C. Wijerathna · P. M. Manage (✉)
Department of Zoology, Centre for Water Quality and Algae Research, University of Sri Jayewardenepura, Gangodawila, Nugegoda, Sri Lanka

Faculty of Applied Sciences, University of Sri Jayewardenepura, Gangodawila, Nugegoda, Sri Lanka

Faculty of Graduate Studies, University of Sri Jayewardenepura, Gangodawila, Nugegoda, Sri Lanka
e-mail: pathmalal@sjp.ac.lk

K. P. P. Udayagee
Faculty of Technology, Department of Bio-systems Technology, University of Sri Jayewardenepura, Gangodawila, Nugegoda, Sri Lanka

F. S. Idroos
Department of Zoology, Centre for Water Quality and Algae Research, University of Sri Jayewardenepura, Gangodawila, Nugegoda, Sri Lanka

Faculty of Applied Sciences, University of Sri Jayewardenepura, Gangodawila, Nugegoda, Sri Lanka

D. B. Pal, A. K. Tiwari (eds.), *Sustainable Valorization of Agriculture & Food Waste Biomass*, Clean Energy Production Technologies,
https://doi.org/10.1007/978-981-99-0526-3_1

producing a range of value-added products from solid waste substances. Moreover, the future trends of biological conversions that use microbial bioconversion generate a number of eco-friendly valorized products like biopesticides, biohydrogen, organic acids, antibiotics, enzymes, food colors, amino acids, and single-cell proteins were discussed. Further, this chapter highlights the multidisciplinary approaches for waste biomass valorization combined with advanced bio-nanotechnology, enzymatic sequent biomass hydrolysis treatments that are becoming popular and research gaps to overcome the challenges of waste biomass valorization by enhancing the process efficiency.

Keywords Biomass valorization · Lignocellulosic biomass · Bio-fertilizers · Bio-energy · Composting

1.1 Introduction

According to a recent UN assessment, the present population of 7.98 billion people worldwide is predicted to increase to 9.7 billion by 2050 (World Health Organization 2019). Consequently, the demand for resources, including water, food, shelter, and energy, will be increased along with human population growth and economic growth. In addition, the amount of waste generated has remarkably increased due to rapid urbanization and industrialization (Adesra et al. 2021). Therefore, waste management is one of the significant environmental concerns. In addition, the different anthropogenic behaviors and changes in lifestyles and consumption patterns have increased solid waste generation (Nanda and Berruti 2021). Approximately 33% of the estimated 2 billion tons of worldwide municipal solid garbage produced annually are not collected by local governments (Gelan 2021). Out of the current generated municipal solid waste (MSW), about 70% is dumped in landfills or other unspecified locations, 19% is recycled, and 11% is used for energy recovery (Aslam et al. 2022; Pheakdey et al. 2022). Nearly 3.5 billion out of the 7.6 billion people in the globe today lack access to essential waste management services (Jebaranjitham et al. 2022). Furthermore, it is estimated that 5.6 billion people will lack sufficient access to basic waste management services by 2050 (Dixit et al. 2022).

Over ten waste commodities can be utilized as biomass resources, including agricultural residues, livestock waste substances, residues of the alcohol industry, sugarcane industry residues, wood waste substances, industrial byproducts, and municipal waste (MSW) (Koul et al. 2022).

Further, the characteristics of waste biomass vary with the income level of a low-to-middle-income population. Organic waste generation is exceptionally high in the lower-income population category, whereas paper, metal, and glass waste generation is more significant in the higher economy-level population group (Mehmood et al. 2021). Table 1.1 describes the major types of solid wastes in the world and their composition.

Table 1.1 Major types of solid waste and their composition

Types of solid wastes	The composition	Global waste generation (per year) MT	Reference
Agricultural waste	Crop residue, straws and husks, olive pits and nut shells, bagasse, rice and wheat straw and husk, cotton stalk, ground nut shell, banana stalk and jute, sisal and vegetable residues	Bagasse—5,752,800 Cotton stick—1,474,693 Rice straw—731,000,000 Wheat straw—354,000,000 Corn straw—204,000,000	Koul et al. (2022) Kaza and Yao (2018)
Municipal solid waste and swage waste	Kitchen wastes, vegetable residues, fruit residues, plastic bottles, paper residuals, glasses, sludge, metals, chemicals	Municipal solid waste—3,760,000,000	Sharma and Jain (2020)
Paper and pulp industry waste	Printed documents, brochures, menus, maps, magazines, newspaper	Pulp waste 117,000,000	Jeswani et al. (2021)
Food industry waste	Fruit and vegetable peelings, flowers and plants, branches, leaves, grass, packing materials, residuals of foods	Food waste—1,300,000,000	Ncube et al. (2021)
Hospital wastes	Cotton, cloths, food residuals, gloves, plastic items, paper waste	Hospital waste—330,000	Chowdhury et al. (2020)
Waste from livestock production	Animal manure, plant residues, feathers, bones	229,000,000	Qi et al. (2021)
Forestry and timber industry waste	Wood particles, leaves, pallets	–	Wang et al. (2021)
Electronic waste	Computer parts, TV	44,700,000	Ilankoon et al. (2018)

At present, municipal councils practice different processes in biomass management such as recycling, combustion, energy conversion, production of compost and vermicompost, and sanitary landfilling (Munir et al. 2021).

The term "biomass" describes the biological matter generated by living processes. For example, solar energy is stored as biomass in organic matter, which is carbon based and made up of a combination of organic molecules combining hydrogen and oxygen. Carbs are the organic molecules that make up biomass (Munir et al. 2021). During photosynthesis, which occurs as trees and plants develop, energy from the sun is used to transform atmospheric CO_2 into carbohydrates. The ability of biomass to efficiently store solar energy makes it unique. It is also the only renewable carbon source and can be converted into functional solid, liquid, and gaseous fuels (Xie et al. 2022).

Lignocellulosic biomass (LB) is one of the most abundant biomass wastes, mainly generated by agricultural and forestry residues. LBs mainly contain three major compounds, including cellulose (35–50%), hemicelluloses (20–35%), and lignin (10–25%). Annually, 181.5 billion tons of lignocellulosic biomass are produced worldwide (Dahmen et al. 2019).

Cellulose constitutes 35–50% of the polysaccharides in lignocellulosic biomass, which gives it mechanical strength and structure (Xie et al. 2022). Hydrogen bonds hold a uniform linear polymeric chain of 1,4-D-glucan together to create paracrystalline microfibrils. About 2000–25,000 glucose residues make up a single unit chain. In addition, hemicellulose accounts for 1F–35% of biomass weight (Wang et al. 2021). Aldopentoses, aldohexoses, and sugar acids make up the polymeric units of hemicellulose. The third primary polymeric unit, lignin, accounts for 10–20% of the total biomass. Since lignin is insoluble and optically inactive under normal circumstances, its breakdown is complex (Roy et al. 2021).

1.2 Significance of Solid Biomass Valorization in the World

Lignocellulosic biomass is regarded as a resource of the ecosystem and a valuable source of organic carbon for the sustainable production of numerous biofuels and other bio-based products with low carbon and sulfur emissions (Mahapatra et al. 2021). In addition, expanding biomass-based businesses in developed and developing countries has produced possibilities in various employment categories since using biomass necessitates several logistics processes, including bulk collection, transportation, and pretreatment (Sangtani et al. 2022).

The world is confronted with an energy crisis and ecological concerns due to the exhaustion of fossil energy and the steady increase in CO_2 gas emissions (Holechek et al. 2022). By 2050, it is anticipated that energy produced from renewable resources will meet 10–40% of the world's energy needs, according to the Global Energy Assessment (GEA) (Holechek et al. 2022). Fossil fuels, the primary cause of the greenhouse effect, are the core of the global energy matrix. Therefore, using biomass as an energy source is crucial and far more inexpensive. Furthermore, lignocellulosic biomass offers more environmental advantages than fuels made from fossil fuels and other processed substances depending on its higher abundance, and renewable potential, environmentally friendly microbial decompositions (Kalair et al. 2021).

Biomass waste can be converted into value-added products, including biogas, bioplastic, biofuel, and bioethanol (Holechek et al. 2022). Further, different physical-chemical, thermochemical, and biochemical methods are used as biomass pretreatment techniques for biomass valorization. Despite the strength and effectiveness of chemical and thermochemical processing techniques, biochemical techniques, including extracellular enzymatic pretreatment techniques, have been popular globally in recent years due to their environmentally friendly nature (Holechek et al. 2022).

The sustainable treatment of these organic wastes has been identified as an urgent concern to prevent harmful environmental and human health consequences. There are some adverse ecological circumstances that can occur due to the accumulation of solid biomass. Some critical concerns are the emission of greenhouse gases, the contamination of surface and groundwater resources, the biomagnification of dangerous chemicals in food chains, acid rain, and the extinction of biological varieties (Holechek et al. 2022). These untreated wastes negatively affect the air, the environment, and human and animal health (Agrawa and Srivastava 2021). In addition, these wastes include various organic components that can be used to create a variety of goods with additional value, lowering the production cost (Agrawa and Srivastava 2021).

1.3 Different Process Mechanisms of Waste Biomass Valorization (Fig. 1.1)

The conversion process of waste biomass to valuable products is referred to as solid waste valorization. There are several different process mechanisms in which waste material is valorized. Most of these valorization techniques are physicochemical processes, and some of these techniques belong to biological treatment methods (Okolie et al. 2021). The physicochemical treatments are mostly combined with the thermochemical treatments. The thermal transformation pathways concentrate on converting biomass wastes into valorized gases, solid products, or liquid biofuels and associated intermediary substances dependent on the process condition (Martinez et al. 2021). Hydrothermal conversion can be divided into three significant

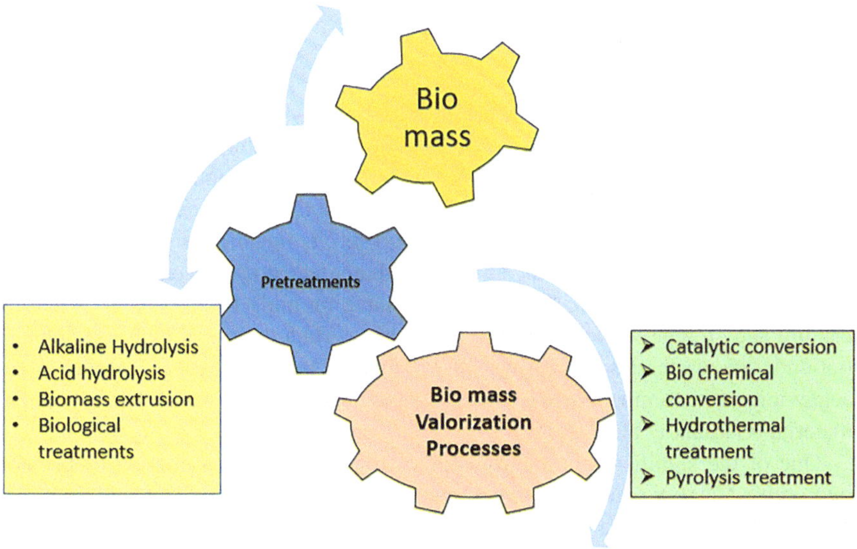

Fig. 1.1 Different process mechanisms of waste biomass valorization

processes: hydrothermal carbonization, hydrothermal liquefaction, and hydrothermal gasification (Okolie et al. 2021; Martinez et al. 2021).

1.3.1 Hydrothermal Carbonization (HTC)

Hydrothermal carbonization (HTC) is a thermochemical process that takes place at pressures of 10–50 bar, residence times of 0.5–8 h, and temperatures usually around 180 and 250 °C (Musa et al. 2022). Most studies on HTC concentrated on the solid product, thoroughly examining its qualities as a biofuel and soil enhancer, even when used in co-combustion with coal (Bevan et al. 2021).

One of HTC's key benefits is that it does not require the raw material to be processed and dried. Instead, hydrogen and oxygen atoms in carbon molecules are released by the action of water at high pressures (about 20 MPa) and temperatures between 180 and 300 °C (Bevan et al. 2021). This procedure can potentially increase the fuel quality of MSW by lowering the water content, removing chlorine, and densifying the energy. As a result, carbon dioxide makes up the bulk of the mixture of gases, including a highly concentrated aqueous solution and an enriched solid carbon (Bevan et al. 2021). The mass of the product concerning the raw material used decreases due to the dehydration and decarboxylation reactions. The HTC process modifies the treated material's ash content and calorific value, two crucial factors for hydrochar energy consumption and international compliance (Garnaik et al. 2021).

1.3.2 Hydrothermal Gasification

A relatively dry solid biomass is transformed primarily into two products during the gasification process: syngas and char. There are various situations where a nonzero amount of aqueous organic liquid can be created, such as updraft gasification. A series of thermochemical reactions occur to carry out the procedure in an atmosphere with low oxygen levels (Okolie et al. 2021). There is much discussion about using biomass gasification syngas as an intermediate feedstock to create power and second-generation biofuels. These methods have multiple benefits such as carbon capture and storage while producing bioenergy. When there is a high ash concentration in the waste, this process has been improved by using pure O_2 instead of air, achieving greater temperatures ($>$1600 °C) and creating a vitrified and inert granular material (Garnaik et al. 2021).

One of the significant problems with gasification is the production of tar as a byproduct. Catalytic cracking, which can work at relatively low temperatures and produce high tar removal efficiency, is acknowledged as the most effective way to reduce tar formation in the gas mixture. It produces more product gas and hydrogen-rich gas by adding a catalyst to lower the gasification temperature, enhance the steam reform, and increase the water-gas exchange processes (Okolie et al. 2021).

1.3.3 Hydrothermal Liquefaction (HTL)

(HTL) offer hope for the generation of bio-oil. However, HTL-derived oils have lower oxygen concentrations and higher heating values. The only thermochemical conversion technology currently in use that can effectively treat biomass with high moisture levels is hydrothermal liquefaction (HTL) (Bevan et al. 2021). As opposed to the pyrolysis and gasification technique, it necessitates costly drying pretreatments to reduce the moisture content 10–20 wt% (Park et al. 2018). HTL eliminates this expensive drying process phase, making it more affordable and desirable. HTL breaks down the complex molecular structures into simpler molecules with combustion capabilities similar to bio-oil (300–400 °C, 15–20 MPa).

It is also feedstock-neutral because it has been demonstrated to transfer a higher range of feed biomass to beneficial outputs. When at subcritical pressure, moisture acts as a nonpolar liquid, some with specific properties like density and viscosity. As a result, organic substances, including fatty acids, carbohydrates, and polymers, are more soluble (Bevan et al. 2021). Moreover, a disproportionately high number of H^+ and OH^- ions are formed due to the rise in the ionic product of water under subcritical circumstances. They encourage processes using an acid-base catalyst, such as the hydrolysis of polymeric larger molecules such as proteins in cellulose and lignin in waste feedstock (Musa et al. 2022).

1.3.4 Anaerobic Digestion (AD)

The anaerobic digestion process occurs in the following four basic steps: (1) hydrolysis, (2) acidogenesis, (3) acetogenesis, and (4) methanogenesis.

1.3.4.1 Hydrolysis

Biomass feedstock comprises long chains of biopolymers like proteins, polysaccharides, and lipids. During the hydrolysis phase, specific fermentative microorganisms produce complicated organic compounds into simpler molecule substances such as simple sugars, amino acids, and fatty acids (Karuppiah and Azariah 2019).

1.3.4.2 Acidogenesis

In this stage, acidogenic bacteria are used to further degrade the hydrolyzed compounds by producing simple acids like acetic acid, butyric acid, and propionic acid. The process produces gases, including NH_3, H_2, CO_2, and H_2S.

1.3.4.3 Acetogenesis

During this process, a particular bacterial group generates acetate from various organic substrates like acetic acid, carbon dioxide, and hydrogen. Further, the methanogenic bacteria utilize the simple forms of compounds produced by acetogenic bacteria as their organic substrate (Amin et al. 2021).

1.3.4.4 Methanogenesis

Methanogens, a genus of bacteria, produce methane from the byproducts of acetogenesis. Methanogens produce the primary byproducts of anaerobic digestion—methane, carbon dioxide, and water (Karuppiah and Azariah 2019) (Fig. 1.2).

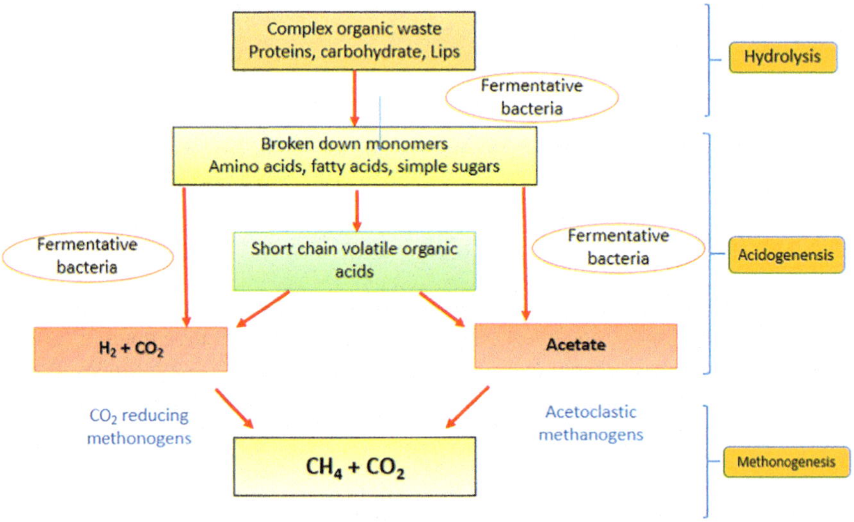

Fig. 1.2 The process mechanism of biomass anaerobic digestion

1.4 Different Products of Solid Waste Valorization and Their Environmental Applications

1.4.1 Production of Biochar

Biomass is increasingly being converted into carbon products to provide a soil amendment that is long lasting. Particularly woody solid materials, including agriculture and forestry waste, can efficiently be utilized as a regenerative resource while also serving as a crucial feedstock for biochar production, which is used in environmental engineering. The solid, carbon-rich byproduct of the thermochemical transformation of solid waste substances, known as biochar, acts as a soil supplement (Tripathi et al. 2019). Traditional uses for oak sawdust as organic solid waste from the forestry sector include solid fuel enhancement or soil fertility restoration. The pyrolysis process reflects the sustainable conversion of biomass to charcoal and biofuels by thermal conversion in an oxygen-restricted atmosphere (Tripathi et al. 2019).

Biochar has many possible uses due to its distinctive physicochemical qualities, which include soil amendments, environmental remediation, waste treatment, tackling climate change, catalysis, activated charcoal with some of these materials, and power generation.

1.4.2 Biopesticides

Biopesticides are biological agents that are used for Integrated Crop Management (ICM). One of the most widely utilized and researched biopesticides is *Bacillus thuringiensis*. As an effective and safe biological insecticide, *Bacillus thuringiensis* (Bt) is effective against lepidopteran, coleopteran, and dipteran insect pests and nematodes. Developing agroecological systems using biopesticides represents a safe alternative that contributes to the reduction of agrochemical use and sustainable agriculture (Sharma and Gaur 2021) (Fig. 1.3).

1.4.3 Biohydrogen Production

Hydrogen is regarded as an emerging fuel source at present. To highlight how far efforts have been achieved to introduce cars powered by bioethanol, it is being sought that biomass is used to produce hydrogen as an energy source for automobiles (Devi et al. 2021).

The need for hydrogen is rising dramatically globally, but it is challenging and difficult to supply the demand. Pretreatment is one of the technological obstacles,

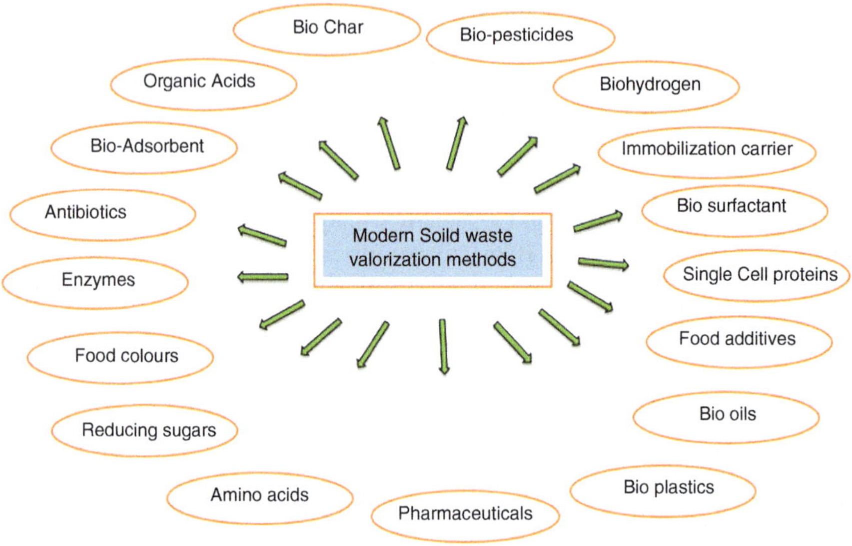

Fig. 1.3 Modern solid waste valorization methods

especially for residues like lignocellulosic biomass that are more resistant to enzymatic breakdowns (Devi et al. 2021).

Hydrogen generation is regarded as a convenient method of manufacturing hydrogen. The biological conversion of waste to hydrogen could be a viable alternative to the well-known chemical and electrolytic processes of producing hydrogen. The processes mentioned above demand high energy and production costs. Specifically, synthesizing biohydrogen from agricultural wastes is regarded as advantageous since agricultural biomasses are readily available and affordable (Hitam and Jalil 2020). The biohydrogen generation through dark fermentation is being popular because of its production efficiency and adaptability to use a variety of biomass types. Some concerns that affect biohydrogen generation include inoculum specificity, waste substrate pretreatment techniques, the temperature of the medium, and pH conditions (Devi et al. 2021).

Either photosynthetic or anaerobic microorganisms could achieve biohydrogen production. Hydrogen and carbon dioxide are produced mainly by photosynthetic and phototrophic microorganisms using carbon dioxide, organic acids, or carbohydrates (Hitam and Jalil 2020). Although this method produces much hydrogen, it is not practical because light energy must be supplied, and efficient photoreactor construction is challenging. Hydrogen is created as a byproduct of the anaerobic acidogenesis process and organic acids. In an anaerobic treatment system, hydrogen will inevitably be created by hydrogen producers (Hitam and Jalil 2020). However, methanogenic consumers will quickly consume hydrogen in the liquid phase. Therefore, methanogenesis must be stopped in the anaerobic treatment system to generate hydrogen gas in large quantities (Hitam and Jalil 2020).

1.4.4 Production of Organic Acids

Fermentation of cellulose is a novel way to produce organic acids. The cost of raw materials is one of many variables that affect how much fermentative acid is produced. When simple sugars are used as the source for manufacturing organic acids, it has proven to be quite expensive. Diverse cultivation techniques have been used to examine the generation of various organic acids, like citric, lactic, and gluconic acid, generated from sugarcane production (Duque et al. 2021). Several approaches for producing organic acids using biomass are given in Table 1.2.

1.4.5 Bio-Adsorbent Production

Removing dangerous substances from aqueous solutions via bio-sorption is an innovative method that includes heavy metals, dyes, and petroleum (gasoline, n-heptane). Bio-sorption is one of the superior approaches for removing a diverse range of contaminants (Rathi and Kumar 2021).

Environmental pollutants such as heavy metal ions, dyes, ammonia, and nitrates can be effectively removed from the environment using adsorbents made from plant wastes as an affordable alternative to expensive conventional procedures (Rathi and Kumar 2021). Owning to their strong adsorption qualities resulting from their ion-exchange capabilities, lignocellulosic agro-wastes are a beneficial resource. Solid biomass can effectively be turned into effective sorbents for removing numerous metals, offering a possibly less expensive option for treatment approaches.

Table 1.2 Production of organic acids using different biomass sources

Acid	Microorganism	Source	Reference
Lactic acid	*Lactobacillus delbrueckii*	Sugarcane bagasse and cassava bagasse	Stylianou et al. (2020)
Gluconic acid	*Aspergillus niger*	Tea waste	Stylianou et al. (2020)
Citric acid	*Aspergillus niger DS 1* *Aspergillus niger PTCC-5010*	Pineapple waste, valencia orange peel, sugarcane bagasse	Cheah et al. (2019)
Ellagic acid	*Aspergillus niger* *Aspergillus niger GH1*	Pomegranate seeds and husk Pomegranate peel	Stylianou et al. (2020)
Oxalic acid	*Agaricus campestris*	Sugarcane bagasse	Cheah et al. (2019)
Gluconomic acid	*Gluconobacter oxydans* *Aspergillus niger* *Pseudomonas savastanoi* *Pseudomonas fluorescens*	Sugarcane bagasse	Stylianou et al. (2020)
Formic acid	*Aspergillus niger*	Sugarcane bagasse	Stylianou et al. (2020)

Chemically altered organic substrates include rice hulls, spent grains, and typical bagasse, altered with formaldehyde in an acidic media, such as NaOH, KOH/K_2CO_3, or acid solution, or they are simply washed with warm water (Castillo et al. 2021). According to scanning electron micrographs with energy spectra, heavy metals can be immobilized in two ways: by cation exchange and adsorption on hypha or by chelating a fungal metabolite (Castillo et al. 2021).

1.4.6 Antibiotics Production

Antibiotics are compounds made by certain microbes that, at shallow doses, destroy or selectively restrict the growth of other germs. Different antibiotics are produced using a variety of agricultural wastes. Antibiotics with agro-industrial waste origin were used in various experiments. Some scientists have used sawdust, rice hulls, and corn cobs as raw materials to make the antibiotic oxytetracycline. Sassu et al. (2018) agro-waste aids in the manufacture of oxytetracycline. Utilizing low-cost carbon sources from various agricultural residues significantly reduced the cost of producing antibiotics.

Agro-industrial waste, known as oil-pressed cake, has been used as a raw material in several studies on the synthesis of an extracellular range of antibiotics with *Amycolatopsis mediterranei*. Two of the various agro-industrial wastes, namely, pulverized nutshell and coconut oil cake, produced the most antibiotics. In addition, the usage of external energy sources allowed for increased antibiotic production.

1.4.7 Enzyme Production

Wastes from the agro-industrial sector have a diverse composition for encouraging microbial metabolism because fermentation creates a variety of valuable enzymes (Mateii et al. 2021). Utilizing these leftover substrates accelerates the growth of fungi by causing the degradation of the lignocellulosic substrate by several enzymes, which results in the creation of a simpler substrate. For example, amylase is one of the key enzymes used in the food industry to break down starch into maltose (Mateii et al. 2021).

Biological processes like the hydrolysis of cellulose to glucose are catalyzed by enzymes made from OFMSW enzymes, which are protein molecules. The enzyme usage is essential since it controls the effectiveness and rate of reactions when manufacturing bulk chemicals and biofuels. However, the substrate cost needed to grow enzymes makes the traditional chemical synthesis of enzymes prohibitively expensive (Mateii et al. 2021). A critical step toward greater cost-efficiency is the production of these enzymes using the waste substrate. For synthesizing enzymes through SFF by microorganisms and fungi, mixed food waste is used as the raw material in numerous research. Enhanced generation of a whole cellulase system

Table 1.3 Different solid waste substrates enzymes produced by micro organisms

Substrate	Enzymes	Related microorganism species	Industrial application	Reference
Papaya waste, orange peel, fruit peel waste, rice bran, wheat bran	A Amylase	*Aspergillus niger* *Bacillus tequilensis* *Lysinibacillus* sp.	Food industry Paper industry Textile industry Food industry	Pham et al. (2021)
2021 Organic Fraction of Municipal Solid waste (OFMS)	Pectinase	*Bacillus megaterium* *Bacillus badius*	Detergent industry Cosmetic industry	Escaramboni et al. (2022)
OFMS	Proteinase	*Bacillus berjingensis* *Bacillus licheniformis*	Paper industry Beverage industry Textile industry	Escaramboni et al. (2022)
OFMS	Cellulase	*Penicillium chrysogenum*	Food industry Detergent industry	Terrone et al. (2018)
Sugar cane bagasse	Xylanase	*Aspergillus niger*	Detergent industry	Escaramboni et al. (2022)
Municipal solid waste	Chitinase	*Aspergillus niger* *Trichoderma harzianum*	Bakery industry Animal food industry Textile industry Agriculture industry	Pham et al. (2021)

using low-cost mixed food waste requires a low-cost and optimized growing medium (Chatterjee et al. 2019). Table 1.3 describes the different studies on producing organic waste using biomass.

1.4.8 Production of Pigments/Food Colors

The production of microbial pigments and food colors is one of the popular biomass valorization methods in the world. Food coloring is a crucial factor that significantly influences how a particular food commodity feels in the mouth. B-carotene and astaxanthin are naturally occurring coloring compounds mostly employed in the food and animal feed sectors as colors. They are produced by microalgae, such as *Haematococcus* sp., and yeast and are still constrained by the expensive production method (Simon et al. 2017). Utilizing waste materials produced during the food production process is a method of reducing manufacturing costs. Because of this, the creation of food colorings utilizing leftovers from agro-food waste is a better alternative (Pham et al. 2021).

Different characteristics of carotenoids may be produced depending on the environmental factors, nutritional factors, and climatic and stress conditions. Carotenoids can be produced from various substrates, including cereals, lipids, glycerol, cellobiose, shrimp waste, organic wastes, and whey milk from various microbial groups (Simon et al. 2017). When cultured on a mixture of green waste and whey, the fungi varieties belonging to the *Penicillium* sp. can produce a wide range of coloring compounds including orange, green, yellow, and red. The antibacterial, antioxidant, and anticancer properties of carotenoids, a group of colored terpenoids, have led to their commercial application as bio pigments in the pharmaceutical sector. One of the most critical nutraceuticals for preventing vitamin A shortage is carotene, one of the provitamin-A carotenoids. Additionally, the global market for biocolors used as food coloring has demonstrated an annual growth rate of 10–15% (Adeleke and Babalola 2020).

1.4.9 Reducing Sugars

Reducing sugars are one of the most valuable products generated from waste biomass worldwide. Reducing sugars are those that transform Tollens reagent into a silver mirror. Reducing sugar can be further converted by microorganisms in a fermentation process to fuel alcohol as a source of biomass energy. As a result, recovering reducing sugars from waste lignocellulosic biomass is also highly interesting. Biomass generated from plants primarily consists of cellulose, a glucose polymer. Cellulose hydrolysis of supercritical water can result in the production of sugars.

Several researchers have looked at the conversion of biomass into reducing sugars and have found certain low-cost biomass products that can produce reducing sugar. One of the less expensive types of biomass is bean dregs, which contain a greater cellulose concentration. A standard procedure for generating reducing sugars using the solid bean mass is by subcritical hydrolysis. Another sort of biomass residue that is produced in significant amounts by the sugar and alcoholic industries in nations like Brazil, India, Cuba, and China is sugarcane bagasse. Approximately 5.4108 t of sugarcane is used yearly, and 1 t of sugarcane typically yields 280 kg of solid waste bagasse. The main components of bagasse are lignin, cellulose, and hemicelluloses. Moreover, some scientists have looked into subcritical hydrolysis to generate reducing sugars from bagasse.

Peanut shell residuals are another most abundant biomass type, producing only 4.5 × 108 t of weight in China. Hemicelluloses and biomass conversion make up about 10.1% of a peanut shell. Zhu et al. (2015) evaluated the influence of different reaction conditions on reducing sugar yield. Reducing sugar extraction from peanut shells was the subject of their research. Reaction time and temperature have the most effects on how easily reducing sugars can be extracted. The reduced sugar yield increased due to the addition of $AlCl_3$.

Corn stalks, coconut meal, ginger bagasse, and wheat straw are other lignocellulosic wastes that are generated in more significant quantities. They are potential bioenergy sources due to their vast abundance and regenerative nature.

1.4.10 Amino Acids

Recovering valuable compounds from waste substrates is becoming popular in the world. Animal protein-rich wastes generally include hog hair, feathers, fish, and shrimp shells. These protein-rich biomass wastes, including some biomass from proteinaceous plants, can be converted into amino acids. About 20% of bean dregs are made up of a high percentage of protein. The primary byproducts of the soybean processing industry are bean dregs. The bean dreg trash produced annually in China has increased to about 80,000 t in recent years.

Some scientists have explored the feasibility of hydrolyzing amino acids from bean dregs in subcritical water. The outcomes demonstrated the ability to generate a range of amino acids. The amino acids include glutamic acid, aspartic acid, cystine, methionine, tryptophan, and threonine, which can be produced using waste biomass. The hydrolysis reaction is greatly influenced by temperature and duration. Therefore, different amino acids are affected differently by reaction temperature and time. The ideal hydrolysis conditions were attained at 200 °C for the process and 20 min. The overall amino acid yield reaches 52.9% under these circumstances. Rice bran is a significant byproduct that makes up 8% of milled rice. De-oiled rice bran can be hydrolyzed successfully using subcritical water to synthesize usable amino acids. The maximum production capacity of amino acids, 8.0 1.6 mg gl, was produced at 200 °C with a 20-minute hydrolysis time (Ziero et al. 2020).

1.4.11 Production of Pharmaceuticals

The waste biomass can be used to create novel green synthetic pathways for synthesizing some active medicinal components and creating novel bioactive molecules (Espro et al. 2021). Mushrooms, which are a significant source of potent novel pharmacological compounds, can develop in an environment that is conducive to biomass. In particular, bioactive substances like antitumor, antiinflammatory, antivirus, and antibacterial polysaccharides can be synthesized using the waste biomass. However, they also contain active ingredients that decrease cholesterol and compounds with immunomodulating capabilities. Prospects for study on the bioactive substances produced by fungus growing on such affordable and common substrates are promising. They may result in advances in developing chemotherapies for antibacterial, antiviral, and anticancer conditions (Espro et al. 2021).

1.4.12 Bioplastics

The word "bioplastics" describes polymers from biomass or renewable sources. Bioplastics are plastics derived from biological sources and made from the renewable feedstock or by different types of microbes, a desirable material. Proteins, starch-based polysaccharides, algae, or wastewater treatment byproducts are used to make bioplastics and other alternative carbon sources like algae. Thermoplastic starch, primarily produced by enzymatic bioreactors, is the most well-known bioplastic available today (Ross et al. 2017).

The same qualities as regular plastics are present in bioplastics derived from plants, and these materials also offer the benefit of a minor environmental impact. Compared to traditional plastics, bioplastics produce fewer greenhouse gases over their lifetime. Consequently, bioplastics help to create a society that is more sustainable. When used on bioplastics derived from soy protein, heat treatment improved mechanical qualities, dehydrothermal treatment boosted superabsorbent capacity, and ultrasounds resulted in a structure with fewer pores (Abe et al. 2021).

As a result, the modified bioplastics might be applied in various ways. Currently, scientists are interested in creating next-generation sophisticated polymers generated from biomass, including poly(ethylene 2,5-furandicarboxylate) (PEF). The best method for disposing FW is to produce bioplastics like PHA. The fact that FW is landfilled and produces unfavorable outcomes like greenhouse gas (GHG) emissions and groundwater contamination is one explanation for this. Creating bioplastics from FW is a renewable, sustainable method that uses carbon-neutral resources to create the materials. Under commercial settings, some bioplastics can be composted and biodegraded (Marichelvam et al. 2019). Table 1.4 explains some studies done on bioplastic production using organic waste.

1.4.13 Production of Bio-Oil

Bio-oils contain a complex mixture of char particles, pyrolytic water, oxygenated hydrocarbons, and moisture from the original biomass. The bio-oil may be a stable emulsion or rapidly split into aqueous and oil phases depending on the content of the biomass and the pyrolysis conditions, albeit the oil phase has a lower heating value due to the substantial amount of dissolved water it contains. The heating value of

Table 1.4 Applications of bioplastics

Bioplastic material	Description	Applications	References
Starch-based bioplastics	Glycerol is added to the starch as a plasticizer	Food packaging, agricultural products, building construction industry, textile industry	Marichelvam et al. (2019)
Cellulose-based bioplastics	Produced by the substitution of hydrogen with acetate and the addition of plasticizer	Packaging material, disposal of household, to produce medical and electronic devices	Abe et al. (2021)
PLA	Produced by polymerization of lactic acid	Food packaging, films	Meereboer et al. (2020)
PHA	PHAs are aliphatic polyesters made up of different polyhydroxyalkanoic acid constituents	Coating, food packaging	Ross et al. (2017) Meereboer et al. (2020)

bio-oil is typically less than 50% of the heating value of diesel fuel since most of the chemicals it has are oxygenated (Kowthaman et al. 2022).

Although bio-oils can be used in boilers, the equipment needs to be adapted; fossil fuel must be used to start the boiler. While particle emissions from bio-oils are typically higher than those from fossil fuels, nitrogen and sulfur oxide emissions are usually lower. In all instances, the bio-oil cannot be utilized as the only fuel; instead, it must be blended with or used as a pilot for another fuel, typically alcohol or diesel. Because bio-oils and fossil fuels generally are not miscible, coffering with fossil fuels necessitates the creation of emulsions, which must be stabilized with pricy surfactants (Broumand et al. 2020).

Although bio-oils have been successfully tested in diesel engines, there are still some issues regarding their flow ability in cold weather, corrosion, the development of gummy deposits, and the requirement for an auxiliary fuel during startup. Less pollution is also released when bio-oils are used in turbines, except for carbon monoxide. Turbine blade rust and fouling are very concerning (Kowthaman et al. 2022).

1.4.14 Food Additives

Waste biomass can be used to make several food additives, such as anthocyanins, tannins, starches, saponins, polypeptides and lectins, polyphenols, lactones, flavones, phenols, pectin, and food gums (Míguez et al. 2021). The *Xanthomonas* species make exopolysaccharides of the type xanthan. Food additives are made from xanthan. As a result, making xanthan from agricultural waste is an intelligent strategy for producing a quality product at a reasonable price (Míguez et al. 2021).

1.4.15 Single-Cell Proteins (SCP)

SCP, which can be utilized in animal and human nutrition, is microorganism biomass or protein extract obtained from microscopic algal, yeast, fungus, or bacterial cultures. SCP is a good substitute for agricultural proteins since its production is more environmentally friendly, uses less water, takes up less space on the land, and has a far more negligible impact on climate change than proteins obtained from agriculture (Sharif et al. 2021). Another benefit of SCP is cultivating SCP-producing microbes using a wide range of biodegradable industrial wastes. Utilizing waste materials can lower production costs and lessen the harmful effects of leftovers on the environment thanks to waste treatment (Sharif et al. 2021).

For the manufacturing of SCP, various raw materials have been investigated as substrates (carbon and energy sources). Before usage, these raw materials were frequently hydrolyzed via physical, chemical, and enzymatic processes (Bratosin et al. 2021). In addition, agriculture wastes, including hemicellulose and cellulose waste from plants, as well as fibrous proteins, like horns, feathers, nails, and hair from animals, are also plentiful sources of waste products for the creation of SCP. These waste products are rich in some growth factors needed by microbes and have been transformed into biomass, protein concentrate, or amino acids utilizing proteases produced by specific microorganisms (Bratosin et al. 2021).

In addition, Khan et al. (2022) investigated how single-cell protein (SCP) could be produced from fruit waste. Using *S. cerevisiae* and submerged fermentation, they produced SCP utilizing cucumber and orange peels as the substrate. They discovered that compared to orange peels, cucumber peels produced a higher amount of protein. It was, therefore, proposed that these fruit wastes could be converted into SCP employing the appropriate bacteria.

1.4.16 Biosurfactant Production

Biosurfactants are amphiphilic compounds produced by a great variety of microorganisms. They can be categorized as glycolipids, particulate surfactants, polymeric surfactants, lipopeptides, phospholipids, and fatty acids (Saranraj et al. 2021). The uses of biosurfactants are comparable to those of synthetic surfactants. However, biosurfactants are particularly well suited for environmental applications because of their biological and environmental safeness for the bioremediation of soils contaminated with crude oil (Saranraj et al. 2021).

The applications of biosurfactants include excellent detergency, emulsification, foaming, dispersing traits, wetting, penetrating, thickening, and microbial growth enhancement. In addition, enhancing biosynthesis efficiency and choosing inexpensive culture media components are necessary to lower the cost of producing microbial biosurfactants because they account for 50% of the overall production expenses (Rath and Srivastava 2021).

The manufacture of biosurfactants can benefit from using food and agricultural wastes as carbon sources, lowering the high production costs of these substances. Some studies have discussed using oily wastes as carbon sources for synthesizing biosurfactants (Rath and Srivastava 2021). However, as far as we know, no studies have discussed the production of biosurfactants utilizing organic waste. The primary carbon source for manufacturing surfactin can be two-phase waste from olive mills. Similar results were obtained with additional bacterial strains and carbon sources for surfactant yield from biomass (Saranraj et al. 2021).

1.4.17 Immobilization Carrier Production

Wastes from the agro-industrial sector can be employed as the carrier for the solid-state fermentation process that immobilizes fungi; among the ten agro-industrial wastes, Sadh et al. (2018) used different organic waste residues for immobilized carrier production and has obtained successful results. They investigated the suitability of several agro-industrial wastes as an SSF immobilization carriers. Before starting the study, they used physiochemical treatment to characterize the agricultural and industrial wastes. These agricultural industry wastes can be utilized in other ways that are advantageous economically and environmentally.

1.5 Traditional Approaches to Waste Biomass Valorization

The biomass valorization is not regarded as a novel concept. The biomass residue can be converted into a wide range of products to fully fill energy requirements and synthesize green products. The typical and most popular method for converting waste biomass from crop plants is using heat energy by combustion. Different types of straws, grass clippings, and wood fragments are examples of biomass that can be utilized in the process. Figure 1.4 illustrates a few of the traditional processes for solid waste valorization.

1.5.1 Biogas Production

One common technique is the decomposition of organic matter, which involves converting organic material directly into biogas, a combination primarily composed of methane and carbon dioxide with trace amounts of certain other gases, including hydrogen sulfide. The main component of biogas utilized for proper gas cleanup is methane, which is also used as fuel in engines, gas turbines, fuel cells, boilers, industrial heaters, and other operations, as well as in the production of chemicals.

Fig. 1.4 Traditional approaches to solid waste valorization

This gaseous biofuel is produced when organic materials are digested without oxygen and consists primarily of methane, a crucial central pillar of biogas, and carbon dioxide (CO_2). In addition, alcohols, carbonyl molecules (aldehydes and ketones), ammonia, terpenes, aromatic compounds, and sulfur compounds are among the other substances that are produced during digestion. Anaerobic digestion involves several coprocesses, with different communities of bacteria catalyzing each step. The mechanisms of acidogenesis, acetogenesis, and methanogenesis are performed in response to the degradation of such complex compounds to monomer units.

1.5.2 Compost and Vermicompost

Composts are a significant added-value product made from solid waste. Compost is an organic fertilizer rich in nutrients and improves the soil. Compost is made from diverse organic materials through the humification process. Vermicomposting, to reiterate, is the biological stabilization of organic materials into the finished product known as vermicompost. Numerous microorganisms support both functions, including bacteria, fungi, earthworms, and others. In these methods, bacterial inocula are used to speed up the composting process, while lignocellulolytic fungal inoculum as pretreatment of organic biomass (Soobhany 2019).

Byproducts from the sugarcane business, particularly PM, are regarded as an excellent fertilizer source when they contain a variety of essential micronutrients in high concentrations, such as calcium, nitrogen, and phosphorus (1.9%, 1.8%, and

0.9%, respectively). In addition, a new *Perionyx sansibaricus* earthworm strain can humify a substrate of sawdust, cow manure, and industrial-grade guar gum waste. Therefore, composting can be viewed as a low-cost technology to transform agricultural biomass into goods with value additions (Soobhany 2019).

1.5.3 Mushroom Production

The regulated growth of edible mushrooms uses cellulose waste and residues from the agro-industrial complex as a bio-conservation. The regulated production of mushrooms that use these wastes as raw material can address the environmental issue with agriculture and agricultural wastes. Mushroom cultivation served as a unique biotechnological approach to valorizing agricultural and industrial waste. By converting agro-based leftovers utilizing a variety of microorganisms, mushroom production demonstrated its strength in ecological and economic terms (Sadh et al. 2018).

1.6 Utilization of Waste Biomass as an Alternative Energy Source

Currently, nearly 87% of the world's energy mix is produced by depleting resources, all of which are carbon-rich fossil fuels like coal (25%), natural gas (21%), and oil (35%), except for nuclear energy (6%). In terms of sustainability, geopolitics, socioeconomics, and the environment, excessive use of fossil fuels has brought up several challenges. In such a situation, the globe must focus on finding alternate resources. As a result, biomass is viewed as a superior option for creating energy due to its abundance of carbon, oxygen, moisture, and ash (Cheah et al. 2019).

A biomass' energy content varies based on its constituent parts. There are four main methods for using biomass energy: anaerobic digestion, pyrolysis, gasification, and direct burning. Heat-related chemical process is a critical component of the development of renewable energy. In this procedure, heat is used to modify the chemical composition of biomass (Cheah et al. 2019).

This method can turn biomass into charcoal or produce gas. The conversion process can occur at temperatures ranging from a few hundred to several hundred degrees centigrade. Additionally, dry biomass can be gasified to create methane, hydrogen, and carbon monoxide, as well as energy, or it can be transformed into liquid fuel (Zhao et al. 2015).

Biodiesel synthesis and bioethanol production are the main techniques used to manufacture liquid fuels. Currently, biomass wastes with high lipid content serve as the primary raw materials for biodiesel manufacturing, while carbohydrate-rich substrates serve as the primary source for bioethanol production. However, using plant leftovers for bioethanol production is difficult since pretreatment is needed to break down the lignocellulose complex and increase the availability of cellulose to the fungi. Pretreatment methods include using acids, thermal steam, and enzymes (Sadh et al. 2018).

Biological process (aerobic and anaerobic): In this process, the treatment of a biological agent causes a shift in the chemical composition of biomass. This method allows for the production of biogas and ethanol from biomass. Anaerobic processes are those that occur when there is insufficient oxygen present in the environment, whereas aerobic processes are those that occur when there is sufficient oxygen present. Biomass (biological process with or without oxygen), biogas, and ethanol make up the process flow (Zhao et al. 2015).

1.7 Multidisciplinary Approaches to Waste Biomass Valorization

Integrating thermodynamic and biological approaches can be effective for feedstock value addition in order to conquer feedstock's diverse and resistant characteristics for a particular substrate biochemical/biofuel generation. An integrated solid waste valorization approaches have a less negative environmental impact. Economically and environmentally, hybrid pyrolysis-anaerobic digestion technologies are promising. The biorefinery concept is used to optimize resource utilization and the production of various valuation commodities. The bio refinery's heat, energy, and material fluxes are planned to reduce the cost of unit operations and remove or lessen the associated adverse environmental impact.

1.7.1 Bio-Nanotechnology

Bio-nanotechnology refers to the combination of nanotechnology and biotechnology. This is an integrated technology for solid waste management. Numerous researchers focus on developing new processes using this bio-nanotechnology for sustainable waste management. The bio-nanotechnology can be classified into two broad categories in solid waste management. The first is to degrade/reduce the specific toxic compounds to minimize their environmental poisoning, and the other is to generate valorized commodities using organic biomass.

Nanomaterials (NMs) are a nanocatalyst, which is key to the catalytic destruction of biomass resources. The small size of the catalyst allows for simple penetration into the plant cell wall and contact with the holocellulose to release fermentable sugars (Roy et al. 2021). NPs are crucial in the production of bioproducts from lignocellulosic biomass. Therefore, a nanotechnological approach is a promising, environmentally benign, economically viable, and effective method. However, since the subject of bioenergy and bioproducts is still in its early stages, an extensive study on the valorization of lignocellulosic biomass is found to be constrained (Roy et al. 2021).

1.8 Biomass Valorization for the Bio-Economy Economy and Meeting Environmental Feasibility

The circular economy has become crucial for creating closed-loop technical and biological cycles. A closed-loop system, often known as the cradle-to-cradle idea, involves either returning materials to the natural ecosystem without impacting the environment or recycling them endlessly without losing any of their original features.

Turning waste biomass into value-added products has several benefits, including reducing our reliance on petroleum-derived goods and opening up new revenue-generating opportunities. In particular, bio refineries can make it possible to realize the circular bio-economy by connecting the streams and loops and enabling the valuing of numerous products. Waste management is unquestionably crucial in sustainability, especially biodegradable waste.

1.9 Challenges of Waste Biomass Valorization

Thermochemical and biological conversion processes can process waste biomass into green fuels and value-added materials. Despite each production method's effectiveness, they all have particular flaws. Due to their high-temperature needs, thermochemical processes are constrained by factors including high processing costs. The high-temperature need for the thermochemical conversion processes makes them expensive and dangerous. The aqueous liquid byproducts of pyrolysis are just one example of the hard-to-dispose-of compounds produced by thermochemical processes (Sikarwar et al. 2021). Contrarily, biological processes face difficulties like limited product yield and extended processing times. For instance, low product yield, poor stability, and lengthy processing times are characteristics of biological processes. An integrated strategy could help to reduce the drawbacks of each method.

In addition, challenges for developing alternative technologies for biomass processing include the heterogeneity of biomass composition, ineffective pretreatment steps, inadequate reactors, dependence on high temperatures and elevated pressures, the need for toxic reagents in large quantities, and calorific power loss due to inorganic contaminants (Lee et al. 2021).

Notably, the creation of biomass energy has several downsides. The first and possibly worst drawback of biomass energy is that it contributes severely to global warming since it generates more greenhouse gases when burned directly as fuel. One another disadvantage of biomass energy is the amount of space it requires. Some biomass commodities take a lot of area and water to develop, and once they are mature, the product needs a lot of storage space before it can be turned into electricity. The high cost of setting up and maintaining the machinery needed to process biomass to produce energy (Jaworek et al. 2021). The cost of producing biomass fuel is also expensive, including paying for the required labor and transportation costs.

1.10 Future Perspective of Waste Biomass Valorization in the Circular Bio-Economy

The ongoing expansion of resource use has served as the impetus for the circular economy, which has recently received much attention as a strategy for meeting current consumption and production demands. The common understanding of CE is that it refers to a sustainability concept that strives to limit or prevent the consumption of virgin resources by enhancing resource utilization, recovery, and recycling. However, studies on the chemical properties of biomass and how they affect its liquefaction are marginal. The most exciting methods in the older research and literature utilize phenol and polyhydric alcohol-based liquefaction processes. Following the discovery of the wood liquefaction phenomenon, studies of various liquefaction parameters were carried out to increase the concentration of biomass in the liquefaction mixture and obtain the accurate liquefaction degree with respect properties to liquefied biomass in organic solvents, comprehension and understanding of the wood liquefaction mechanism, and further application.

1.11 Conclusions

Over the past several years, biomass waste has been utilized in manufacturing processes and as an alternative energy substitute for fossil fuels. However, utilizing waste biomass for energy and other commodities has substantial adverse environmental effects. Waste biomass can be valorized into different products using different mechanisms such as hydrothermal treatments, anaerobic fermentation, bi-refinery techniques, biotechnology, and bio-nanotechnology to produce a range of value-added products using waste biomass. Further, valorized products such as enzymes, amino acids, pharmaceuticals, single-cell proteins, food additives, coloring compounds, and compost can be generated using the waste biomass.

Moreover, bioethanol, biogas, and bio-oil, like different energy products, can serve as alternative energy supply for the world's current energy crisis. Biomass crop leftovers have a heating value of roughly 3106 kcal Mg^{-1}, making them a vital alternative energy resource. Different physicochemical and biological processes are used to produce valorized products from the biomass. However, the integrated biomass valorization approaches linked with nanobiotechnology are becoming popular worldwide due to their greater conversion efficiency. Therefore, the effective usage of waste biomass plays a crucial role in the modern circular bio-economy and environmental sustainability.

References

Abe M, Adamson P, Borcean M, Bortoletto D, Bridges K, Carman SP, Chattopadhyay S, Coleman J, Curfman NM, DeRose K, Deshpande T (2021) Matter-wave atomic gradiometer interferometric sensor (MAGIS-100). Quantum Sci Technol 6(4):044003

Adeleke BS, Babalola OO (2020) Oilseed crop sunflower (Helianthus annuus) as a source of food: nutritional and health benefits. Food Sci Nutr 8(9):4666–4684

Adesra A, Srivastava VK, Varjani S (2021) Valorization of dairy wastes: integrative approaches for value added products. Indian J Microbiol 61(3):270–278

Agrawa P, Srivastava AK (2021) Agricultural domestic waste management in India. Int J Mod Agric 10(2):2145–2154

Amin FR, Khalid H, El-Mashad HM, Chen C, Liu G, Zhang R (2021) Functions of bacteria and archaea participating in the bioconversion of organic waste for methane production. Sci Total Environ 763:143007

Aslam B, Maqsoom A, Tahir MD, Ullah F, Rehman MSU, Albattah M (2022) Identifying and ranking landfill sites for municipal solid waste management: an integrated remote sensing and GIS approach. Buildings 12(5):605

Bevan E, Fu J, Luberti M, Zheng Y (2021) Challenges and opportunities of hydrothermal carbonisation in the UK; case study in Chirnside. RSC Adv 11(55):34870–34897

Bratosin BC, Darjan S, Vodnar DC (2021) Single cell protein: a potential substitute in human and animal nutrition. Sustain For 13(16):9284

Broumand M, Albert-Green S, Yun S, Hong Z, Thomson MJ (2020) Spray combustion of fast pyrolysis bio-oils: applications, challenges, and potential solutions. Prog Energy Combust Sci 79:100834

Castillo NET, Sierra JSO, Oyervides-Muñoz MA, Sosa-Hernández JE, Iqbal HM, Parra-Saldívar R, Melchor-Martínez EM (2021) Exploring the potential of coffee husk as caffeine bio-adsorbent–a mini-review. Case Stud Chem Environ Eng 3:100070

Chatterjee B, Goswami S, Mazumder D (2019) Enzymatic application in anaerobic digestion (AD) of organic fraction of the municipal solid waste (OFMSW). In: Advances in waste management. Springer, Singapore, pp 289–301

Cheah YK, Vidal-Antich C, Dosta J, Mata-Álvarez J (2019) Volatile fatty acid production from mesophilic acidogenic fermentation of organic fraction of municipal solid waste and food waste under acidic and alkaline pH. Environ Sci Pollut Res 26(35):35509–35522

Chowdhury R, Heng K, Shawon MSR, Goh G, Okonofua D, Ochoa-Rosales C, Gonzalez-Jaramillo V, Bhuiya A, Reidpath D, Prathapan S, Shahzad S (2020) Dynamic interventions to control COVID-19 pandemic: a multivariate prediction modelling study comparing 16 worldwide countries. Eur J Epidemiol 35(5):389–399

Dahmen N, Lewandowski I, Zibek S, Weidtmann A (2019) Integrated lignocellulosic value chains in a growing bioeconomy: status quo and perspectives. GCB Bioenergy 11(1):107–117

Devi A, Singh A, Bajar S, Pant D, Din ZU (2021) Ethanol from lignocellulosic biomass: an in-depth analysis of pre-treatment methods, fermentation approaches and detoxification processes. J Environ Chem Eng 9(5):105798

Dixit A, Singh D, Shukla SK (2022) Changing scenario of municipal solid waste management in Kanpur city, India. J Mater Cycl Waste Manage 24:1648–1662

Duque A, Álvarez C, Doménech P, Manzanares P, Moreno AD (2021) Advanced bioethanol production: from novel raw materials to integrated biorefineries. PRO 9(2):206

Escaramboni B, Garnica BC, Abe MM, Palmieri DA, Fernández Núñez EG, de Oliva Neto P (2022) Food waste as a feedstock for fungal biosynthesis of amylases and proteases. Waste Biomass Valoriz 13(1):213–226

Espro C, Paone E, Mauriello F, Gotti R, Uliassi E, Bolognesi ML, Rodríguez-Padrón D, Luque R (2021) Sustainable production of pharmaceutical, nutraceutical and bioactive compounds from biomass and waste. Chem Soc Rev 50:11191–11207

Garnaik PP, Pradhan RR, Chiang YW, Dutta A (2021) Biomass-based CO_2 adsorbents for biogas upgradation with pressure swing adsorption. In: Climate change and green chemistry of CO_2 sequestration. Springer, Singapore, pp 231–262

Gelan E (2021) Municipal solid waste management practices for achieving green architecture concepts in Addis Ababa, Ethiopia. Technologies 9(3):48

Hitam CNC, Jalil AA (2020) A review on biohydrogen production through photo-fermentation of lignocellulosic biomass. Biomass Conv Biorefinery 2020:1–19

Holechek JL, Geli HM, Sawalhah MN, Valdez R (2022) A global assessment: can renewable energy replace fossil fuels by 2050? Sustain For 14(8):4792

Ilankoon IMSK, Ghorbani Y, Chong MN, Herath G, Moyo T, Petersen J (2018) E-waste in the international context–a review of trade flows, regulations, hazards, waste management strategies and technologies for value recovery. Waste Manag 82:258–275

Jaworek A, Sobczyk AT, Marchewicz A, Krupa A, Czech T (2021) Particulate matter emission control from small residential boilers after biomass combustion. A review. Renew Sust Energ Rev 137:110446

Jebaranjitham JN, Christyraj JDS, Prasannan A, Rajagopalan K, Chelladurai KS, Gnanaraja JKJS (2022) Current scenario of solid waste management techniques and challenges in Covid-19-a review. Heliyon 2022:e09855

Jeswani H, Krüger C, Russ M, Horlacher M, Antony F, Hann S, Azapagic A (2021) Life cycle environmental impacts of chemical recycling via pyrolysis of mixed plastic waste in comparison with mechanical recycling and energy recovery. Sci Total Environ 769:144483

Kalair A, Abas N, Saleem MS, Kalair AR, Khan N (2021) Role of energy storage systems in energy transition from fossil fuels to renewables. Energy Storage 3(1):e135

Karuppiah T, Azariah VE (2019) Biomass pretreatment for enhancement of biogas production. Anaerobic Dig 150:111509

Kaza S, Yao L (2018) At a glance: a global picture of solid waste management. World Bank, Washington

Khan MKI, Asif M, Razzaq ZU, Nazir A, Maan AA (2022) Sustainable food industrial waste management through single cell protein production and characterization of protein enriched bread. Food Biosci 46:101406

Koul B, Yakoob M, Shah MP (2022) Agricultural waste management strategies for environmental sustainability. Environ Res 206:112285

Kowthaman CN, Kumar PS, Selvan VAM, Ganesh D (2022) A comprehensive insight from microalgae production process to characterization of biofuel for the sustainable energy. Fuel 310:122320

Lee CS, Conradie AV, Lester E (2021) Review of supercritical water gasification with lignocellulosic real biomass as the feedstocks: process parameters, biomass composition, catalyst development, reactor design and its challenges. Chem Eng J 415:128837

Mahapatra S, Kumar D, Singh B, Sachan PK (2021) Biofuels and their sources of production: a review on cleaner sustainable alternative against conventional fuel, in the framework of the food and energy nexus. Energy Nexus 4:100036

Marichelvam MK, Jawaid M, Asim M (2019) Corn and rice starch-based bio-plastics as alternative packaging materials. Fibers 7(4):32

Martinez CLM, Saari J, Melo Y, Cardoso M, de Almeida GM, Vakkilainen E (2021) Evaluation of thermochemical routes for the valorization of solid coffee residues to produce biofuels: A Brazilian case. Renew Sust Energ Rev 137:110585

Mateii JC, dos Santos OliveiraI JA, PamphileI JA, PolonioI JC (2021) Agro-industrial wastes for biotechnological production as potential substrates to obtain fungal enzymes. Ciên Nat 43:e72

Meereboer KW, Misra M, Mohanty AK (2020) Review of recent advances in the biodegradability of polyhydroxyalkanoate (PHA) bioplastics and their composites. Green Chem 22(17): 5519–5558

Mehmood T, Nadeem F, Qamar SA, Bilal M, Iqbal H (2021) Bioconversion of agro-industrial waste into value-added compounds. In: Sustainable bioconversion of waste to value added products. Springer, Cham, pp 349–368

Míguez JL, Porteiro J, Behrendt F, Blanco D, Patiño D, Dieguez-Alonso A (2021) Review of the use of additives to mitigate operational problems associated with the combustion of biomass with high content in ash-forming species. Renew Sust Energ Rev 141:110502

Munir MT, Mohaddespour A, Nasr AT, Carter S (2021) Municipal solid waste-to-energy processing for a circular economy in New Zealand. Renew Sust Energ Rev 145:111080

Musa U, Castro-Díaz M, Uguna CN, Snape CE (2022) Effect of process variables on producing biocoals by hydrothermal carbonisation of pine Kraft lignin at low temperatures. Fuel 325: 124784

Nanda S, Berruti F (2021) Municipal solid waste management and landfilling technologies: a review. Environ Chem Lett 19(2):1433–1456

Ncube LK, Ude AU, Ogunmuyiwa EN, Zulkifli R, Beas IN (2021) An overview of plastic waste generation and management in food packaging industries. Recycling 6(1):12

Okolie JA, Epelle EI, Nanda S, Castello D, Dalai AK, Kozinski JA (2021) Modeling and process optimization of hydrothermal gasification for hydrogen production: a comprehensive review. J Supercrit Fluids 173:105199

Park CS, Roy PS, Kim SH (2018) Current developments in thermochemical conversion of biomass to fuels and chemicals. IntechOpen, London, pp 19–41

Pham VHT, Kim J, Shim J, Chang S, Chung W (2021) Purification and characterization of strong simultaneous enzyme production of protease and α-Amylase from an extremophile-Bacillus sp. FW2 and its possibility in food waste degradation. Fermentation 8(1):12

Pheakdey DV, Quan NV, Khanh TD, Xuan TD (2022) Challenges and priorities of municipal solid waste management in Cambodia. Int J Environ Res Public Health 19(14):8458

Qi D, Lai W, Roe BE (2021) Food waste declined more in rural Chinese households with livestock. Food Policy 98:101893

Rath S, Srivastava RK (2021) Biosurfactants production and their commercial importance. In: Environmental and agricultural microbiology: applications for sustainability. Wiley, Hoboken, pp 197–218

Rathi BS, Kumar PS (2021) Application of adsorption process for effective removal of emerging contaminants from water and wastewater. Environ Pollut 280:116995

Ross G, Ross S, Tighe BJ (2017) Bioplastics: new routes, new products. In: Brydson's plastics materials. Elsevier, Amsterdam, pp 631–652

Roy S, Dikshit PK, Sherpa KC, Singh A, Jacob S, Rajak RC (2021) Recent nanobiotechnological advancements in lignocellulosic biomass valorization: a review. J Environ Manag 297:113422

Sadh PK, Duhan S, Duhan JS (2018) Agro-industrial wastes and their utilization using solid state fermentation: a review. Bioresour Bioprocess 5(1):1–15

Sangtani R, Sonbhadra S, Nogueira R, Kiran B (2022) Algal biorefinery: a paradigm to sustainable circular bioeconomy. In: Zero waste biorefinery. Springer, Singapore, pp 295–323

Saranraj P, Sayyed RZ, Sivasakthivelan P, Devi MD, Al Tawaha ARM, Sivasakthi S (2021) Microbial biosurfactants sources, classification, properties and mechanism of interaction. In: Microbial surfactants. CRC Press, Boca Raton, pp 243–265

Sassu EL, Bossé JT, Tobias TJ, Gottschalk M, Langford PR, Hennig-Pauka I (2018) Update on Actinobacillus pleuropneumoniae—knowledge, gaps and challenges. Transbound Emerg Dis 65:72–90

Sharif M, Zafar MH, Aqib AI, Saeed M, Farag MR, Alagawany M (2021) Single cell protein: sources, mechanism of production, nutritional value and its uses in aquaculture nutrition. Aquaculture 531:735885

Sharma P, Gaur N (2021) Microbial biopesticides use in insect-pest management: an overview. In: Microbial biotechnology in crop protection. Springer, New York, pp 123–145

Sharma KD, Jain S (2020) Municipal solid waste generation, composition, and management: the global scenario. Soc Responsib J 16(6):917–948

Sikarwar VS, Pohořelý M, Meers E, Skoblia S, Moško J, Jeremiáš M (2021) Potential of coupling anaerobic digestion with thermochemical technologies for waste valorization. Fuel 294:120533

Simon JE, Decker EA, Ferruzzi MG, Giusti MM, Mejia CD, Goldschmidt M, Talcott ST (2017) Establishing standards on colors from natural sources. J Food Sci 82(11):2539–2553

Soobhany N (2019) Insight into the recovery of nutrients from organic solid waste through biochemical conversion processes for fertilizer production: a review. J Clean Prod 241:118413

Stylianou E, Pateraki C, Ladakis D, Cruz-Fernández M, Latorre-Sánchez M, Coll C, Koutinas A (2020) Evaluation of organic fractions of municipal solid waste as renewable feedstock for succinic acid production. Biotechnol Biofuels 13(1):1–16

Terrone CC, de Freitas C, Terrasan CRF, de Almeida AF, Carmona EC (2018) Agroindustrial biomass for xylanase production by Penicillium chrysogenum: purification, biochemical properties and hydrolysis of hemicelluloses. Electron J Biotechnol 33:39–45

Tripathi N, Hills CD, Singh RS, Atkinson CJ (2019) Biomass waste utilisation in low-carbon products: harnessing a major potential resource. NPJ Clim Atmosp Sci 2(1):1–10

Wang H, Bi X, Clift R (2021) Utilization of forestry waste materials in British Columbia: options and strategies. Renew Sust Energ Rev 150:111480

World Health Organization (2019) National systems to support drinking-water: sanitation and hygiene: global status report 2019: UN-Water global analysis and assessment of sanitation and drinking-water: GLAAS 2019 report

Xie Y, Khoo KS, Chew KW, Devadas VV, Phang SJ, Lim HR, Rajendran S, Show PL (2022) Advancement of renewable energy technologies via artificial and microalgae photosynthesis. Bioresour Technol 363:127830

Zhao S, Zhu L, Li D (2015) Microplastic in three urban estuaries, China. Environ Pollut 206:597–604

Zhu G, Zhu X, Xiao Z, Zhou R, Feng N, Niu Y (2015) A review of amino acids extraction from animal waste biomass and reducing sugars extraction from plant waste biomass by a clean method. Biomass Conv Biorefinery 5(3):309–320

Ziero HDD, Buller LS, Mudhoo A, Ampese LC, Mussatto SI, Carneiro TF (2020) An overview of subcritical and supercritical water treatment of different biomasses for protein and amino acids production and recovery. J Environ Chem Eng 8(5):104406

Chapter 2
Biomass Valorization as Energy Production Using Waste Biomass

Amit Kumar Tiwari, Piyush R. Chauhan, Dan Bahadur Pal, and Sumit Kumar Jana

Abstract Humans have been using biomass energy since primitive man, biomass energy can be obtained from living things that have been burned using wood to keep food warm. Biomass is being utilized in power generators and other machines. Organic stuff generated from living objects or organisms such as plants and animals is referred to as biomass. The most common biomass materials commonly used for energy production are plants, wood, and waste. They are known as raw materials. Biomass energy can be seen as a unique source of energy production. Biomass includes energy that can be first obtained from the sun. For example, plants use photosynthesis to absorb the energy of the sun and convert carbon dioxide and water into nutrients or simple carbohydrates. The energy obtained or obtained organically or from these organisms can be converted and can be used directly or indirectly in the appropriate method through direct or indirect methods or means. Biomass can be burned to cause direct heat, can be transformed directly into electricity, and can be treated as an indirect biofuel. Biomass can be used for burning energy by a method called thermal conversion. Thermal conversion involves heating a biomass feedstock for combustion, dehydration, and stabilization. The most common forms of biomass feedstock that can be used for thermal conversion are feedstock such as municipal solid waste and by-products from paper mills or sawmills. Biomass is an important component of the carbon cycle on Earth. The carbon system is the series of carbon being transferred across all of the Earth's layers, including the atmosphere, hydrosphere, biosphere, and lithosphere. The carbon cycle can utilize different shapes. Carbon also aids in the regulation of solar radiation entering the Earth's climate. Photosynthesis, degradation, respiration, and human activity all contribute to the exchange. For example, when plants release carbon-based nutrients into the biosphere through photosynthesis, the carbon absorbed by the soil as organisms

A. K. Tiwari · P. R. Chauhan · S. K. Jana (✉)
Department of Chemical Engineering, Birla Institute of Technology, Ranchi, India
e-mail: sumitkrjana@bitmesra.ac.in

D. B. Pal
Department of Chemical Engineering, Harcourt Butler Technical University, Kanpur, India

D. B. Pal, A. K. Tiwari (eds.), *Sustainable Valorization of Agriculture & Food Waste Biomass*, Clean Energy Production Technologies,
https://doi.org/10.1007/978-981-99-0526-3_2

decompose can be recycled. Decomposing organisms can be transformed into peat, coal, or oil under the right conditions before being restored by nature or human intervention.

Keywords Biomass · Combustion · Biofuels · Energy production · Alcohols · Feedstock

2.1 Introduction

Energy has become a very important entity or asset which acts as a very important driving force to carry out our day-to-day activities. From toy airplane launches to mega GSLV launches, some or a major amount of energy is for sure required to carry out the business. Without energy, it's very much difficult to imagine a life. According to the industrial raw materials survey of 2019, approximately 70% of the energy-generating companies functioning within India uses fossil fuels (majorly coal) as raw material to produce energy. So, to remove the dependency of the overall earth on fossil fuels, the biomass energy production technique will be a perfect alternative. Biomass is a type of organic matter that comes from plants and recently living organisms (Walker et al. 2010). For example, the use of wood in fireplaces is a good example of biomass most familiar to most people (Panwar et al. 2011). There are many ways in which biomass can be used to produce energy (Srinivasnaik et al. 2015). In the case of energy production using biomass, carbon dioxide is generated and harmful to the human body, but on the other hand, the same amount of energy is used for plant regeneration, creating a balanced atmosphere (Srinivasnaik et al. 2015). Biomass can be used for various purposes in our daily life, both for personal and commercial use. Pyrolysis is the heat-based breakdown of biomass in anoxic environments. It's the fundamental chemical process that happens during the first two seconds and is the foundation of both combustion and gasification. By-products of biomass pyrolysis include biochar, bio-oil, and gases such as methane, hydrogen, carbon monoxide, and carbon dioxide. Pyrolysis produces mostly biochar at low temperatures, less than 450 °C or lower when the warming rate is slow, and primarily gases at high temperatures, higher than 800 °C, when the heating rate is quick, based on the thermal conditions and the ultimate temperature. The major product is bio-oil, which is produced at a medium temperature and quite high heating rates.

2.2 Fossil Fuel Formation

About millions of years ago, some or many parts of the earth (land and oceans) were covered with plants and vegetation. In the oceans, major amounts of algae grow under the shallow waters; hence, plants grew and died over thousands of years and leaving many layers of different plants and algae materials on either land regions or the seafloor. Over time (about millions of years), these plant deposits of matter or algae were buried underground by earthquakes, volcanos, landslides, silting from flooding rivers, eroded soils, or any other natural cause or calamity (Perea-Moreno et al. 2019). After millions of years, the pressure trapped over the top of these deposited materials and heat from the Earth's core caused them to change into fossilized versions such as different forms of coal (Peat, Lignite) and oil of different kinds. Peat and coal can be found near the Earth's surface and immediately underneath the ground level (Perea-Moreno et al. 2019). From the plant materials, oil and gas materials are separated, as these materials pass through the cracks in the rock around it and get collected in pools which are very much underground in most cases. Many times, gases having less density than denser oils get deposited over the pool (Tripathi et al. 2016).

2.3 Types of Renewable Energy Resources

2.3.1 Astral Energy

Better known as sun or solar energy is among the resources that can be obtained freely after paying some equipment and setup costs (Eisenberg and Nocera 2005). The quantity of solar energy coming to the Earth's face is very much sufficient to fulfill our requirement, and according to a survey, the quantity of solar energy coming to the Earth's face in only one hour is much greater as compared to the energy required to completely run a power plant for a whole year (Srinivasnaik et al. 2015). After reading this, it seems that it is a perfect renewable energy source to fulfill all the requirements. But the amount of solar energy we use depends directly on several factors like weather, season, humidity, etc. Overall, the amount of solar energy received at Shimla is way less than that of Jaipur (Eisenberg and Nocera 2005). The important benefit of using solar energy is because of its abundant availability. Rather than fossil fuels, if we try to use this alternative energy source in the mainframe, it will contribute largely to overcoming pollution caused by the overutilization of fossil fuels, hence improving environmental conditions (Eisenberg and Nocera 2005).

2.3.2 Wind Energy

Wind power is a non-destructible energy source. In coastal areas of countries, there were wind farms where windmills are fitted in larger areas, and they fulfill the need of the people and can be stored when not in use. In most of the cases, wind energy is supplied to the National Electricity Board too, which can be used for the country's development process and supplied to those areas where there is a kind of electricity problem. Not every place is suitable for wind turbines; it differs according to place to place (Arora and Chaudhry 2015). Also, wind energy is one of the energies of the clean source that does not pollute the air and environment. Also, it does not release carbon dioxide and other harmful gases that affect human health.

2.3.3 Hydro Energy

If we are talking about renewable energy, then hydro energy is very much popular for the commercial process used by people; it is also one of the best resources for providing electricity to people and to fulfill their day-to-day or daily needs. With the help of a dam or by constructing the dam, the water will flow in the desired manner, which will rotate the turbine and produce electricity that can be stored for furthermore use. And it is considered more reliable when compared to solar energy and wind energy. And it does not pollute the environment; therefore, it gets credit for being the primary environmentally friendly energy source.

2.3.4 Geothermal Energy

Geothermal energy resource is a type of energy resource in which heat is stored or trapped which was obtained from the Earth's shell (crust) during the creation of the Earth 4.5 billion years back and the radioactive materials stored in the Earth's crust. A sometimes large amount of heat comes out together in the form of volcanic eruptions (Perea-Moreno et al. 2019). Then, this heat is stored and utilized to produce geothermal energy utilizing the vapor obtained from the high-temperature water from the Earth's shell, which later can be used for the turbine to produce electricity and used commercially. Although geothermal energy is not the same as other types of renewable resources, it has a different potential for the supply of energy when compared to other sources (Bugaje 2006).

2.3.5 Ocean Energy

This type of energy can possess multiple types of sub-energy units: one is thermic, and another is monotonous. Ocean thermic energy majorly depends on heated water collected with surface temperatures to obtain energy from a variety of sources (Bugaje 2006). Ocean monotonous energy plays on the flow of tides to induce electricity, and tides are created by a joint operation of the Earth's rotation and the gravitational force of the moon. Ocean energy is calculable regarding the amount of energy going to be produced as compared to other energy forms. The most populated are mostly located near the ocean or harbor which makes it easier to use the energy produced by the wave for their local population.

2.4 Environment Impacts on Renewable Energy

Renewable energy projects play a vital role in enhancing environmental appearances, such as reductions in carbon dioxide and other harmful (inflammable, toxic to inhale, hard for lungs to digest) fumes, reviving society about the ambiance change in the area (Panwar et al. 2011). Recent research revealed that there was very little of an effect on local residents, tourism, energy delivery costs, or educational outcomes (Arora & Chaudhary et al. 2015). But, on the other hand, changes were seen in the improvement of living standards, social bond creation, and community evolvement (Strachan et al. 2015).

2.4.1 Laws for Renewable Energy

Energy services are needed by all societies to meet fundamental human requirements and to serve gainful processes. For society's development, energy services must be delivered with minimal environmental impact. Pyrolysis produces mostly biochar at low temperatures, less than 450 °C or lower, when the warming rate is slow, and primary gases at high temperatures, higher than 800 °C, when the heating rate is quick, based on the thermal conditions and the ultimate temperature. The major product is bio-oil, which is produced at a medium temperature and has quite high heating rates (Strachan et al. 2015). Future energy demand and supply are predicted to grow at a compound annual growth rate (CAGR) of 7% over the 12th and 13th plan periods (Bugaje 2006). Relentless on the occupation-as-usual for the evolution of the generation of electricity from fossil fuels over long-term use had some limitations because of various factors, such as the limited availability of fossil fuel resources in nature, risks to the security of supply of imported fuel, macroeconomic constraints such as the disruptive balance of payments issues and high current account losses, the external effects of fossil production, international pressures relating to climate mitigation due to this, constraints of water availability for thermal cooling, etc. (Verbruggen et al. 2010). Reliance on fossil fuel imports exposes India to variable pricing. There is a significant fluctuation in the foreign currency rate, which is also a tremendous risk for the economy; there is a rivalry with other importers; and the source countries' internal demands. In a business-as-usual scenario, the approach employed for a cost-effective energy system would indeed be cost, which would become the sole determining factor for the supply of energy to meet these requirements and work (Verbruggen et al. 2010). The appeal of a certain or specific type of energy supply choice will be influenced by the larger economic as well as the society's and country's environmental and social factors (Sweeney et al. 2020). When the environmental and social externalities of usable power generation are computed or calculated for the pricing of fossil fuels used in power generation for the society and country's smooth functioning, RE-based power becomes competitive or even cheaper than fossil fuel-based power generation methods (Verbruggen et al.

2010). Furthermore, renewable energy sources are the only free mechanically generated sources in society that compete with the high cost of fossil fuels (Srinivasnaik et al. 2015). A portfolio of renewable energy sources has a lower risk-adjusted cost of generation or development than a fossil fuel portfolio (Verbruggen et al. 2010).

2.5 Types of Biofuels

2.5.1 Ethanol

Ethanol is a renewable resource that may well be created from a variety of plant components, giving it the name biomass. Ethanol is an alcohol that can be used to boost the octane of gasoline while lowering carbon monoxide and other emissions. E10 is the most common ethanol blend that is readily available. Some vehicles, known as flexible-fuel vehicles, are built to run on E80, an option for E10 fuel that contains significantly more ethanol than conventional gasoline (Jahirul et al. 2012). Plant starches and sugars are used to make the majority of ethanol. Fermentation is the most frequent process for turning biomass into ethanol. Microorganisms digest plant carbohydrates and generate ethanol throughout this process.

2.5.2 Biodiesel

Biodiesel is a type of liquid fuel created from renewable resources such as fresh and old vegetable oils and animal fats that burns cleaner than petroleum-based diesel. In nature, biodiesel is nontoxic and biodegradable. It's manufactured by combining recycled vegetable oils, animal fat, or frying grease with alcohol (Sattanathan 2015). Biodiesel is used to power compression-ignited engines, sometimes known as diesel engines, much as petroleum-derived diesel. Biodiesels, such as B100 (pure diesel) and B20 (the most common mixture), may be mixed with petroleum diesel in any ratio (a blend containing 20% biodiesel and 80% petroleum diesel) (Sattanathan 2015).

2.5.3 Methanol

Methanol, like ethanol, is a kind of alcohol that is utilized as a clean fuel to power vehicle engines across the world, particularly racing automobiles. In chemical composition, methanol is very similar to methane; the main difference is that methane is a gas, whereas methanol is a liquid. Biomass is then easily transformed into methanol via the gasification process, which takes place at extremely high temperatures but with the help of a catalyst.

2.5.4 *Butanol*

This is a form of, or must we say another class of, alcohol that is used as a biofuel. Butanol is a liquid that is created during the fermentation process and has a greater energy per unit content than ethanol and methanol. It has a chemical composition and performance similar to gasoline, but it is very hard to manufacture. It is generated mostly from plants that produce high-energy grains, such as wheat and sorghum. It may be injected directly into gasoline engines without modification due to its high energy content and longer hydrogen link.

2.6 The Process of Pyrolysis

Biomass was just the first form of energy utilized by mankind, and many impoverished countries still use it today. Earlier, charcoal which was generally obtained from biomass was firmly used for heating purposes (Thewes et al. 2012). But this process had several disadvantages such as excessive pollution and slow and low energy yield (Jahirul et al. 2012). To overcome such issues, several reformations have been made in the processes, and in the end nowadays mainly, three processes were used, combustion, gasification, and pyrolysis (Balat 2008).

2.6.1 *Combustion*

It is the total oxidation of biomass to convert it to heat (Thewes et al. 2012). Alas, the efficiency of this process is just 10%, and it produces a significant quantity of pollutants (Malik et al. 2014).

2.6.2 *Gasification*

It is a partially oxidizing process that turns solid biomass fuel straight into gaseous form (Malik et al. 2014). Pyrolysis is the result of combining the two processes previously discussed. When organic matter is heated in a nonreactive environment, it is an extremely complicated process that includes simultaneous and consecutive reactions. At 350–550 °C, components begin to decompose thermally. The larger and longer chains of carbon, hydrogen, oxygen, and carbon compounds were split into smaller particles, and the gaseous form was generated (Jahirul et al. 2012).

The extent and rate of decomposition of biomass depend on the following mentioned factors:

- Pyrolysis reactor type
- Temperature
- Biomass feed rate
- Biomass heating rate
- Pressure
- Feedstock type

2.7 Classification of Pyrolysis

2.7.1 Slow Pyrolysis

This method has been used to boost char production for a long time, and it does so at a low simmer and with a slow cooking rate. This procedure takes something from 5 to 30 min to complete. Because of the length of time, the components in the vapor phase react with one another, resulting in solid char and various liquid by-products (Balat 2008). However, it has several negative implications, including low-quality and low-yield biofuel production. Furthermore, due to the extended residence time and restricted heat transmission, larger energy input is required. As a result, this method of pyrolysis is no longer used (Jahirul et al. 2012).

2.7.2 Fast Pyrolysis

In this process, the biomass components were quickly fried at a high temperature in the absence of oxygen. This process produces 60–75% oily products (often a mixture of oil and other liquid products), 15–25% solids (biochar), and 10–20% gaseous phase. The liquids were derived from biomass in general at low temperatures, with a high heating rate and a short residence period. Instantaneous cooling of aerosols and other vapors leads to high bio-oil yield. This technology is very much famous for producing liquid fuels and various other specialty chemicals. Liquid products are easy to store and transported. This process required relatively low investment for industrial setup as compared to the other technologies. Also, it provides more energy output, especially on a small scale (Jahirul et al. 2012).

2.7.3 *Flash Pyrolysis*

Solid, liquid, and gaseous fuel from biomass components is the end product of this process, with a yield percentage of 74%. Quick devolatilization in an absence of air, a rapid heating rate, an increasing reaction temperature between 451 and 1001 °C, and a shorter gas contact time are all features of this process. However, it has several drawbacks, such as lower heat stability, soil corrosivity, and increased viscosity due to char's catalytic effect (Jahirul et al. 2012).

2.8 Pyrolysis Reactor Type

A reactor is a central component of any reaction carried out. Only one chamber was used for the majority of the critical operations and conversions. Reactors have recently become one of the most popular study subjects in the chemical and energy industries. To boost the bio-oil yield, pyrolysis reactor designers previously used strategies such as tiny biomass particle size (around 1 mm) and short residence time. However, the following scientific studies have shown that this strategy produces distinct results (Balat 2008). The composition of bio-oil is only affected by particle size and residence period. Over time, many reactor models have been created to optimize performance characteristics and produce the upgraded quality of bio-oil as well as rigorous research and studies to push pyrolysis technology to new heights. The following are the most common and widely utilized reactor designs among the various reactor designs (Jahirul et al. 2012).

2.9 Different Reactor Designs

2.9.1 *Fixed Bed Reactor*

This pyrolysis system includes a reactor with a gas cleaning and cooling system. This reactor's prototype is especially effective for fuels with a low small particle percentage and a relatively homogeneous size distribution. In this type of reactor, solid particles sink through a vertical shaft and come into contact with a counter-current rising product gas stream (Aysu 2015). A conventional fixed bed reactor is made up of firebricks, concrete with a fuel-feeding device, steel, an ash remover unit, and a gas outlet (Aysu 2015). When there is a high level of carbon conservation, a long residence time, and a low gas velocity, these beds work. A majority of these beds were used in small-scale power applications. The removal of tar is one of the reactor's major drawbacks, which is why it isn't widely used in large-scale companies (Jahirul et al. 2012).

2.9.2 *Fluidized Bed Reactor*

It is made up of a solid-liquid combination with solid characteristics. This is easily accomplished by applying pressure to the fluid via solid particle matter. These reactors are commonly employed for fast pyrolysis because they allow for quick heat transfer, as well as good thermal transport within the system, strong solid-liquid contact per unit volume, and a high relative velocity between the solid and liquid phases (Jahirul et al. 2012). Fluidized bed reactors are of two types. They bubble fluidized bed reactors and circulating fluidized bed reactors (Aysu 2015).

2.9.3 *Ablative Reactor*

Ablative pyrolysis differs from liquid bed procedures in that the heat transmission system in the reactor heat exceeds the melted layer and there is no exhaust gas. The biomass is pressed into the heated reactor's wall using mechanical pressure (Balat 2008). The substance that comes into contact with the wall "dissolves," and the leftover oil dissipates like pyrolysis vapor as it is removed. Ablative reactors have the advantage of not requiring substantial feed grinding and allowing for larger biomass particle sizes compared to the other types of pyrolysis reactors. Those reactors could use crystallite sizes up to 20 mm, as opposed to the 2 mm required for water-filling bed systems. However, because of the nature of the mechanical process, this setup is a little more complicated (Aysu 2015). Because this system operates in a controlled environment, measurement is a function of the heat transmission line. As a result, unlike other reactor designs, ablative reactors may not profit from economies of scale. Ablative vortex and ablative rotating disk reactors are the two most popular forms of ablative reactors, as explained in the following sections (Jahirul et al. 2012).

2.9.4 *Vortex Reactor*

In a vortex reactor, biomass particles are pyrolyzed in a flow of hot inert gas (vapor or N_2) and incorporated into the reactor tube. The biomass particles were driven along the reactor's wall at a rapid rate due to the high centrifugal force. The particles are kept at a temperature of around 625 °C before being planted on the heated wall of the reactor, resulting in fluid film biology. A specific circuit of a solid is used to process the particles that have not responded. Couples produced on the reactor's wall are promptly eliminated by gas in the range of 50–100 ms. This design fits the requirements for rapid pyrolysis and has yielded a 65% bio-oil output (Jahirul et al. 2012).

2.9.5 Rotary Disk Reactor

The rotary disk reactor of the rotating plate forcibly slides the supply stock of the biomass on a hot rotation disk. The pressure and heat exchanger soften and come into contact with the rotating disk on the high-temperature surface of the biomass, causing the pyrolysis reaction. The fundamental advantage of this reactor is that it does not require an inactive gas environment, reducing the size of the technical equipment. However, because this method is based on the surface area, scaling larger objects may be tricky (Jahirul et al. 2012).

2.9.6 Vacuum Pyrolysis Reactor

This reactor performs a slower pyrolysis process at a lower heat transfer rate to reduce the yield of 35–50% compared to 75% by weight, and 75% by weight of the fluidized bed compared to 35–50%. In a vacuum reactor, the pyrolysis process is mechanically complicated and costly to set up and operate. The biomass is transported into a high-temperature vacuum chamber through a moving metal belt. A mechanical stirrer is used to mix the biomass on the belt at regular intervals. An induction heater with molten salt as a heater and a burner is used to heat the biomass. Because pyrolysis reactors function in a vacuum, unique equipment for feeding and discharging solid particles is always required to maintain good air tightness. Vacuum reactors have the advantage of being able to handle larger biomass particles than fluid bed reactors (up to 2–5 cm) (Jahirul et al. 2012).

2.9.7 Rotary Cone Type Reactor

The most efficient technique to give heating to petitioned biomass is to use a tight combination of biomass and high-temperature inert particles. Mixing the fluidized bed, on the other hand, mandates the use of wasteful inert gas. The reaction is allowed to perform during the mechanical mixing of biomass and hot sand in a rotating conical reactor, and there is no inert gas used. The centrifugal force moves to the lips of the conical lips as business and sand biomass are introduced from the base of the conical basin during rotation. The pair of pairs will be directed to the condenser when a solid is different from the ribs of the cone. The coal and sand are delivered to the combustion chamber, where the sand is reheated and returned to the bottom of the furnace. The coal and sand are delivered to the combustion chamber, where the sand is reheated and replaced with fresh feed biomass at the bottom of the cone. Despite the rotating conical reactor's complicated construction, it produces excellent quantities of bio-oil (Jahirul et al. 2012).

2.9.8 *Plasma Pyrolysis Reactor*

A cylindrical quartz tube is normally flanked by two copper electrodes. An uneven-speed screw feeder placed at the tube's peak feeds the biomass particles into the tube's center. To provide thermal energy for the gas flow via the tube, the electrodes are coupled to an electrical source. The inert gas contained in the reactor removes oxygen. This inert gas is also employed in plasma manufacturing as a working gas. A variable-speed vacuum pump is used to remove the pair of pairs pairing from the reactor (Jahirul et al. 2012). Only the plasma reactor, which consumes a lot of electricity and has a high running cost, gives a unique benefit from biomass pyrolysis as compared to other reactors. The elevated energy density and temperature obtained through plasma pyrolysis correspond to a rapid response, which could be a solution to problems resulting from sluggish pyrolyzes, such as heavy node connection and the occurrence of low-performance synthesis gas. The influence of a highly active plasma environment with diverse molecules of varying electrons, ionicness, atoms, and activated molecules is used to remove the resin production in this type of reactor. However, the thermal plasma emits a considerable proportion of its energy to radiation and a conductive environment (Jahirul et al. 2012).

2.9.9 *Microwave Reactor*

The most recent examination in the area of pyrolysis is the microwave response, which shows that energy is delivered through the interaction of molecules or atoms using microwave heads. The oven microwave company receives biomass that has been dried and pyrolyzed. The inert gas circulates continuously through the reactor, generating a glass oxygen environment and serving as a backup gas (Balat 2008). Microwave atoms offer certain advantages over the system's slow pyrolysis in terms of restoring valuable compounds in biomass. These benefits include improved heat transfer, exponential management, and chemical reactivity of the heating process, as well as a reduction in the creation of undesirable species. In addition, at microwave reactors, unexpected physical phenomena such as "hot spots" appear to increase the yield of syn-gas (Jahirul et al. 2012). Therefore, various biomass and industrial wastes can be processed in a microwave reactor with desired products such as syn-gas and bio-oil in high yield. Biomass pyrolyzed using a microwave reactor is sewage sludge, coffee hulls, glycerin), rice straw, corn stalks, waste automobile oil, woodblock, and sawdust (Balat 2008).

2.9.10 *Solar Reactor*

Pyrolysis using solar reactors is a convenient approach to storing solar energy as chemical energy. Quartz tubes with opaque outside walls are used in these reactors, which are subjected to concentrated solar radiation. The sun radiation is focused using parabolic solar concentrators, which are mounted to the reactor. High temperatures (>700 °C) can be generated in pyrolysis reactors by concentrated solar radiation. Solar reactors, on the other hand, have significant advantages over sluggish reactors. A portion of the raw material is used to generate process heat in slow pyrolysis. As a result, it reduces the amount of usable raw materials while also causing pollution (Jahirul et al. 2012). As a result, pyrolysis using solar energy maximizes the amount of usable raw material and eliminates the build-up problem. Solar reactors can also start and stop more quickly than sluggish reactors.

2.10 Process Description

For effective heat transmission and efficient functioning, all pyrolysis reactors have size constraints on feed materials. The particle size requirement for a fluidized bed pyrolysis reactor, for example, is generally 2–6 mm. As a result, biomass should be processed by operational and crushing processes. When there is no dry stuff, such as straw, the biomass material must be dried with a moisture level of 10% by weight or even less (Rashid et al. 2015). Drying is essential to avoid the negative effects of flooding on pyrolysis aspects of sustainability, pH, and other properties. The output of fluid is increased by polishing and drying raw materials, but this also raises manufacturing expenses (Rashid et al. 2015). The biomass is dehydrated and crushed before being fed into the reactor for pyrolysis. Since these droplets generated in the reactor act as a catalyst for steam cracking, charcoal cleaning cyclones are utilized to recover charcoal from the reactor directly after pyrolysis. And on the other hand, little ball particles are generally transported via cyclones and mingled with fluid items (Rashid et al. 2015). After splitting the solid, the pair and gas should always be quickly shut off to avoid further shattering of the organic molecules (char pyrolysis liquid condensers, which cool the vapors directly with bio-oil or hydrocarbon liquids, are the most prevalent approach for vapor reduction (Jahirul et al. 2012).

2.10.1 *Feed Preparation*

The pyrolysis process and calorie value have a detrimental effect on the high amount of biomass. If the raw material's humidity level is too high, you'll produce nanoparticles with a lot of wetness, which will lower the calorie count. In theory, the proper pyrolysis process necessitates a moisture level of less than 5% by weight. As a result, before pyrolysis, biomass must undergo a thorough drying process to minimize humidity. The drying process, on the other hand, is likely to cause a thermal oxidation reaction, leading to an inadequate condensation mechanism and superior heat of the biomass compound. Temperature and drying time affect the features, chemical content, and yield of pyrolysis products. Woody biomass dehydrates at temperatures of 199 °C and 241 °C for 46 and 89 min, respectively. According to their findings, increasing the temperature range influences the yield of pyrolysis products, whereas increasing the drying time has no effect. It improves the efficacy of bio-oil by reducing the volume of water and acid in it through a drying process (Rashid et al. 2015). After that, it is crushed to a particle size of 2–6 mm, resulting in tight granules that react swiftly in the pyrolysis reactor.

2.10.2 Catalysts Selection

Catalysts accelerate a chemical reaction without being consumed or modified in the process. It is commonly utilized in biomass pyrolysis processes and plays an essential role (Wei et al. 2019). Catalysts are most commonly employed to speed up the pyrolysis process by splitting down large molecules into smaller hydrocarbon products (Balat 2008). Various accelerators, on the other hand, yield different product distributions under different operating conditions. Pyrolysis catalysts are divided into three categories based on their use. The first group is added to the biomass before it is fed into the reactor. In the reactor, a second group is added. As a result, steam interaction is permitted. Resinous and hard (Wei et al. 2019), the third group is housed in the second reactor, which is adjacent to the pyrolysis reactor. However, Khan and Kim divided the catalysts into four groups according to their composition. These include dolomite catalysts, nickel-based catalysts, alkali metal-based catalysts, and novel metal-based catalysts. The dolomite catalyst, you will greatly reduce the formation of the resin in a rich product because it attracts a lot of attention. As a result, stone dolomite is widely scarce in different reactors such as fixed beds and fluidized bed reactors. He and others used a fixed reactor for the pyrolysis of municipal solid waste using dolomite that was calcined as a catalyst for the feed laboratory (Balat 2008). In this study, the presence of a fired white body showed that oil product yields reduced oil and quality yields, resulting in a significant impact on product yield through significant production increases. However, the fired white catalyst is difficult to access or exceed a slight limit, including low melting points, and is inefficient in heavy cracks and to achieve or exceed the transition of 90–95% (Wei et al. 2019). However, to increase productivity, we studied several different types of catalysts in biomass pyrolysis. Some of them are Ni, CeO_2, Al_2O_3, aluminum oxide aluminum, sodium inspection, CeO_2, RH, SiO_2, Li, Na, K carbonate, Na_2CO_3, K_2CO_3, $ZnCl_2$, Ni/SiO_2N, zeolite, and ZrO_2 (Wei et al. 2019). These Ni and alkali metal catalysts are effective for the removal of severe resins and achieving more than 99% of the destruction of the resin. But it was found to be disabled by carbon deposition. Most of them have been used only to improve gas production for research purposes (Wei et al. 2019).

2.10.3 *Biomass Heating*

One of the most essential concerns is biomass heating or heat transmission in pyrolysis reactors. Heat transmission to the reactor's heat transfer medium (solids and gases in a fluidized bed reactor or the wall in a molten reactor) and heat transfer medium into pyrolysis biomass are two critical needs for pyrolysis reactors. These heat transfer techniques can be solid or solid in gas form, with heat transferred by convection from a hot gas to pyrolysis biomass particles, or solid with heat transferred by conduction. In fluidized bed reactors, conduction contributes to around 90% of heat transfer, with convection accounting for up to 10% of heat transfer due to the use of an appropriate solid combination. In addition to convection and conduction, radiant heat transfer occurs in all types of reactors. Various heating techniques are utilized in different pyrolysis reactors to accomplish the efficient conversion of biomass to liquid fuel (Valle et al. 2019). A significant feature of heat transport is biomass' low thermal conductivity, which is strongly reliant on temperature. The biomass particle size must be very small to meet the needs of fast heating to produce high bio-oil output, which is dependent on gas-solid heat transfer (Jahirul et al. 2012). Char is a medium solid residual water formed in the reactor during the pyrolysis process. The symbol is separated from the vapor quickly and efficiently, which is essential since it leads to the synthesis of polycyclic aromatic hydrocarbons (PAH) in the pyrolysis process, especially at low temperatures. The symbol is separated from the reactor using a cyclone approach in this scenario. However, this process is limited to the passage of small particles via a cyclone, which is collected from the liquid product, speeding up aging and producing instability (Valle et al. 2019). Although many other approaches have been used to overcome this limitation, this method, which includes the distribution of filtering, filtration, and rotating particles in the vapor, is also complicated, involving an interaction between a helicopter and a helicopter that forms a cologne phase that quickly blocks due to filter difficulty. An attempt is made to resolve this problem by changing the liquid microstructure using a solvent such as methanol or ethanol. However, this causes the liquid product to be diluted with the solvent, increasing the process cost. As a result, research into effective mechanisms for separating coal in the pyrolysis process continues.

2.10.4 *Liquid Collection*

Gaseous pyrolysis products are typically available in real vapors, aerosols, and non-condensable gases. Aerosols require coalescence or agglomeration, while these require quick cooling to limit secondary reactions and condense genuine vapors (Valle et al. 2019). Due to the thick dirt on the recuperators, heat recovery at high temperatures is not recommended. Cooling with a simple heat exchanger, quenching with product oils or immiscible hydrocarbon solvents, or employing traditional aerosol separators like detectors and electrostatic precipitators are all common methods for this purpose. Except for electrostatic precipitators, which are probably the preferred technology for fluid separation in the pilot plant, none of these are highly effective (Valle et al. 2019).

2.10.5 Products

Charcoal, perpetual gas, and coupled condensed into dark brown sticky substance fluid at room temperature are the three primary outputs of biomass pyrolysis. Temperatures between 350 °C and 500 °C produce the most fluid (de Paula Gomes and de Araújo 2009). This is because the pyrolysis process produces a wide range of reactions at varying degrees. As a consequence, the particles and leftover particulates in the solution are degraded at extreme temperatures, resulting in a reduced particle that boosts the gas proportion. The following steps can be taken to maximize the productivity of the biomass pyrolysis product: (1) charcoal (low temperature, low heating speed), (2) liquid product (low temperature, high combustion speed, brief time), and (3) fuel gas (high temperature, low heating speed, long gas accommodation). The relative humidity of the biomass appears to have a significant impact on the pyrolysis products, resulting in massive amounts of liquid water gaseous hydrocarbons (Valle et al. 2019). This makes it easier to remove water-soluble chemicals from gaseous and resinous phases, resulting in less gaseous and solid waste (Vasudevan et al. 2005).

Pyrolysis bio-oil is a liquid that results from the condensation of pyrolysis vapor. It will almost certainly be used to replace gasoline. The warming value of bio-oil is between 40% and 50% of that of hydrocarbon fuels. Bio-retirium or biologist are frequently mentioned in pyrolysis oil, it mentions biologists. It consists of complex mixtures of oxygen compounds (de Paula Gomes and de Araújo 2009). Carbonyl is one of the chemical functionalities found in bio-oil. Carboxyl and phenolic compounds have both promise and drawbacks in their utilization. However, pyrolysis biology's thermophysical characteristics are influenced by several unknown elements (Valle et al. 2019).

2.11 Social Importance of Producing Energy from Biomass

Since the time of cavemen, when they used firewood and burnt it to maintain their feasts warm, consumers have been utilizing biomass energy, or energy produced from living things. Biomass is now utilized to fuel electric generators and other devices (Srinivasnaik et al. 2015). Biomass is biological, meaning it is composed of living organisms such as plants and animals (Walker et al. 2010). The three most common biomass resources used for fuel are plants, timber, and garbage. Fuel sources are what they're called. Biomass energy can also be considered a nonrenewable source of energy. Plants, for example, use photosynthesis to convert carbon dioxide and water into nutrients, which are commonly referred to as carbohydrates. Biomass may be utilized for fuel after being burnt through a technique known as thermal conversion (Demirbas and Balat 2006). Even during the thermal upgrading process, the biomass feedstock is roasted to combust, dehydrate, and set. The most common form of biomass feedstock that may be used for thermal decomposition is raw materials such as sewage sludge and leftovers from paper or wood mills (Walker et al. 2010). Biomass is an essential component of the Earth's carbon cycle. The carbon cycle is defined as the transfer of carbon between the atmosphere, hydrosphere, biosphere, and lithosphere, as well as all other interiors of the globe. The carbon cycle can take many different shapes. Carbon also helps to modulate the amount of solar energy that enters the Earth's climate. When a plant employs photosynthesis to deliver carbon-based nourishment into the environment, the carbon absorbed by the ground as soon as an organism decomposes, for example, may be reused. The decomposing organisms can change into peat, coal, or petroleum under the correct or perfect parameters, which can subsequently be extracted by nature or people. During transfer periods, chemical energy or carbon is stored. The carbon in fossil fuels has been sequestered for long periods. When fossil fuels are harvested and used for energy, the carbon contained in them is released into the atmosphere. Fossil fuels do not reabsorb carbon. Biomass, unlike fossil fuels, is made from living organisms (Walker et al. 2010). The carbon in biomass may be exchanged forever throughout the carbon cycle. Humans have relied on biomass energy from the beginning of time, which may be obtained from living creatures that have been burnt with wood to keep the food warm. Biomass is being utilized in power generators and other machines (Dagar and Shah 2013). Organic stuff generated from living objects or organisms such as plants and animals is referred to as biomass. Plants, wood, and garbage are the most frequent biomass resources utilized for energy generation. Raw commodities are what they're called. Biomass energy may be thought of as a one-of-a-kind source of energy. Biomass contains energy that is derived initially from the solar (Srinivasnaik et al. 2015). For example, photosynthesis is the process through which plants receive the sun's energy and transform carbon dioxide and water into nutrients or simple carbohydrates. The energy generated biologically or from those creatures can be transformed and also used actively or passively in the proper fashion via explicit or implicit techniques or procedures (Surriya et al. 2015). Biomass can be torched to provide warmth, converted directly into electrical energy, or seen as an indirect biofuel. Thermal conversion can be used to convert biomass into fuel and energy. Thermal conversion involves heating a biomass feedstock for combustion, dehydration, and stabilization. Heating a biomass feedstock for combustion, dehydration, and stabilization is known as thermal conversion (Demirbas and Balat 2006). Sludge and by-products from pulp mills and sawmills seem to be the most frequent type of biomass feedstock that can be used for thermal conversion. Biomass is an essential component of the carbon cycle on the planet. The carbon cycle is a sequence of carbon being exchanged across all of the Earth's layers, including the atmosphere, hydrosphere, biosphere, and lithosphere. The carbon cycle can take many different shapes. Carbon also aids in the regulation of the quantity of sunlight entering the Earth's atmosphere. Photosynthesis, degradation, respiration, and human activity all contribute to the exchange. For example, when plants release carbon-based nutrients into the biosphere through photosynthesis, the carbon acquired by the soil as organisms decompose may be recycled. Decomposing organisms can be transformed into peat, coal, or oil under the right conditions before being restored by nature or human intervention. Carbon is kept around between transfers.

2.12 Policies and Laws Regarding Biomass Production

The rising attention on bioenergy policy and legislation in communities and nations is due to some issues (Amann et al. 2011). These include the political importance of weather change's lengthy impacts and their linked influence or implications on world energy consumption and the global financial system (Da Pillay and Da Silva 2009). Another aspect is the agriculture sector problem, which has been characterized by massive supplies of crop production and dwindling worldwide market prospects, particularly in industrialized nations (Stąsiek and Szkodo 2020). It's hardly astonishing, therefore, that biofuels have already been touted as a way to save the World Trade Organization's (WTO) Doha Round of farm contract talks. Many nations are currently seeking new and innovative strategies for disseminating and boosting the bioenergy industry as a result of political, sociological, financial, and environmental challenges (Surriya et al. 2015).

2.12.1 The International Context

There are various national policies that have been built and followed by the country (especially their legislative initiatives) to encourage or promote bioenergy production as their demands are enormously high in the international markets (Pillay and Da Silva 2009). More than 30 nations have already implemented or are in the process of implementing biofuel regulations, including gasoline-ethanol schemes (Jebaraj and Iniyan 2006). In comparison to the ethanol business, the international market for biodiesel is significantly younger, with European nations leading the way in biodiesel synthesis (Pillay and Da Silva 2009). With considerable initiatives now being made in the biodiesel business in Australia, Brazil, India, Indonesia, Malaysia, and the United States, these nations are expected to become big suppliers of these biofuels (Perea-Moreno et al. 2019).

2.12.2 Climate Change Mitigation

Climate change has significant implications or consequences for poor countries (Das et al. 2012). The weakest sections of the globe are most prone to natural disasters, droughts, and epidemics; thus, they are expected to face the brunt of global warming's impacts. In many areas of the planet, crop output is expected to be challenged by erratic weather conditions and the unknown consequences or challenges of global warming, but the hazards are greatest in poor countries that rely on small-scale farming (Surriya et al. 2015). As per current forecasts, food yields in sub-Saharan Africa might drop by 20%, and global warming-related famine could force more than 250 million people to relocate by 2050 (Pillay and Da Silva 2009). These forecasts have prompted new initiatives around the planet to impose or coerce the implementation of climate policy (Das et al. 2012). They are indeed driving legislation and other legislative efforts to shift away from fossil fuels and toward carbon-free bioenergy sources (Stąsiek and Szkodo 2020).

2.12.3 International Trade

Biomass production for international markets poses several obstacles for developing countries (Jebaraj and Iniyan 2006). Tariff changes and manufacturing quality requirements can have an impact on these emerging countries' revenue and dignity in global biomass overseas markets (Shahbaz et al. 2020). The development of the framework was only at raising or enhancing production in developed nations, as well as guarded deal measures designed at limiting market access, which may limit potential trade prospects (Stąsiek and Szkodo 2020). Therefore, tariff barriers such as the 6.5% ad valorem tax on biodiesel imports into the European Union and the 54 cents/gallon tariff on most imported ethanol to the United States are like barriers when dealing with others. Attractive developing countries are one of the most important consumer markets for bioenergy (Perea-Moreno et al. 2019). In recent years, agreements have been developed with many priority EU-US trade promotion initiatives to address these concerns and global demand for biomass benefits in developing countries (Surriya et al. 2015). Numerous advantages can be obtained from biomass (Shahbaz et al. 2020).

2.12.4 National Policies and Legislation

It is also one of the most important as well as burning topics for government officials to understand which type of regulatory tools or steps or necessary methods need to be taken by the national governments and their agencies to improvise the popularity of bioenergy worldwide (Stąsiek and Szkodo 2020). Government officials enacted policies and legislation aimed at encouraging and publicizing private investment in bioenergy industries as well as providing appropriate or legitimate financial advice and approaches to both companies and countries from national, bilateral, and multilateral sources for capital-intensive bioenergy projects. Before the end of the 1990s, Brazil became the first country to establish Proalcool, a substantial government-sponsored ethanol initiative (Shahbaz et al. 2020). It was created to boost sugarcane ethanol output to meet expanding energy demands in transportation fuels at a period when global energy commodity prices were sky-high (Perea-Moreno et al. 2019). Hence, one of the chief advantages of this program was to, therefore, to reduce the national energy bill and minimize the overall dependency of the country on energy over fossil fuels (Shahbaz et al. 2020). It also increased hard currency revenues and fostered energy independence. Since then, Brazil has been promoting the use of ethanol and has become a major ethanol producer and exporter for different countries.

2.13 Conclusion

After carrying out all the experiments and research, we concluded that energy production from biomass will soon become a very much popular practice in the coming era. After processing the biomass components through drying and grinding chambers (to remove moisture and keep the particle size around 5 mm), among all the pyrolysis (combination of combustion and gasification) methods, fast pyrolysis is the most suitable process because of its minimal cost of setup and normal CO_2 balance, the high energy density of product fuel generated, and the storability and transportability of end products. The reactor is used in fluidized bed reactors because of its maximum output efficiency of by-product generation (60–75%) among all the other reactors. The final product majorly includes pyrolysis bio-oil. The advantage of liquid pyrolysis fuel is that it has a favorable CO_2 balance, can be used in both large power plants and small electrical systems, can be stored and transported, and has a higher density than biomass gasification fuel. It may also be used in existing power plants.

References

Amann E, Baer W, Coes D (eds) (2011) Energy, biofuels, and development: comparing Brazil and the United States, vol 87. Taylor & Francis, Milton Park

Arora P, Chaudhry S (2015) Dealing with climate change: concerns and options. Int J Sci Res 4: 847–853

Aysu T (2015) Catalytic pyrolysis of Alcea pallida stems in a fixed-bed reactor for the production of liquid biofuels. Bioresour Technol 191:253–262

Balat M (2008) Global trends on the processing of biofuels. Int J Green Energy 5(3):212–238

Bugaje IM (2006) Renewable energy for sustainable development in Africa: a review. Renew Sust Energ Rev 10(6):603–612
Dagar N, Shah IH (2013) Experimental investigation on two cylinder diesel engine using biodiesel and diesel as fuel with EGR technique. Int J Sci Res 2(7):333–336
Das B, Khara U, Bandyopadhyay A (2012) Main physical causes of climate change and global warming-a general overview. Int J Sci Res 1:119–132
de Paula Gomes MS, de Araújo MSM (2009) Bio-fuels production and environmental indicators. Renew Sust Energ Rev 13(8):2201–2204
Demirbas MF, Balat M (2006) Recent advances on the production and utilization trends of bio-fuels: a global perspective. Energy Convers Manag 47(15-16):2371–2381
Eisenberg R, Nocera DG (2005) Preface: overview of the forum on solar and renewable energy. Inorg Chem 44(20):6799–6801
Jahirul M, Rasul M, Chowdhury A, Ashwath N (2012) Biofuels production through biomass pyrolysis —a technological review. Energies 5(12):4952–5001
Jebaraj S, Iniyan S (2006) Renewable energy programs in India. Int J Global Energy Issues 26(3-4): 232–257
Malik D, Singh S, Thakur J, Kishore R, Kapur A, Nijhawan S (2014) Microbial fuel cell: harnessing bioenergy from Yamuna water. Int J Sci Res 3(6):1076–1081
Panwar NL, Kaushik SC, Kothari S (2011) Role of renewable energy sources in environmental protection: a review. Renew Sust Energ Rev 15(3):1513–1524
Perea-Moreno M-A, Samerón-Manzano E, Perea-Moreno A-J (2019) Biomass as renewable energy: worldwide research trends. Sustain For 11(3):863
Pillay D, Da Silva EJ (2009) Sustainable development and bioeconomic prosperity in Africa: bio-fuels and the South African gateway. Afr J Biotechnol 8:11
Rashid IM, Atiya MA, Hameed BH (2015) Production of biodiesel from waste cooking oil using CaO-eggshell waste derived heterogeneous catalyst. Int J Sci Res 6(11):94–103
Sattanathan R (2015) Production of biodiesel from castor oil with its performance and emission test. Int J Sci Res 4(1):273–279
Shahbaz M, Raghutla C, Chittedi KR, Jiao Z, Vo XV (2020) The effect of renewable energy consumption on economic growth: evidence from the renewable energy country attractive index. Energy 207:118162
Srinivasnaik M, Sudhakar TVV, Balunaik B (2015) Bio-fuels as alternative fuels for internal combustion engines. Int J Sci Res Publ 5(12):2250–3153
Stąsiek J, Szkodo M (2020) Thermochemical conversion of biomass and municipal waste into useful energy using advanced HiTAG/HiTSG technology. Energies 13(16):4218
Strachan PA, Cowell R, Ellis G, Sherry-Brennan F, Toke D (2015) Promoting community renewable energy in a corporate energy world. Sustain Dev 23(2):96–109
Surriya O, Saleem SS, Waqar K, Gul Kazi A, Öztürk M (2015) Bio-fuels: a blessing in disguise. In: Phytoremediation for green energy. Springer, Dordrecht, pp 11–54
Sweeney C, Bessa RJ, Browell J, Pinson P (2020) The future of forecasting for renewable energy. Wiley Interdiscip Rev Energy Environ 9(2):e365
Thewes M, Müther M, Brassat A, Pischinger S, Sehr A (2012) Analysis of the effect of bio-fuels on the combustion in a downsized DI SI engine. SAE Int J Fuels Lubr 5(1):274–288
Tripathi L, Mishra AK, Dubey AK, Tripathi CB, Baredar P (2016) Renewable energy: an overview on its contribution in current energy scenario of India. Renew Sust Energ Rev 60:226–233
Valle B, Remiro A, García-Gómez N, Gayubo AG, Bilbao J (2019) Recent research progress on bio-oil conversion into bio-fuels and raw chemicals: a review. J Chem Technol Biotechnol 94(3):670–689

Vasudevan P, Sharma S, Kumar A (2005) Liquid fuel from biomass: an overview. J Sci Ind Res 64: 9

Verbruggen A, Fischedick M, Moomaw W, Weir T, Nadaï A, Nilsson LJ, Sathaye J (2010) Renewable energy costs, potentials, barriers: conceptual issues. Energy Policy 38(2):850–861

Walker G, Devine-Wright P, Hunter S, High H, Evans B (2010) Trust and community: exploring the meanings, contexts and dynamics of community renewable energy. Energy Policy 38(6): 2655–2663

Wei J, Wang T, Cao X, Liu H, Tang X, Sun Y, Lin L (2019) A flexible Cu-based catalyst system for the transformation of fructose to furanyl ethers as potential bio-fuels. Appl Catal B Environ 258: 117793

Chapter 3
Volatile Organic Compounds Impacts on Environment: Biofiltration as an Effective Control Method

Rahul, Reena Saxena, Shravan Kumar, and Dan Bahadur Pal

Abstract Volatile organic compounds (VOCs) can be toxic for human beings. At the present scenario, air pollution, which has been generated by various types of human activities, has become one of the major problems all around the world, including most parts of India. Due to the hazardous effect of pollutants, the lives of millions of people have been adversely affected, and due to that, great economic damage to ecosystems and society has been found. The passage of Clean Air Act Abatement (CAAA) and Occupational Safety and Health Administration's (OSHA) regulations has enforced stringent emission standards whose compliance presents a major challenge to the chemical industry. The biofilters are most economical and effective if the contaminant concentration is less and treated gases should not contain secondary pollution. The manufacturing and designing of biofilters has tremendous potential in the coming years. According to on the results obtained for incremental reactivities and correlations of VOC concentrations with smog, various types of biofilters are proposed depending upon the required elimination capacities. Reviewing the papers, it was found that fixed entities such as the packing, microorganism cultured, and type of immobilization used have been studied less in comparison to variable parameters such as nutrient solution circulation rate, superficial velocity, pH, temperature, loading rate, concentration of VOC, flow rate, and water content. Several successful bioengineering pathways including redesigning the metabolic pathway, display of complex dioxygenases on a single microbial entity by tailoring selected enzymes, crinoids' manipulation, etc., were identified which would enable us to degrade a complex mixture of pollutants.

Rahul
Department of Paint Technology, Government Polytechnic Bindki, Fatehpur, UP, India

R. Saxena (✉)
School of Applied Science, Suresh Gyan Vihar University, Jaipur, Rajasthan, India

S. Kumar
Department of Biochemical Engineering, Harcourt Butler Technical University, Kanpur, India

D. B. Pal
Department of Chemical Engineering, Harcourt Butler Technical University, Kanpur, India

D. B. Pal, A. K. Tiwari (eds.), *Sustainable Valorization of Agriculture & Food Waste Biomass*, Clean Energy Production Technologies,
https://doi.org/10.1007/978-981-99-0526-3_3

Keywords Volatile organic compounds · Toxic · Pollutants · Biofiltration · Clean air · Control

3.1 Introduction

Without air, life cannot be possible, but when it gets polluted, it becomes a bitter enemy of life. Environmental changes due to air pollution are one of the most important concerns of today. A large quantity of volatile organic compounds (VOCs) is measured in the atmosphere, and these VOCs are produced by anthropogenic and biogenic emissions and their photochemical degradation. At the present scenario, air pollution in urban area is considered as a major effect worldwide (Burnett et al. 2018), including developing countries in South America. The rapid incensement in population has created industrialization, urbanization, consumption, traffic, and utilization of energy in many of the urban areas (González et al. 2017). Out of these VOCs, some (such as benzene, toluene, ethylbenzene, and xylene) are toxic and potentially cause harmful effects to the human health, whereas others can greatly affect the environmental conditions at different scales with the change in oxidizing capacity of the atmosphere. Both natural and anthropogenic sources are the major reason for the production of VOCs. There are various kinds of chemicals including aromatic and aliphatic hydrocarbons, ketones, alcohols, aldehydes, esters, and halogenated compounds. Urbanization and industrialization are the major causes for anthropogenic sources. Fuel combustion, industrial processes, consumer products, pesticides and insecticides, dry cleaning operations, pharmaceutical manufacturing, refrigeration, gas filling, etc., are some of the sources of VOCs. When VOCs mix with the atmospheric components, they encounter other kind of compounds, which are already present in the environment, such as nitrogen oxides (NO) found in the air, and play major role in atmospheric photochemical reactions (Carbonell 2014). Other than these, solvents such as m/p-xylene, butadiene, toluene, butene, and acetone are considered as one of the major contributors for the ozone formation (Liu and Zheng 2020); out of these, some solvents are ingredients which are used in paint industry. According to many researchers, it has been observed that the rate of reactivity for different VOCs was found to be different in the atmosphere and is responsible for the production of different quantity of ozone at a particular time (Jacob et al. 1995). Ground-level ozone has been created by the interactions of gas-phase volatile organic compounds (VOCs) and also oxides of nitrogen (NO_x) in the presence of sunlight (National Research Council (NRC) 1991). For the reduction of ozone, development of various effective controlling strategies is required to model predicatively the effects of changing VOCs and NO_x emissions on ozone (Carter 1994). Based on this model, one can not only be able to identify appropriately the effects of variations in the total amount of VOC emissions but also represent appropriately the effects of variations in the chemical content of the emissions (Carter 1995).

3.2 Harmful Effects of Tropospheric Ozone

Smog contains major component as ozone (O_3). The upper atmospheric layer of O_3 is useful for the Earth because it protects the Earth from the sun's harmful ultraviolet radiation, but at the ground-level (tropospheric) area, high concentrations of O_3 were found to be a major issue for health and environmental concern. The reaction rate of O_3 is so high that it can create health problems by damaging the lung tissue, reducing functions of the lungs, and also exposing the lungs to other irritants (Lippman 1993). According to the obtained scientific data of previous years, it has been noticed that ambient levels of O_3 not only affect people's respiratory systems, like asthmatics, but also it creates many threads for adults and children of healthy stage. When someone gets exposed to O_3 for several hours at relatively low concentrations, their lung's function has been found to be significantly reduced, and due to that, some respiratory inflammation were induced in normal, healthy people at the time of exercise (Tampion and Tampion 1987). The ozone is considered as a potential greenhouse gas, and it is not usually emitted directly but is formed through complex chemical reactions in the atmosphere. According to some previous studies, it has been reported that due to some inconsistent evidence between formaldehyde and VOCs, the risk of asthma (Hussain et al. 2019; Mitha et al. 2013) can be created. The risk of asthma during childhood stage, and interest in these exposures (e.g. particulates increases oxidative stress and inflammation in the lungs (Mir 2007)), has led to a number of prior reviews investigating elevated VOCs and childhood asthma (Kelly and Fussell 2015; Patelarou et al. 2015; Dick et al. 2014; Al-daghri et al. 2013). There are some precursor compounds like VOC and NO_x, which play a major role in the formation of O_3 with the help of sunlight. Ultraviolet radiation and temperature are the important source of these reactions, so large quantity of O_3 produced typically during the warmer days as well as warmer period of the year (Cooper et al. 1996).

Some key points about the ozone:

1. Component of the ozone (the principal component of smog is ground-level ozone):
   a. Precursor compounds and source: Obtained when some pollutants VOCs and NO_x chemically react to each other.
   b. Impact on health: Certain issues related to breathing, reduction in the lung's function, asthma, irritation in the eyes, stuffy nose, decreased resistance to colds, and many other infections may speed up aging of the lung tissue.
   c. Impact on environment: Living organisms (plants and trees) can be damaged by ozone (Fruekilde et al. 1998); visibility also can be reduced due to smog.
   d. Damage of property: Damages some industrialist material like rubber, fabrics, etc.

3.3 Environmental Regulations Dealing with Ozone Reduction by VOC Control

In 1993, within the framework of the UNECE Convention on Long Range Transboundary Atmospheric Pollution, a protocol was passed committing signatories (including 14 EU Member States) to reduce VOC emissions by 30% by 1999 as compared to 1990 (Liebscher 2000). Precursors of ozone which are listed in the European Union Air Quality Directive 2002/3/EC are NO*x* and VOCs including hydrocarbons and carbonyls (Duane et al. 2002). According to the ozone reduction of the US Environmental Protection Agency (EPA), all emissions of VOCs are considered to be harmful if they are more reactive than ethane (Chang and Rudy 1990). VOCs act both as primary and secondary pollutants. The current US ambient standard for ozone is 125 ppb, and 82 ppb is the maximum acceptable level in Canada (Jiang et al. 1997). Data provided by the member states of the European Union to the European Commission, in accordance with the provisions of Directive 92/72/EEC, indicate that during summer season, the effect of ozone level for the safety of human health (110 mg/m^3, expressed as an average value over 8 h) exceeded in all member states of the European Union (Vautard et al. 2005). The problem of tropospheric ozone formation was the driver for drafting an EU VOC Directive (Sillman and Samson 1995; Simpson 1995; Simpson et al. 1995), which came into force in March 1999 (Liebscher 2000).

The US EPA Standard 40 CFR Part 63 has established an emission limit of 10 g TOC/m^3. The German TA-Luft Standard has set an emission limit to 150 mg TOC (except methane) per cubic meter of loaded product (0.15 g TOC/m^3) (USEPA 1991; Khan and Ghoshal 2000). Occupational Safety and Health Administration (OSHA) demands even more stringent compliance to protect worker health safety. As we can see, emissions of VOCs which are reactive can be regulated according to their mass. Research has shown that the impact of VOC is dependent upon the reactivity level of the VOC compounds with respect to hydroxyl group (OH) rather than to the total amount of VOCs generation. To account for differences in reactivity level of strategies used for ozone control, the State of California has begun using a set of maximum incremental reactivity (MIR) factors developed by Carter (Carter et al. 1995). Few number of researchers (McNair et al. 1994) have demonstrated that the applicability of MIR scale would be higher for the urban core conditions, where controlling system of VOC is most effective as compared to rural area.

3.4 Relation Between VOC, NO*x*, and Ozone

There are several factors, which affect the relationship between ozone precursors and photochemical smog. It is necessary to identify the specificity of these relationships to develop an effective policy response and ozone control directive. Like several developing countries, India is still to pass regulations to control VOCs. Cities like

Mumbai, Kanpur, and Kolkata are bogged down by chronic pollution of the air. Ozone isopleth plots show the relationships between the concentration of ozone precursors and ozone. The isopleth plot depends upon the VOC/ NO*x* ratio (Hirsch et al. 1996). A high ratio means NO*x* sensitive, and a low ratio indicates VOC-sensitive chemistry according to common sense (Hanna et al. 1996). It has been observed that as the air parcel ages, the VOC/NO*x* ratio increases thereby causing a shift from VOC- to NO*x*-sensitive chemistry. This is due to NO*x* titration as NO is a major constituent of initial emissions (Azzi et al. 1995). In the morning, the characterization of pollutant ratio of VOC/NO*x* was observed lower than 10 and was equationalized with VOC-sensitive peak ozone, and ratios of VOC/NO*x* greater than 20 correspond to NO*x*-sensitive peak ozone (Sillman 1999). For reducing ozone levels, some extent parameter, which approximately reveals the relative merit of VOC and NO*x* control, has been calculated (Chang et al. 1997). From a 24-hour perspective, the total ozone accumulation is a function of the cumulative NO*x* emitted and is widely subjected to interference by meteorological factors and topography. On the other hand, immediate pollution may be affected by VOC emissions, sunlight intensity, and humidity (Lam et al. 2005). The amount of redicals which was generated during the process of photochemical evolution increases regularly with increased sunlight, VOC, and humidity; both are considered as a big reason for the switch from NO*x*-sensitive to VOC-sensitive conditions (Kleinman 1991). By expressing the smog particulation (SP) term as a function of cumulative sunlight exposure (Johnson 1983; Johnson and Quigley 1998) in the integrated empirical rate (IER) model, it was shown that the production of ozone can be considered as a two-phase process. In the first phase, a "light-limited regime" or VOC-sensitive regime is considered where the smog production rate is due to the intensity of sunlight and the reactivity as well as concentrations of the VOCs emitted into the air. The second phase is the shift to NO*x* chemistry as the plume moves downwind. By observing the isopleths, it has been found that a decrease in amount of NO*x* will be effective only if sensitive chemistry of NO*x*- predominates and may actually increase the ozone level in VOC-sensitive areas. This trend has been confirmed recently in the Po valley (Milan region) by GAMES mesoscale simulation (Gabusi and Volta 2005) and in the Pearl River Delta region (Wang et al. 2005).

3.4.1 VOC Reactivity

VOC reactivity along with total emission determines the ozone sensitivity to changes in concentration of NO*x* and VOC (Carter and Atkinson 1987; Carter and Atkinson 1989). The OZIPR program was combined with CALGRID 3D chemical mechanism (Jiang et al. 1997) to determine sensitivity values for the Canadian Lower Fraser Valley. In the relative terms, the base case ARO_2, which contains most of the isomers of xylene and trimethyl benzene emissions, contribute 41.48% of the maximum ozone at base case level in LFV. The ARO_2 group emissions are emitted from gasoline fuels. Some amount of VOCs, which are emitted into the air such as

Table 3.1 Carter's MIR factors

VOC	Carter MIR factor (ppm O_3/ppmC)	Δ VOC (ppmC)
Formaldehyde	4.47	0.08
Acetaldehyde	2.53	0.15
t-2 Butene	2.91	0.13
Ethane	2.13	0.17
Propene	2.75	0.13
p-Xylene	1.83	0.20
Toluene	0.749	0.49
Ethylbenzene	0.745	0.5
n-Octane	0.183	2
Butane	0.310	1.2
Benzene	0.115	3.2
Carbon monoxide	0.0318	12
Methane	0.005	74

methylglyoxal, have very high specific value for reactivity, while the quantity emitted in the LFV is observed very few. Therefore, no significant reduction achieved in ozone level and average sensitivities are also found very low by the process of controlling the emissions of these compounds in the LFV (Jiang et al. 1997). In this case, controlling these emissions alone will help in solving the environmental problem without taking recourse to expensive measures. The method of "propylene-equivalent carbon" was developed by (Chameides et al. 1992) in which the weighted reaction rate with OH is used to determine the contribution of an individual compound in the ozone formation. VOCs with highest mechanistic reactivity were formaldehyde, methanol, the alkenes, and the alkyl benzenes. Chamber experiments for smog (Derwent et al. 1998) were designed to measure increasing level of reactivates for a wide range of individual compounds (Kelly and Chang 1999), as shown in Table 3.1.

3.4.2 Contribution of Biogenic VOCs

The role of biogenic compounds as potential pollutants has been highlighted recently (Solberg et al. 2001). For the biogenic VOCs, some evaluations have been performed (specially isoprene), and their oxidative products at atmospheric level are used for the estimations of their potential effect for measurement of quality of air comparatively with VOCs found in anthropogenic sources (Vogel et al. 1995; Duane et al. 2002). It has been found that biogenic emission of hydrocarbons in the USA during summer season is estimated to be equal to the total emission of hydrocarbon which is emitted by manmade sources (Geron et al. 1994). The ratios of HCHO that can be attributed to isoprene oxidation vary over the year and reach values above around 50% in the summer months with the highest biogenic activity in Insubria

(Duane et al. 2002). HCHO is a very reactive VOC with one of the highest incremental reactivity and is formed in the process of photooxidation of isoprene. During summer season, OH reactivity index for isoprene, methylacrolein, and methyl vinyl ketone on monthly basis exceeds up to the maximum limit as compared to other VOCs determined in the air (Duane et al. 2002), and the incremental reactivity measured by Bowman and Seinfeld 1994, was also on the higher side. The isoprene was found with just 1.5 ppb average concentration, which represent less than 1% of total VOC, but still it accounts for 25% of total VOC reactivity (Chameides et al. 1992) for a region near Los Angeles (Winner et al. 1995). Other group of compounds whose origin is biogenic are determined in ambient air and counted as the monoterpenes. Vegetations such as conifers, aromatic scrubs, deciduous trees, and rubber plantations are not located along close proximity to Indian Metros, which also have either a continental or tropical climate. The biogenic emissions can be safely neglected in the case of India for the purpose of policy making. In major urban areas such as in the eastern USA, the predicted chemistry for peak ozone would shift from VOC sensitive to NO*x* sensitive (Kleinman et al. 1994) if replacement of the older estimation emission rate of biogenic emissions can be done by new inventories (Sillman 1999). Concentrations of ozone (O_3) have been reduced significantly over the past 40 years in California, and the reason of this decrement is adaptation of stringent emission controls, but the quantity of emitted residual ozone (O_3) continues to create harmful effect on public health (Faloona et al. 2020).

3.5 Contribution of Various Sectors in Smog Formation

The transportation sector is found to be the largest source for the emission of NO*x* and VOC and is found to be responsible for about 49.2% and 68.0% of the total emissions produced by anthropogenic sources, respectively. In the case of only NO*x*, the contributions of industrial and power sectors are about 15% and 20%, respectively (Cardellino and Chameides 1990). For the production of O_3, the contribution of the transportation sector is found to be larger, accounting for 17.8%, as compared with industry level sources and power generators, which are 6.4% and 6.7%, respectively, in the Guangdong province as analyzed with the help of STEM-2K1 and MM5 (*The NCAR/Penn State Fifth-Generation Mesoscale Model model*) (Wang et al. 2005). Emissions from on-road vehicles were found to be 56% of the total Taiwanese emissions of NO*x*, 32% of VOC, and 82% of carbon monoxide (CO) (Lucher et al. 1992). The optimization of different combinations for different objectives, the reductions of maximum MIR and VOC, and the minimum level of ozone marginal cost have been analyzed and compared. At the end, the performance of the abatement strategies across a variety of control costs and marginal cost for MIR and VOCs should be given priority in the order of their results.

When aromatics content get controlled in gasoline, fuels could have a significant impact on ozone formation as they are the ones having higher incremental relativities

Table 3.2 VOC emission sectorwise with chronological variations (in thousand tons) (National Emissions Standards for Hazardous Air Pollutants 2003)

Source/year	1970	1980	1990	1995	2000	2003
Fuel combustion elec. util	30	45	47	44	62	56
Fuel comb industrial	150	157	182	206	173	170
Fuel comb other	541	848	776	823	949	878
Chemical and allied product mfg	1341	1595	634	660	254	218
Metals processing	394	273	122	125	67	72
Petroleum and related industries	1194	1440	611	642	428	380
Other industrial processes	270	237	401	450	454	412
Solvent utilization	7174	6584	5750	6183	4831	4562
Storage and transport	1954	1975	1490	1652	1176	1178
Waste disposal and recycling	1984	758	986	1067	415	427
Highway vehicles	16,910	13,869	9388	6749	5325	4428
Off highway	1616	2192	2662	2890	2644	2572
Miscellaneous	1101	1134	1059	551	733	704
Total	34,659	31,107	24,108	22,042	17,511	16,057

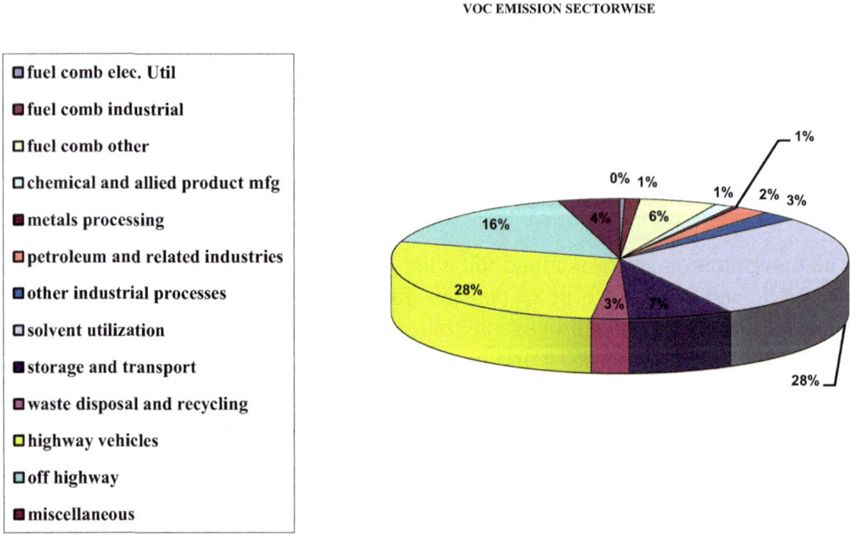

Fig. 3.1 Pie chart denoting the relative roles of various sectors of economy toward VOC emissions

(Jiang et al. 1997). The data from US EPA Report 2003 has been taken, which shows the VOC emission from the various sectors of industry, as shown in Table 3.2 and Fig. 3.1.

It is imperative that more R&D funds are granted for VOC removal in automobiles as this particular problem has failed to attract attention though its contribution to VOC emission is almost equal to that from solvent utilization industries. Though steps have been taken to keep a check on VOC emissions and treatment units have been established in most process industries, still 16,056 thousand tons of VOC waste was released in the environment in 2003 alone. Though VOC emissions have been brought down to a laudable 50% in the 1970s, but the technology used is very energy intensive, the merits of which will be discussed later in the paper.

3.6 Requirement of Monitoring and Setting Up of Air Quality Standards of VOC and NO*x* and Suggestions About Control Technology

It is strange that India does not have any VOC regulations although the NOx regulations are being strictly enforced. So far, not much attention has been paid toward monitoring of air toxics especially trace organic like BTEX, VOC, and HAPs. The Central Pollution Control Board (CPCB) (n.d.), Government of India, has initiated monitoring of BTEX and HAPs. Unfortunately, State Pollution Control Boards have not given much through of measurement of these VOCs and HAPs. There is a strict need for reviewing and setting the NO*x* and VOC standard as the technology makes it possible to achieve a higher degree of removal. VOCs not only cause primary pollution but also secondary photochemical pollution, which cannot be neglected for long. The presence of a pollution cloud over India for a long time can have severe adverse effects. The reduced sunlight might stagnate the agricultural growth which is crucial if we have to clock a GDP growth of excess of 10%. New NO*x* control technologies are being used as selective noncatalytic reduction (urea injection), selective catalytic reduction (ammonia injection) (World Bank Group 1998), Exxon Thermal DeNO (Kovacs et al. 1998), and catalytic reduction.

3.7 Future NO*x* Treatment Technologies

*3.7.1 NO*x *Reduction as an In Situ Process*

Recent research in the destruction of pollutants by photocatalysts has paved the way for a revolutionary technology. Location advantages of photocatalytic concrete blocks laid on the roads were found as they can help in tackling the pollution for the transportation sector, which causes almost 44% (EPRI 2003) of the pollution. The relative pollution caused on the Indian roads is shown by the Table 3.3.

Table 3.3 The relative pollution caused on the Indian roads

Type of cars[a]	NO*x* emission coefficient (g/km)
Passenger car	1.255
Light passenger car	0.267
General truck	5.454
Small truck	2.427
Light truck	0.524
Bus	6.158

[a] Average of Japanese cars (Source: Central Pollution Control Board (CPCB) (n.d.) (Government of India)

3.7.2 *Control of NO*x *in Stack Gases with the Help of Ozone*

The process for controlling NOx is based on the ozonation of nitrogen oxides in hot flue gases, (mostly nitrogen oxide (NO) and nitrogen dioxide (NO_2)) to N_2O_5, which is highly soluble in water, and is found to be easily removed as nitric acid. Heat exchangers are used for the reduction of the flue gas temperature below the range of 350 °F. Unlike *urea injection*, *ammonia injection,* or precombustion NO*x* mitigation techniques, the oxidation process at low temperature recovers majority of the useful heat. The process has been examined in the capacity of 400 hp gas-fired boiler and has achieved reduction of NO*x* from a range of 144 ppm to below 2 ppm (EPRI 2003).

3.8 VOC Control Technology in Relation to the Problem of Photochemical Smog

For the reduction process of photochemical smog, a necessary control on the emissions of the primary pollutants is required (Milford et al. 1989; Kumar and Mohan 2002), which are majorly NO and VOCs in the troposphere region (Logan 1989). This is found as a problem, which is not easily soluble. This reduction process requires cooperation between different organizations like government, industries, and individuals. Alternative methods for controlling industrial and automotive emissions (Chang et al. 1999), which are the main sources of these pollutants, include conservation, use of alternate fuels, and also development of new technologies. Some of the suitable methods to control VOCs and NO*x* sectorwise are discussed below:

3.8.1 *VOC Control in Transport Sector*

VOCs from the transport sector are most pernicious from the emission volumes. A concentration of 1000 ppbc of VOC is sufficient to produce ozone at the rate of

30 ppb/h. Recently source contribution estimates for Mumbai Air were determined (Srivastava et al. 2004). Clearly the evaporative emissions seem to dominate the total emissions in Mumbai, and this is very much expected in other cities of India too.

3.8.2 Dealing with Vehicular VOC Emissions

Controlling process of some contents which are related to aromatics in gasoline fuels could have played an important role on the formation of ozone (Russell 1990). With the modification of certain parameters related to fuel, like reduction of benzene content in petrol as well as reduction of VOC contents in the atmosphere, the photochemical smog problem could be mitigated. The contents of benzene found in petrol were about 3% in the year 2001 in Mumbai (Srivastava 2004). Emissions of isomers of xylene (41.5%), trimethylbenzenes (26.7%), and 10-carbon aromatics (9.2%) in the LFV from light-duty gasoline vehicles (LDGV) (54.3%), gasoline marketing (22.2%), and evaporative emissions from the LDGV (10.2%) contribute most to the ozone formation as shown by model calculations (Jiang et al. 1997).

Corona discharge technology was found as the base of products and is assumed for the reduction of CO emissions by more than 80% and NOx emissions by more than 50% (EPRI 2003). The knowledge how of the process is shown in Fig. 3.2.

3.8.3 Dealing with Evaporative VOC Emissions

Due to high volatility, higher water solubility, and high mobility of VOCs, *losses* in refueling can cause a major reason for the entry of huge quantity of hydrocarbon vapor into the environment (Lamb et al. 1985). A vehicle tank has an empty space inside it which is filled with hydrocarbon gases, and when the tank is completely filled, these gases are released out into the atmosphere. Due to this effect in addition, a huge loss can be observed from further evaporation and fuel spillage. Biofiltration is a cost-effective and versatile pollutants and odor treatment technology and has earn much acceptance in recent years for the treatment of VOCs (Leson and Winer

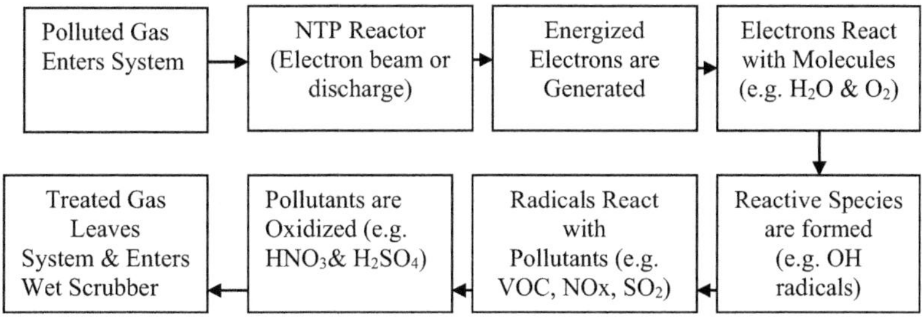

Fig. 3.2 Process of NTP technology

1991). With the help of biofilters, removal of VOCs from contaminated air streams has been observed on a large scale due to their lower cost, easy availability, operational simplicity, and lack generation of harmful by-products, although biofilter has been accepted for the control of industrial VOCs and for odor control (Fritsche and Lechner 1992).

3.8.3.1 Rooftop Biofilters

A rooftop biofilter might just be an ideal apparatus to control VOC loss from filling stations. As these biofilters have lower cost of capital and operating system, biofiltration may offer advantages economically in an application where the stream of air has VOC concentrations of about 1000 ppm, and that is exactly what happens at a petrol pump. Microbes degrading pollutants emitted from gasoline and other fuels have been listed in the Table 3.4.

Table 3.4 Some microorganisms used to degrade various compounds

Degrading strains	Compound	Reference
1. *Rhodococcus* sp. *strain DK17* 2. *Rhodococcus* sp. *strain B3* 3. *Pseudomonas stutzeri strain OX1* 4. *Pseudomonas putida (arvilla) mt-2* 5. *Azoarcus* spp. 6. *Bacillus sphaericus*	Xylenes	Kim et al. (2002) Worsey and Williams (1975) Hess et al. (1997)
1. *Azoarcus and Dechloromonas* 2. *Geobacter chapellei, Geothrix fermentans, Azoarcus evansii* 3. *Pseudomonas putidaBTE1* 4. *Bacillus sphaericus*	Benzene	Mancini et al. (2003) Attaway et al. (2001)
1. *Pseudomonas putida (ATCC 23973)* 2. *Pseudomonas picketer PKO1* 3. *Pseudomonas fluorescens* 4. *Microbacterium laevaniformans* 5. *Agrobacterium rubi* 6. *Bacillus sphaericus*	Toluene	Harding et al. (2003) Olsen et al. (1994) Seipke and Salzwedel (2002)
1. Soil microflorae not successful 2. *Pseudomonas fluorescens, Alcaligenes xylosoxidans* 3. Consortium in compost 4. *Pseudomonas putida TX1* 5. *Pseudomonas putida Idaho* 6. *Ps. aeruginosa CM323* 7. *Ps. putida CM337* 8. *Ps. aureofaciens CM332*	1,2,4-trimethylbenzene Mainly found in gasoline emissions	Serena et al. (1998) Journal of Industrial Microbiology and Biotechnology 1998; 20: 101–108 Delhomenie et al. (2003) Attaway et al. (2001) Cruden et al. (1992) Cavalca et al. (2000)
1. *Ps. oleovorans ATCC 29347* 2. *Ps. aureofaciens RWTH 529* 3. *E. coli GEC137* 4. *Acinetobacter calcoaceticus TM-31*	Decane	Vomberg and Klinner (2000) Pleshakova et al. (2001)

3.8.4 The Full Potential of Biofiltration Can Be Realized

Methods to enhance attachment

- Though inert packings can be grafted and made more conducible to hosting microorganisms, organic packings are the preferred choice for microorganisms. Microbial adsorption was compared as 248 and 2 mg/g on wood chips and inorganic silica, respectively (Gemeiner et al. 1994).

3.8.5 Selection of Suitable Microorganisms for Degradation

Selection of the proper microbial strain is the single most important parameter (Juliette et al. 1999), that affects the biodegradation rate to the greatest extent (Cohen 2001). The choice of microorganism can make a great difference to elimination capacities and removal efficiency (Hwang and Wu 2002). The predominant microorganisms treating VOCs in biofilters are heterotrophs (Kolot 1988). For influent gases which contain major inorganic constituents, the microorganisms are found to be chemoautotrophs, which use CO_2 as a carbon source, in either case as mesophiles and thermophiles predominate. The phenomenon of stratification in type and number also takes place in a normal biofilter but to a lesser extent compared to that of rotating drum biofilters. The development of higher densities was found near the influent end where the readily biodegradable substrates enter. As they reach deeper into the bed, smaller populations of various types of organisms adapted to low concentrations of more complex substrates will exist (Veiga and Kennes 2001). There are certain well-established principles which one should keep in mind in designing the microbes.

1. Incorporating different degradation pathways into a suitable microbe can make biodegradation more industrially implementable. Research has shown that degradation of certain VOCs is suppressed in the presence of certain others (Liu et al. 2005; Hwang et al. 2003). For example, a strain of hybrids which has the capability of mineralizing certain components of a benzene, toluene, and p-xylene mixture simultaneously can be constructed by rearrangement of metabolic pathway of *Pseudomonas putida 13* (Chen et al. 2005).
2. Display of complex dioxygenases or monooxygenases can endow a single-host microbial entity to degrade a set of VOCs. Active expression of soluble methane monooxygenase from *Methylosinus trichosporium* in *Pseudomonas putida* led to pollutant TCE being degraded, whereas it was not degrading previously (Shim et al. 2000).
3. Instead of scouting for new microbes, it is generally better to use already acclimatized microbes and then design them for the desired traits.
4. Unique features of rhizoremediation with phytoremediation can be combined (Chen et al. 2005) with metabolic engineering which might yield results which are still cheaper than biofiltration and practically need no energy whatsoever.

5. Enzyme engineering tools can help in tailoring selected enzymes to specific purposes. Several protein engineering software are in the market which gives appreciable results.

3.8.6 Combination of Natural Materials to Form an Ideal Packing

Organic packings serve the purpose very well, but for long time satisfactory operation, the packing has to be customized and tailored according to the VOC being degraded. A good natural packing might well be composed of peat, compost, sludge, or manure as the primary ingredient. Limestone or marl might be added to augment the buffering capacity (Zhuqiu et al. 2019). A good adsorbent like activated carbon might be added to increase adsorption of VOC, water, and nutrient solution. The adsorbent acts as a reservoir for nutrients and prevents from both starvation and shock loading. Variations in VOC supply and nutrients are easily mitigated in this way.

3.9 Conclusions

There is no need to have expensive VOC control apparatus in rural areas in the summer months for the sake of ozone prevention. Only NOx control consideration would be effective while deciding a policy framework. It has been conclusively proved that most of the rural areas appear to be majorly NOx sensitive as mentioned earlier in the paper. Clearly for formulating a policy for urban areas in India for the summer period, VOC emissions are not that important. Also the control devices can be adsorption, absorption, and condensation based. Cost-effectiveness should be affected rather than to have 100% efficiency. So biofilters are the best devices for these conditions. In the cold countries near cities, VOC chemistry predominates, so it is essential to completely remove VOCs. Technology, which would give 100% efficiency, should be preferred. Catalytic oxidation and high-temperature plasma processing might be suitable in these circumstances. Both have around 99% efficiency. Cost of high-temperature plasma processing is $100–200 per ton of waste. It removes recalcitrant VOCs, best suited when combined with a VOC accumulation technology like GAC adsorption. In summer biogenic emissions are greatly exceeded, so there should be different policies for NOx and VOC emissions in winter and summer in certain areas like the area of Insubria and Northern Italy where frequently smog formation occur. In particular, our main concern is the problem related to the production of photochemical smog which is increasing with the rapid increment in atmospheric pollutants. Therefore, it is an urgent requirement to understand and realize the basic balance between the environmental constituents and to practice science and technology in an ethical way for the protection of environment.

References

Al-daghri NM, Alokail MS, Abd-Alrahman SH, Draz HM, Yakout SM, Clerici M (2013) Polycyclic aromatic hydrocarbon exposure and pediatric asthma in children: a case-control study. Environ Health Glob Access Sci Sour 12:1

Attaway H, Gooding CH, Schmidt MG (2001) Biodegradation of BTEX vapors in a silicone membrane bioreactor system. J Ind Microbiol Biotechnol 26:316–325

Azzi M, Johnson GM, Hyde R, Young M (1995) Prediction of NO_2 and O_3 concentrations for NO, plumes photochemically reacting in urban air. Math Comput Model 21(9):39–48

Bowman FM, Seinfeld JH (1994) Fundamental basis on incremental reactivities of organics in ozone formation in VOC/NOx mixtures. Atmos Environ 28:3359–3368

Burnett R, Chen H, Szyszkowicz M, Fann N, Hubbell B, Arden C, Apte JS, Brauer M, Cohen A, Weichenthal S, Coggins J, Di Q, Brunekreef B, Frostad J, Lim SS, Kan H, Walker KD, Thurston GD, Hayes RB, Performed JVS (2018) Global estimates of mortality associated with long-term exposure to outdoor fine particulate matter contributed new reagents/analytic tools. Proc Natl Acad Sci U S A 115:38. https://doi.org/10.1073/pnas.1803222115

Carbonell JC (2014) Pinturas y barnices, Tecnologíab´asica. In: Ediciones Díaz de Santos. Pinturas y barnices: Tecnologíab´asica - Jordi CalvoCarbonell - Google Libros.

Cardellino CA, Chameides WL (1990) Natural hydrocarbons, urbanization, and urban ozone. J Geophys Res 95(13):971–979

Carter WPL (1994) Development of ozone reactivity scales for volatile organic compounds. J Air Waste Manage Assoc 44:881–899

Carter WPL (1995) Computer modeling of environmental chamber studies of maximum incremental reactivities of volatile organic compounds. Atmos Environ 29(18):2513

Carter WPL, Atkinson R (1987) An experimental study of incremental hydrocarbon reactivity. Environ Sci Technol 21:670–679

Carter WPL, Atkinson R (1989) A Atmospheric reactions of modeling study of incremental hydrocarbon reactivity. Environ Sci Technol 23:864–880

Carter WPL, Pierce JA, Luo D, Malkina IL (1995) Environmental chamber study of maximum incremental reactivities of volatile organic compounds. Atmos Environ 29(18):2499–2511

Cavalca L, Confalonieri A, Larcher S, Andreoni V (2000) Evolution of a degradative bacterial consortium during the enrichment of naphtha solvent. J Appl Microbiol 88:1009–1018

Central Pollution Control Board (CPCB) (n.d.) Government of India

Chameides WL, Fehsenfeld F, Rodgers MO, Cardellino C, Martinez J, Parrish D, Lonneman W, Lawson DR, Rasmussen RA, Zimmerman P, Greenberg J, Middleton P, Wang T (1992) Ozone precursor relationships in the ambient atmosphere. J Geophys Res 97:6037–6056

Chang TY, Rudy SJ (1990) Ozone-forming potential of organic emissions from alternative-fueled vehicles. Atmos Environ 24A:2421–2430

Chang TY, Chock DP, Nance BI, Winkler SL (1997) A photochemical extent parameter to aid ozone air quality management. Atmos Environ 31(17):2787–2794

Chang TY, Nance BI, Kelly NA (1999) Atmospheric modeling smog chamber measurements of incremental reactivities of volatile organic compounds. Environment 33:4695–4708

Chen W, Mulchandani A, Deshusses MA (2005) Environmental biotechnology: challenges and opportunities for chemical engineers. AICHE J 51:3

Cohen Y (2001) Biofiltration of the treatment of fluids by microorganisms immobilized into the biofilter bedding material: a review. Bioresour Technol 77:257–274

Cooper DA, Peterson K, Simpson D (1996) Hydrocarbon, PAH and PCB emissions from ferries: a case study in the Skagerak-Kattegat-Oresund region. Atmos Environ 30(14):2463–2473

Cruden DL, Wolfram JH, Rogers RD, Gibson DT (1992) Physiological properties of a *Pseudomonas* strain which grows with *p*-xylene in a two-phase (organic-aqueous) medium. Appl Environ Microbiol 58(9):2723–2729

Delhomenie MC, Bibeau L, Gendron J, Brzezinski R, Heitz M (2003) Degradation of toluene, xylene, and trimethylbenzene vapors by biofiltration, a comparison. J Air Waste Manage Assoc 53:217–226

Derwent RG, Jenkin ME, Saunders SM, Pilling MJ (1998) Photochemical ozone creation potentials for organic compounds in northwestern Europe calculated with a master chemical mechanism. Atmos Environ 32:2429–2441

Dick S, Friend A, Dynes K, Alkandari F, Doust E, Cowie H, Ayres JG, Turner SW (2014) A systematic review of associations between environmental exposures and development of asthma in children aged up to 9 years. BMJ Open 4:e006554

Duane M, Poma B, Rembges D, Astorga C, Larsen BR (2002) Isoprene and its degradation products as strong ozone precursors in Insubria. Atmos Environ 36:3867–3879

EPRI (2003) An analysis of new treatment technologies for the California Energy Commission. California Energy Commission, Sacramento

Faloona IC, Chiao S, Eiserloh AJ, Alvarez RJ, Kirgis G, Langford AO, Senff CJ, Caputi D, Hu A, Iraci LT, Yates EL, Marrero JE, Ryoo JM, Conley S, Tanrikulu S, Xu J, Kuwayama T (2020) The California baseline ozone transport study (CABOTS). Bull Am Meteorol Soc 101(4):E427–E445. https://doi.org/10.1175/BAMS-D-18-0302.1

Fritsche K, Lechner U (1992) Biotechniques for air pollution abatement and odour control policies. Elsevier, London, pp 287–292

Fruekilde P, Hjorth J, Jensen NR, Kotzias D, Larsen B (1998) Ozonolysis at vegetation surface a source of acetone, 4-oxopentanal, 6-methyl-5-heptene-2-one and geranyl acetone in the troposphere. Atmos Environ 32:1893–1902

Gabusi V, Volta M (2005) Seasonal modelling assessment of ozone sensitivity to precursors in northern Italy. Atmos Environ 39:2795–2804

Gemeiner P, Rexova L, Svec F, Norrlow O (1994) Natural and synthetic carriers suitable for immobilization of viable cells, active organelles and molecules. In: Veliky IA, McLean RJC (eds) Immobilized biosystems. Chapman & Hall, London

Geron CD, Guenther AB, Pierce T (1994) An improved model for estimating emissions of volatile organic compounds from forests in the eastern United States. J Geophys Res 99(12):773–791

González CM, Gómez CD, Rojas NY, Acevedo H, Aristizábal BH (2017) Relative impact of on-road vehicular and point-source industrial emissions of air pollutants in a medium-sized Andean city. Atmos Environ 152:279–289. https://doi.org/10.1016/j.atmosenv.2016.12.048

Hanna SR, Moore GE, Fernau ME (1996) Evaluation of photochemical grid models (UAM-IV, UAM-V, and the ROM/UAM-IV couple) using data from the Lake Michigan

Harding RC, Hill GA, Lin YH (2003) Bioremediation of toluene-contaminated air using an external loop airlift bioreactor. J Chem Technol Biotechnol 78(4):406–411

Hess A, Zarda B, Hahn D, Haner A, Stax D, Hohener P, Zeyer J (1997) In situ analysis of denitrifying toluene- and m-xylene-degrading bacteria in a diesel fuel-contaminated laboratory aquifer column. Appl Environ Microbiol 63(6):2136–2141

Hirsch AI, Munger JW, Jacob DJ, Horowitz LW, Goldstein AH (1996) Seasonal variation of the ozone production efficiency per unit NO*x* at Harvard Forest, Massachusetts. J Geophys Res 101(12):659–666

Hussain S, Parker S, Edwards K, Finch J, Jeanjean A, Leigh R, Gonem S (2019) Effects of indoor particulate matter exposure on daily asthma control. Ann Allergy Asthma Immunol 123:375–380

Hwang SCJ, Wu SJ (2002) Water transformation in the media of biofilters controlled by *Rhodococcus fascians* in treating an ethyl acetate-contaminated airstream. J Air Waste Manage Assoc 52:511–520

Hwang SJ, Lee CM, Lee HC, Pua HF (2003) Biofiltration of waste gases containing both ethyl acetate and toluene using different combinations of bacterial cultures. J Biotechnol 105:83–94

Jacob DJ, Heikes BG, Dickerson RR, Artz RS, Keene WC (1995) Evidence for a seasonal transition from NO*x*- to hydrocarbon-limited ozone production at Shenandoah National Park, Virginia. J Geophys Res 100:9315–9324

Jiang W, Singleton DL, Hedley M, Mclaren R (1997) Sensitivity of ozone concentrations to VOC and NOx emissions in the Canadian Lower Fraser Valley. Atmos Environ 31(4):627–638

Johnson GM (1983) An empirical model of photochemical smog formation. In: Proc. 15th world congress on air quality, vol 1. IUAPPA, Paris, pp 25–32

Johnson GM, Quigley SM (1998) A universal monitor for photochemical smog. Air Waste Manage Assoc 8:89–29

Juliette N, Anderson R, Fraga JL, Ringelberg D, Lovley DR (1999) Microbial communities associated with anaerobic benzene degradation in a petroleum-contaminated aquifer. Appl Environ Microbiol 65:3056–3063

Kelly NA, Chang TY (1999) An experimental investigation of incremental reactivities of volatile organic compounds. Atmos Environ 33:2101–2110

Kelly FJ, Fussell JC (2015) Air pollution and public health: emerging hazards and improved understanding of risk. Environ Geochem Health 37:631–649

Khan FI, Ghoshal AK (2000) Removal of volatile organic compounds from polluted air. J Loss Prev Process Ind 13:527–545

Kim DK, Kim YS, Kim SK, Zylstra SW, Kim GJ, Kim E (2002) Monocyclic aromatic hydrocarbon degradation by *Rhodococcus* sp. strain DK17. Appl Environ Microbiol 68(7):3270–3278

Kleinman LI (1991) Seasonal dependence of boundary layer peroxide concentration: the low and high NO*x* regimes. J Geophys Res 96(20):721–734

Kleinman LI, Lee YN, Springston SR, Nunnermacker L, Zhou X, Brown R, Hallock K, Klotz P, Leahy D, Lee JH, Newman L (1994) Ozone formation at a rural site in the southeastern United States. J Geophys Res 99:3469–3482

Kolot FB (1988) Immobilized microbial systems: principles, techniques and industrial applications. Krieger, New York

Kovacs J, Egyhazy T, Tac PC (1998) Selective catalytic reduction of NOx by NH_3 in an adiabatic reverse flow reactor. In: Proceedings of the 4th Pannonian international symposium on catalysis. Smolenice, Slovak Republic

Kumar P, Mohan D (2002) Photochemical smog: mechanism, ill effects and control. TERI Inf Digest Energy Environ 3:445–456

Lam KS, Wang TJ, Wu CL, Li YS (2005) Study on an ozone episode in hot season in Hong Kong and transboundary air pollution over Pearl River Delta region of China. Atmos Environ 39: 1967–1977

Lamb B, Westberg H, Allwine G, Quarles T (1985) Biogenic hydrocarbon emissions from deciduous and coniferous trees in the United States. J Geophys Res 90:2380

Leson G, Winer AM (1991) Bio-filtration: an innovative air pollution control technology for VOC emissions. J Air Waste Manag Assoc 41(8):1045

Liebscher H (2000) Economic solutions for compliance to the new European VOC directive. Prog Org Coat 40:75–83

Lippman M (1993) Health effects of tropospheric ozone: review of recent research findings and their implications to ambient air quality standards. J Expo Anal Environ Epidemiol 3:103–128

Liu J, Zheng G (2020) Emission of volatile organic compounds from a small-scale municipal solid waste transfer station: ozone-formation potential and health risk assessment. Waste Manag 106: 193–202. https://doi.org/10.1016/j.wasman.2020.03.031

Liu Y, Quan X, Zhao Y, Chen S, Zhao H (2005) Removal of ternary VOCs in air streams at high loads using a compost-based biofilter. Biochem Eng J 23:85–95

Logan JA (1989) Ozone in rural areas of the United States. J Geophys Res 94:8511–8532

Lucher DR, Jackson MD, Walsh MP, Hsu CCF (1992) The role of alternative fuels in reducing emissions from mobile sources in Taiwan, ROC. In: Alternatives fuels in transportation. Society of Automotive Engineers, Warrendale

Mancini SA, Ulrich AC, Couloume GL, Sleep B, Edwards EA, Lollar BS (2003) Carbon and hydrogen isotopic fractionation during anaerobic biodegradation of benzene. Appl Environ Microbiol 69(1):191–198

McNair LA, Russell AG, Odman MT, Croes BE, Kao L (1994) Air shed model evaluation of reactivity adjustment factors calculated with the maximum incremental reactivity scale for transitional-low emissions vehicles. J Air Waste Manage Assoc 44:900–907
Milford J, Russell AG, McRae GJ (1989) A new approach to photochemical pollution control: implications of spatial patterns in pollutant responses to reductions in nitrogen oxides and reactive organic gas emissions. Environ Sci Technol 23:1290–1301
Mir L (2007) Indoor ultrafine particles and childhood asthma: exploring a potential public health concern. Environment 6:256
Mitha N, Levy J, Annesi-Maesano I, Lafortune J, Magnier A, Ibanez G (2013) Indoor air quality and adult asthma. Rev Mal Respir 30:374–413
National Emissions Standards for Hazardous Air Pollutants (2003) Reinforced plastic composites production, federal register. Environ Prot Agency 68:19375–19443
National Research Council (NRC) (1991) Committee on tropospheric ozone formation and measurement. Rethinking the ozone problem in urban and regional air pollution. National Academy Press, Washington
Olsen RH, Kukor JJ, Kaphammer B (1994) A novel toluene-3-monooxygenase pathway cloned from *Pseudomonas pickettii* PKO1. J Bacteriol 176:3749–3756
Patelarou E, Tzanakis N, Kelly FJ (2015) Exposure to indoor pollutants and Wheeze and asthma development during early childhood. Int J Environ Res Public Health 12:3993–4017
Pleshakova EV, Muratova AY, Turkovskaya OV (2001) Degradation of mineral oil with a strain of *Acinetobacter calcoaceticus*. Appl Biochem Microbiol 37(4):342–347
Russell A G (1990) Air quality modeling of alternative fuel use in Los Angeles, CA: sensitivity of pollutant formation to individual pollutant compounds. The AWMA 83rd
Seipke RF, Salzwedel JL (2002) Isolation and plasmid characterization of eight toluene-degrading bacteria from soil. Independent study of soil bioremediation of petroleum
Serena FS, Marchal R, Blanchet D, Vandecasteele JP (1998) Selection of microbial populations degrading recalcitrant hydrocarbons of gasoline by monitoring of culture-headspace composition. Biodegradation 9:319–326
Shim H, Ryoo D, Barbieri P, Wood TK (2000) Aerobic degradation of tetrachloroethylene by toluene-*o*-xylene monooxygenase of *Pseudomonas stutzeri* OX1. Nat Biotechnol 18:775
Sillman S (1999) The relation between ozone, NOx and hydrocarbons in urban and polluted rural environments. Atmos Environ 33:1821–1845
Sillman S, Samson PJ (1995) The impact of temperature on oxidant formation in urban, polluted rural and remote environments. J Geophys Res 100(11):497–508
Simpson D (1995) Biogenic emissions in Europe 2 Implications for ozone control strategies. J Geophys Res 100(22):891–906
Simpson D, Guenther A, Hewitt CN, Steinbrecher R (1995) Biogenic emissions in Europe 1. Estimates and uncertainties. J Geophys Res 100(22):875–890
Solberg S, Dye C, Walker SE, Simpson D (2001) Long term measurements and model calculations of formaldehyde at rural European monitoring sites. Atmos Environ 35:195–207
Srivastava A (2004) Source apportionment of ambient VOCS in Mumbai city. Atmos Environ 38: 6829–6843
Srivastava A, Joseph AE, Wachasunder SD (2004) Qualitative detection of volatile organic compounds in outdoor and indoor air. Environ Monit Assess 96:263–271
Tampion J, Tampion MD (1987) Immobilized cells: principles and application. Cambridge University Press, Cambridge.to growth. Appl Environ Microbiol 59:2991–2997
USEPA (1991) EPA handbook: control technologies for hazardous air pollutants. Washington, USEPA: Office of Research and Development
Vautard R, Honore C, Beekmann M, Rouil L (2005) Simulation of ozone during the August (2003) heat wave and emission control scenarios. Atmos Environ 39:2957–2967
Veiga MC, Kennes C (2001) Parameters affecting performance and modeling of biofilters treating alkylbenzene-polluted air. Appl Microbiol Biotechnol 55:254–258

Vogel B, Fiedler F, Vogel H (1995) Influence of topography and biogenic volatile organic compounds emission in the state of Baden-Württemberg on ozone concentrations during episodes of high air temperatures. J Geophys Res 100(11):22907–22928

Vomberg A, Klinner U (2000) Distribution of *alkB* genes within n-alkane-degrading bacteria. J Appl Microbiol 89(2):339–348

Wang X, Carmichael G, Chen D, Tang Y, Wang T (2005) Impacts of different emission sources on air quality during March 2001 in the Pearl River Delta (PRD) region. Atmos Environ 39:5227–5241

Winner DA, Cass GR, Harley RA (1995) Effect of alternative boundary conditions on predicted ozone control strategy performance: a case study in the Los Angeles area. Atmos Environ 29:3451–3464

World Bank Group (1998) Nitrogen oxides: pollution prevention and control pollution prevention and abatement handbook. World Bank, Washington, pp 245–249

Worsey MJ, Williams PA (1975) Metabolism of toluene and xylenes by *Pseudomonas putida*(*arvilla*) [sic] mt-2: evidence for a new function of the TOL plasmid. J Bacteriol 124:7–13

Zhuqiu G, Ding C, Xi J, Lichao L, Yang B (2019) Enhancing biofilm formation in biofilters for benzene, toluene, ethylbenzene, and xylene removal by modifying the packing material surface. Bioresour Technol 296:122335

Chapter 4
Utilization of Waste Biomass for Producing Useful Chemicals

Harsh Singh, Swapnajeet Pandey, Nirupama Prasad, Dan Bahadur Pal, and Sumit Kumar Jana

Abstract Biomass is a very important renewable feedstock. Biorefineries from biomass are a good alternative for the replacement of various petrochemical products. Biorefineries need to be upgraded in order to compete with the presently well-established fuel industries. Refineries can be upgraded to integrated biorefineries. Chemicals that derived from biomasses include compounds like 5-hydroxymethylfurfural (HMF-5), levulinic acid, furfurals, sugar alcohols, drinkable acids, carboxylic acid, and phenols. These chemicals are also known as "platform chemicals." These chemicals are used to assemble an outsized range of chemicals on an industrial scale. This chapter provides information concerning the various chemicals obtained from biomasses.

Keywords Biomass · Lignocellulosic · Dehydration · Hydrogenation · Hydrodeoxygenation

4.1 Introduction

Biomass feedstocks are gaining attention due to numerous reasons such as high dependence on fossil-based energy fuels, depleting crude oils, and the pledge to cut the present levels of harmful carbon emissions to safeguard the surroundings. Scientists and researchers intended to develop alternative sources of various crude-based products. Biomass like residues of agricultural activities and wood make up for a low-cost feedstock. The main chemicals that may be obtained from the wood

H. Singh · S. Pandey · S. K. Jana (✉)
Department of Chemical Engineering, Birla Institute of Technology, Ranchi, India
e-mail: sumitkrjana@bitmesra.ac.in

N. Prasad
Department of Chemical Engineering, Birsa Institute of Technology, Sindri, Dhanbad, India

D. B. Pal
Department of Chemical Engineering, Harcourt Butler Technical University, Kanpur, India

D. B. Pal, A. K. Tiwari (eds.), *Sustainable Valorization of Agriculture & Food Waste Biomass*, Clean Energy Production Technologies,
https://doi.org/10.1007/978-981-99-0526-3_4

chips and agricultural residue include organic acids such as levulinic acid and carboxylic acid, different alcohol compounds, and furanic compounds like 5-hydroxymethylfurfural or HMF-5 and alternative furfurals. These chemicals further regenerate into vital chemical merchandise and derivatives that have a variety of applications in several solvent-based primary industries (Wirth et al. 2003). Many studies have been conducted in the past regarding the use of biomass and the different chemicals derived from them, along with the various engineering problems related to their synthesis. This chapter puts together all the significant chemicals, their derivation process, and other chemicals that can be derived from them together.

Lignocellulosic biomass is primarily composed of polysaccharide (34–52 wt%), polymer (12–28 wt%), and hemicellulose (20–32 wt%). The significant and essential part is polysaccharide that is given 1/2 the organic carbon. Polysaccharide consists of anhydrous aldohexose units, whereas hemicellulose, another significant part, consists primarily of various C-5 sugar monomeric chains. Lignin, the third principal part, is a complicated compound with a 3-D structure and plenty of alternative crisscrossed biopolymers with compounds like phenylpropane units with hydrophobic and aromatic characteristics.

4.2 Common Methods of Biomass Velarization as Various Chemicals

Some common mechanisms and reactions that are concerned within the process of chemicals from the biomass include dehydration, chemical change reactions, and hydrodeoxygenation. This mechanism and reaction sequence are described in the following section (Liu et al. 2018).

4.2.1 Dehydration

In this dehydration reaction, a water molecule is liberated from the substrate, particularly alcohol. The catalysts that are used during this reaction include chemicals like Bronsted acids because the cluster could also be an inferior exploit group (Abednatanzi et al. 2019; Brown et al. 2015). In the presence of Bronsted acids, first the cluster gets protonated and forms $R\text{-}H_2O^+$. Due to this, the catalyst bost the reaction mechanism and removed water molecule. This results improved graphene bond with a pair of carbon atoms (C=C) that being designed as per Zaitsev rule. Lewis acid is electrophilic, and due to this property, it lowers the density of electrons. This results in the formation of compounds like Lewis acids, olefin compounds, and hydroxide species (Rotstein et al. 2014). Upon reaction with the free Beta Protons, the Lewis acid hydroxide forms water alongside the initial catalyst species (Bermejo-Deval et al. 2014). Dehydration reactions unit of measurement is the foremost common variety of reactions that take place. This reacts in a vital way as it converts the biomass to different base chemicals (Iglesias-Juez et al. 2022).

4.2.2 Hydrogenation

Hydrogenation is a reaction within which element atoms are additional to associate degree unsaturated matter, which reduces the quantity of double bonds and triple bonds (Ockerman 1996). The addition of an element to chemical species needs a catalyst, and therefore this reaction is catalyzed by heterogeneous catalyst. This successively will increase the feasibility of the reaction (Colantonio et al. 2020). The catalysts used are heterogeneous catalysts that include solid metals, chemical process catalysts, and typically H_2 molecules and also used as catalysts for conversion reactions (Dhakshinamoorthy et al. 2011). The mechanism of chemical process with the assistance of heterogeneous catalysts is as follows:

- Firstly, surface assimilation of the element molecule happens on the catalyst's surface.
- Secondly, it is then followed by the formation of the H–H bond that ends up forming two absorbed element atoms.
- Thirdly, sorption of the unsaturated chemical on the catalyst surface takes place.
- Finally, the covalent bond opens when the third step, and currently the transfer of the element atoms, takes place on the catalyst surface, and this step is reversible.

Hydrogenation is one of the foremost common reactions in chemistry, and the chemical process of the present biomass-derived monosaccharides in lignocellulosic biomass. Biomass helps to form sugar compounds. Later, the chemicals obtained are used as a solvent in various industries, such as monomers and biofuels.

4.2.3 Hydrodeoxygenation (HDO)

In this hydrodeoxygenation reaction, the associate degree chemical element atom is far from the compound in the element's presence (Grochala and Edwards 2004). The removal of the elemental atoms can cause various mechanisms such as direct reaction, dehydration, chemical change, and decarbonylation (Pujro et al. 2021). This mechanism additionally needs catalysts that are selective and includes noble metals. The mechanism depends on the conditions of the reaction, and that catalyst is getting used (He and Wang 2012). A general order of the reactivities of the HDO compounds is as follows:

Alcohols > Ketones > radical Ethers > carboxylic Acids

Generally, HDO reaction is carried out at conditions of air mass and high temperature, and this then ends up in the formation of various products that are shaped through the cleavage of the carbon-oxygen bond (Richardson and Gorton 2003; Liu et al. 2017).

4.3 Some of the Most Important Chemicals That Are Obtained from the Biomass

4.3.1 1,4-Diacids (Four Carbon Compounds Like Succinic, Fumaric, and Malic Acids)

1,4-Diacids contain four carbon chemical group acid's function which is an essential element for several necessary chemicals (Kolb et al. 2001). From the analysis, it is found that carboxylic acid behavior is analogous to that of malic acid that springs from the fossil fuel product (Berger 1984). Carboxylic acid is derived from a range of several biochemicals by exploiting some designed types of enzymes like succinic produce and *Escherichia coli* strains developed by the US DOE labs.

Generally, the fermentation path needed for synthesizing carboxylic acid is extensive, and due to this, it becomes a challenge and necessity to develop some economically possible fermentation routes for its development (Rodriguez et al. 2014).

4.3.1.1 Chemicals Derived from Succinic Acid

- Primarily, the carboxylic acid is employed as a reducer within the selective reduction to provide some important chemicals like butanediol, tetrahydrofuran, and gamma-butyrolactone group chemicals. Carboxylic acid undergoes a chemical reaction to present these compounds (Shinde 2007).
- Further, different compounds are also derived from gamma-butyrolactone, like Pyrrolidinones, a vital element utilized in the solvent market (Varadarajan and Miller 1999).
- Succinic acid is directly regenerated to pyrrolidinone, which is then used to manufacture diammonium succinate (Cheng et al. 2012).

4.3.2 *Succinic Acid*

The name succinic acid has been derived from the Latin word *sucsuccum*, which in English means amber. A succinate anion is an ion that is formed by succinic acid. Research has identified multiple biological roles of the succinate anion. Succinate ion finds its use as a metabolic intermediate, and when it undergoes reasoning the electron transport chain in the presence of succinate formed in the presence of an enzyme, the succinate ion transforms (Cecchi and Carolis 2021). This chain synthesizes adenosine triphosphate as a signaling compound to reflect the cellular metabolic state. The food additive, known commonly as E363, is also derived from succinic acid. The tricarboxylic acid cycle generates a succinate compound in the cell's mitochondria. Succinate compounds begin to function in the protoplasm and the intracellular space of the cell upon exiting the chondriosome array of the cell. In doing so, it changes the gene expression patterns, modulates the epigenetic landscape, and demonstrates hormone-like signaling. As such, succinate compounds aid in metabolism along with the regulation of cell functions. In some of the identified genetic mitochondrial diseases, it has been found that dysl desolation, the synthesis of the succinate anion, occurs. Leigh syndrome and Melas syndrome are two common diseases associated with this defect. The degradation of the succinate anion can further lead to severe morbid conditions such as hostile changes, redness, and tissue injury.

4.3.2.1 Physical Properties of Succinic Acid

Succinic acid is an odorless solid compound that is white. Succinic acid readily ionizes and leads to forming its conjugate base, which is called succinate, when mixed in the aqueous solution. Succinic acid is a diprotic acid. It means that the succinic acid undergoes deprotonation reactions in two subsequent orders.

$$(CH_2)_2(CHOH)_2 \rightarrow (CH_2)_2(CHOH)(CO_2)^- + H^+ \tag{4.1}$$

$$(CH_2)_2(CHOH)(CO_2)^- \rightarrow (CH_2)_2(CO_2)_2^{2-} + H^+ \tag{4.2}$$

The pK_a value calculated for process 1 is 4.3, and for those 2, it comes out to be approximately 5.6. The anions formed in both of the above reactions are colorless in nature, and they can be separated easily as salts like $Na(CH_2)_2(CO_2H)(CO_2)$, which is obtained from the compound formed in reaction 1, and $Na_2(CH_2)_2(CO_2)_2$, which is formed from the compound formed in reaction 2. Succinic acid is found in its basic form, called succinate, in the organisms. A radical group, succinic acid, exists in the form of an ion called a succinate group. Succinic acid is not very harmful like most simple least monocarboxylic and dicarboxylic acids, but prolonged and continuous exposure to succinic can lead to irritation to the skin and eyes and may turn out to be harmful.

4.3.2.2 Methodology for the Production of Succinic Acid

Succinic acid can be derived from biomass-based compounds using following techniques:

1. Furfural as the feedstock for producing succinic acid: This process uses Bronsted acid-base catalysts. One of the catalysts used for this reaction is an Amberlyst-15 resin-based catalyst. The reaction is performed in the presence of toluene, and the catalyst used is an Amberlyst-15 resin-based catalyst along with hydrogen peroxide (H_2O_2). Toluene is used for this reaction because succinic acid is insoluble in it, resulting in an easier separation when furfural undergoes oxidation. When the Amberlyst-15 resin catalyst is used, it results in a yield of approximately 52%. This yield can be further improved by using a carbon-based catalyst such as SO_3H-CD and then employing ultrasound in the presence of H_2O_2. Furfural reacts with the Amberlyst-15 resin catalyst in hydrogen peroxide to form 2-hydroxy furan, further forming Furan-2(3H)-one. Furan-2(3H)-one then reacts with-cationic medium and again with peroxide to finally form succinic acid.
2. Levulinic acid as the feedstock for the production of succinic acid: This is the traditional method for the production of succinic acid using levulinic acid as the feedstock. It involves some environmental concerns as it forms toxic salts of mercury and also the use of nitric acid as the oxidant is hazardous. The most suitable mechanism involves using methyl levulinate, is by transesterification with DMC, and then it undergoes deprotonation with the help of a base which leads to the formation of a nucleophile to add to the carbonyl carbon of DMC. It leads to the formation of a species that can remove the methyl acetate and yield dimethyl succinate. Further reaction of the dimethyl succinate with DMC allows for the formation of 2-methyl dimethyl succinate, which then leads to the formation of succinic acid.
3. Commercial production of succinic acid: According to the historical data, succinic acid was synthesized by distillation from amber. Hence, succinic acid got the spirit of amber. Some typical industrial applications of succinic acid include hydrolysis of maleic acid, corrosion of 1,4-butanediol, and carbonylation of ethylene glycol. Succinate, the base of succinic acid, is synthesized from butane with the help of maleic anhydride. Global succinic acid production every year is estimated to be around 16,000–30,000 tons, and the calculated annual growth rate amounts to about 10% per year.

4.3.2.3 Applications of Succinic Acid

4.3.2.3.1 Use of Succinic Acid as a Precursor

Succinic acid is an antecedent to some standard polyesters and an essential component of some alkyd resins (Islam et al. 2014). 1,4-Butanediol (BDO) is synthesized from succinic acid, where succinic acid is employed as an antecedent. BDO is used heavily in automotive and electronics industries. Succinic acid is also used as the starting and base compound for some of the essential biodegradable polymers, which are of great importance and regularly find application in tissue engineering (Lu and Mikos 1996). A vaccination reaction is the type of reaction that involves the acylation reaction with succinic acid. When more than one succinate compound gets added to a substrate, over-succinate occurs, which is optional for the proper vaccination reaction.

4.3.2.3.2 Succinic Acid as Food and Nutritional Supplement

According to the various departments involved in Food and Drug Management, succinic acid is recognized as safe to use (Patel and Goyal 2015). In the food and beverage industry, succinic acid is regularly used as a pH control agent due to its property of forming essential compounds. Succinic acid is a flavoring medium with a sour, astringent, umami taste (Laffitte et al. 2016). Succinic acid is also used as a bulking agent in many pharmaceutical products due to its ability to control acidity and is utilized as a counter ion.

4.3.3 Fumaric Acid

Fumaric acid ($HO_2CCH{=}CHCO_2H$) is a biologically natural compound. Physically, fumaric acid is a white solid compound readily available in nature. Fumaric acid has a characteristic fruit-like taste. Like succinic acid, fumaric acid is also widely used as a nourishment additive. Fumarates are the salts and esters of the fumaric. In the solution form, fumarate exists as $C_4H_2O^{2-}{}_4$.

4.3.3.1 Methodology for the Production of Fumaric Acid

Fumaric acid is produced as the side product in the cleavage of the C–C bond of the HMF in the reaction for the production of maleic acid. The primary method for producing fumaric acid includes chemical isomerization and microbial fermentation. However, among the two processes, chemical isomerization is more beneficial as it has better efficiency. In eukaryotic organisms, fumaric acid is synthesized from the succinate eaten (the anion of the succinic acid). The carboxylic acid batch exists as trans (E) in the fumaric acid, and maleic acid and fumaric acid exist as cis (Z) (Hronska et al. 2017). Fumaric acid is also found in many other compounds and items. The intermediate formed in the citric acid cycle is known as fumarate. In the cells, fumaric acid finds application in the production of energy. From food, energy is produced through adenosine triphosphate (ATP) (Yang et al. 2011). Oxidation of succinate compound leads to Adenosine Triphosphate or ATP with the help of enzymes succinate dehydrogenase (Pollard et al. 2003). Then with the help of the enzyme fumarase, fumarate is converted to malate. When the skin is exposed to sunlight, fumaric acid is produced naturally in the human skin. In the urea cycle, fumaric acid is also formed as a byproduct.

4.3.3.2 Applications and Uses of Fumaric Acid

1. Food: For a very long time, fumaric acid has been used as a subsistence acidulant. Fumaric acid has been used as a subsistence additive. It is also an acidity regulator (Liao et al. 2008). Fumaric acids are used in various kinds of beverages and baking powders. Fumaric acid has been used as a food preservative in making wheat tortillas and as the acid in the leavening of the tortillas. Fumaric acid, in many cases, is used as an alternative to tartaric acid, and sometimes it is also occasionally used to add sourness to the food in place citric acid. Fumaric acid is an essential component of some artificial vinegar flavors. Generally, fumaric acid is a nontoxic acid, but if anyone is exposed to high doses of it, it becomes nephrotoxic after long-term use (Romero et al. 1993).
2. Medicine: Fumaric acid is also used as a medicine, and it is generally used to cure the autoimmune condition known as psoriasis. Later, the primary ester, dimethyl fumarate, was synthesized from fumaric acid, which is used as a medicine for treating multiple sclerosis (Meissner et al. 2012).
3. Other uses: Fumaric acid has been used in manufacturing polyester resins and polyhydric alcohols. Fumaric acid is an essential ingredient in dyes and used as a mordant for dyes (Giménez-Arnau et al. 2009).

4.3.4 Maleic Acid

Malic acid is a natural compound with the formula $C_4H_6O_5$. Malic acid is a dicarboxylic acid that is synthesized by all kinds of living organisms. Malic acid is the reason for the sour taste of fruits. Malic acid is utilized as a nutrient additive in many other edible compounds (Safari et al. 2021). Malic acid exists in two isomeric spatial forms: L-enantiomers and D-enantiomers. Only the L-isomer exists in nature from the two forms of the enantiomers. The salts and esters of malic acid are called malates and are formed as an intermediate in the citric acid cycle (Safari et al. 2021).

4.3.4.1 Biochemistry of Maleic Acid

There are two forms of malic acid. The form of the malic acid that occurs naturally is called L-Malic acid. A mixture of L- and D-malic acid is generated by synthetic methods. Malate is crucial in the biochemistry and properties of malic acid. The intermediate formed in the citric acid sequence is called S-malate (Cammas et al. 1993). The inclusion of an –OH group on the face of fumarate leads to the formation of S-malate. Malate assists the potassium cations during the uptake of solutes into the guard cells when it behaves as a double anion. It helps to maintain the balance of electrical properties in the cells of the plants. The stockpiling of these solutes within the guard cells decreases due to the presence of solute potential. Due to this, water makes its way to the plant's cell and, thus, helps promote the aperture of the stomata (Lawson et al. 2014).

4.3.4.2 Methodology for the Production of Maleic Acid

Some of the steps for the synthesis of maleic acid are as follows:

1. Oxidation of Furfural to Maleic Acid: Furfural is very cheap cost-wise as it is produced from agricultural waste, which is available in abundance. The process can be of two types—either gas phase or liquid phase. The catalyst used for this process is Hydrogen Peroxide (H_2O_2). When the vanadium compound comes in contact with hydrogen peroxide, it forms a hydroxyl radical. One of the hydrogen atoms gets removed, leading to the generation of an intermediate called furfural radical intermediate. Then, the electron transfer of the furfural radical transforms it into a furfural cation through an electron transfer, and after this, the compound formed undergoes 1,4-rearrangement. The hydrolysis of the compound thus formed takes place along with the oxidation of the compound formed in the previous step, which finally leads to the formation of maleic acid.
2. Oxidation of 5-Hydroxymethylfurfural (HMF) to Maleic Acid: The catalyst used for this process is a graphene-oxide-supported vanadium catalyst. Protic solvents, like acetic acid, are also used for easy breaking of the C–C bond of the HMF. The cleavage of the C–C bond between the furan ring and the hydroxymethyl group of the HMF is the starting step for producing maleic acid through this process. This step is catalyzed by the vanadium cation and leads to forming a compound with formic acid as the side product. The compound formed in the first step then undergoes 1,4-rearrangement to yield a compound, and this compound then undergoes removal of the hydrogen atom followed by decarboxylation to form a compound known as 5-hydroxy-furan-2-one. Then in the final step, further oxidation in the presence of V^{5+} species leads to the formation of maleic acid.

On an industrial scale, double hydration of maleic anhydride leads to malic acid production. The chiral resolution of the racemic mixture technique is used to separate the enantiomeric mixture, which is formed during malic acid production. When fumaric acid undergoes fermentation, it leads to the production of S-Malic acid (Crosby 1991). Self-condensation of malic acid in the existence of fuming sulfuric acid results in the synthesis of pyrone coumaric acid. Maleic acid plays a vital role in the finding and study of the Walden inversion phenomenon and the Walden cycle. In this process, in the presence of phosphorus pentachloride, (−)-malic acid is first converted into (+)-chlorosuccinic acid. (+)-malic acid is formed when the wet silver oxide reacts with chlorine compounds. (+)-malic acid, formed in the previous step, then undergoes a reaction with PCl_5 to form a compound commonly called the (−)-chlorosuccinic acid. When silver oxide takes (−)-chlorosulfonic acid back to (−)-malic acid, this marks the completion of the Walden cycle.

4.3.5 2,5-Furan Di-Carboxylic Acid (FDCA)

These compounds are shaped either by reduction or by direct chemical change. FDCA is a painfully necessary cluster of acids as these are highly utilized in several polyesters like synthetic resin terephthalate (PET), and butylene terephthalate (PBT) as FDCA will replace terephthalic acid (Harmsen et al. 2014).

4.3.5.1 Methodology for the Production of FDCA from Biomass

Biomass is first converted to HMF, and then from the HMF, FDCA is synthesized. This process involves the oxidation of an aldehyde group present in the HMF along with the oxidation of an alcohol group. There are two pathways available for this reaction. In the first type of mechanism, an aldehyde group's first oxidation occurs, forming carboxylic acid and HMFCA. Then, after the formation of HMFCA, the alcohol group present in HMFCA is oxidized to the aldehyde group and then to carboxylic acid, resulting in the formation of FDCA. In the second type of mechanism, the reaction begins with the oxidation of the hydroxyl group of the HMF to aldehyde, which results in the generation of DFF. This second mechanism then involves the oxidation of the aldehyde group, which occurs in two steps, and finally, FDCA is formed. In the second mechanism, FFCA is formed as an intermediate, but later, this FFCA is also oxidized like the aldehyde group, forming FDCA.

4.3.5.2 Properties and Conversion of FDCA

FDCA is a very safe and secure compound. Chemically it has a very high melting point (around 342 °C). It has a high level of insolubility character in most of the solvents. This fact indicates the presence of intermolecular hydrogen bonding (Zhang and Deng 2015). Despite being a stable chemical, FDCA undergoes reactions typical for carboxylic acids. These include reactions like halogen substitution, which gives carboxylic dihalides, di-ester formation, and amides formation reactions. The amalgamation of diethyl ester and dimethyl ester, and the amidation reactions, are some of the most common modifications in the reactions involving FDCA. FDCA is a very versatile chemical, and due to its flexible and versatile nature, it is found in several derivatives. These chemical derivatives are available in many simple chemical changes and reactions (Serrano-Ruiz et al. 2011). Selective reduction of FDCA leads to the formation of partially hydrogenated products. These include compounds like 2,5-hydroxymethyl furan and some completely esterified materials like 2,5-bis(hydroxymethyl)tetrahydrofuran.

4.3.5.3 Chemicals Derived from FDCA

- Selective reduction of FDCA results in forming part-modified products like 2,5-hydroxymethyl furan and conjointly modified materials like 2,5-bis (hydroxymethyl)tetrahydrofuran. Also, with the mixture of FDCA, these chemicals conjointly type various biomass-derived products.
- FDCA may be used as a base chemical to assemble carboxylic acid and connected chemical compounds (Zhou et al. 2019).
  FDCA may be a vital substance, and several technical hurdles need to be overcome for its synthesis. A number of these embrace the evolution and management of esterification reactions, conjointly with the reactivity of the compound made from FDCA (Banerjee et al. 2017).
  The use of FDCA as a part of the production of PET and PBT reveals new markets, and these new ways got to be developed for the selective chemical reaction and dehydration of FDCA for its use effectively (Sajid et al. 2018).

4.3.5.4 Applications and Uses of FDCA

Polymerization of the FDCA to form various chemicals is often considered the most critical group. The most common example of the compound formed through the depolymerization of FDCA is polyethene 2,5-furan-dicarboxylate (Papadopoulos et al. 2021). The polymerization of FDCA also leads to the production of many other compounds, such as polyesters, polyamides, polyurethanes, and many other essential compounds. FDCA has many applications in the field of pharmacology. Anilids which are formed chemically from FDCA possess robust antibacterial characteristics. FDCA is also a potent complexing agent, which finds applications in many medicines used to remove kidney stones. Metabolism of FDCA in mammals, including humans, leads to the formation of HMF. When FDCA is mixed with tetrahydrofuran in diluted form, it is utilized for preparing artificial veins, which are then used for transplantation (Ozcelik et al. 2014). FDCA is a very stable chemical, and due to its highly durable nature, the polycarboxylic acids derived from it are used as a base constituent for the production of fire foams. These fire foams help to put out the fires brought about by polar and nonpolar solvents. FDCA has a strong perspective as a replacement for terephthalic acid. Terephthalic acid is widely used in various polyesters, such as polyethene terephthalate (PET) and polybutylene terephthalate (PBT). Partial hydrogenation of the FDCA leads to the synthesis of compounds and products like 2,5-hydroxymethyl furan and completely esterified materials like 2,5-bis(hydroxymethyl)tetrahydrofuran (Joanna et al. 2018). When these materials are combined with FDCA, then it leads to the formation of an entirely new family. FDCA is also used as the base chemical compound for making succinic acid.

4.3.6 3-Hydroxypropionic Acid (3-HPA)

This acid is primarily developed from reactions like reduction and dehydration of production or base chemicals of the biomass. This acid acts as a crucial building block for each artifact. The development of low-price fermentation pathways for the synthesis of 3-HPA may be a significant challenge (Delidovich et al. 2016).

4.3.6.1 Chemicals Derived from 3-HPA

- Selective or direct reduction of 3-Hydroxypropionic acid leads to forming one or three humectants. This chemical is the production of Sorona fibers.
- Selective dehydration of 3-HPA with no aspect reactions results in the formation of the propenoate family of chemical compounds. For this reaction, heterogeneous catalysts are used (Andanson and Baiker 2010).

4.3.7 Aspartic Acid

The different reactions for converting aspartic acid to completely different chemicals include the formation of acid in the presence of ammonia, primarily involving the chemical transformation of aspartic acid. For the biotransformation of aspartic acid, once salt undergoes conversion hysteria within the organic process, and it results in the formation of aspartic acid. This mechanism involves either fermentation conversion or protein conversion (Krautkramer et al. 2021). The fermentation of aspartic acid provides an occasional yield; therefore, raising the yield simultaneously and keeping the price effective remains a severe challenge.

4.3.7.1 Chemicals Derived from Aspartic Acid

- Reduction of aspartic acid results in paraffin butanediol, paraffin tetrahydrofuran, and paraffin butyrolactone. These chemicals are similar in characteristics and properties to natural resin and malic acids. Synthesis of those chemicals involves selective reduction at mild conditions like gas pressure and cold. Tolerance to the catalyst and its poisoning remains a serious challenge.
- When aspartic acid undergoes selective dehydration with no aspect reactions occurring, then it results in the formation of aspartic chemical compounds. Liquid

catalysts were used; however, heterogeneous catalysts have replaced them antecedently.

- When aspartic acid undergoes selective esterification, it results in the formation of polyaspartics. Branching management is additionally done to regulate the organic compound shape's relative molecular mass and properties (Kricheldorf 2006).

4.3.8 Glucaric Acid

This acid comes from a family of compounds called oxidized sugars. Chemical transformation (one step aqua fortis chemical reaction) involves the selective chemical reaction of alcohols (ROH) and carboxylic acids (RCOOH) (Litter 1999).

The second technique involves a chemical reaction of starch in bleach. Some of the most critical challenges involving glucaric acid production include avoiding expensive oxidants and exploiting air for the chemical reaction, oxygen, dilute peroxide, and alternative chemicals. Further, developing optimum heterogeneous catalysts conjointly creates a severe challenge (Anastassiadis and Morgunov 2007).

4.3.8.1 Methodology for the Production of Glucaric Acid

The starting step for producing glucaric acid from biomass involves the production of gluconic acid. Then, further oxidation of the hydroxyl group present on the terminal of the gluconic acid leads to the formation of glucuronic acid. The isomerization of glucuronic acid takes place in the reacting medium and forms 5-keto-gluconic acid. The additional cleavage of the C–C bond following the oxidation reaction leads to the formation of organic acids of smaller molecular forms and the main product, glucaric acid.

4.3.8.2 Chemicals Derived from Glucaric Acid

- Lactones: once glucaric acid undergoes selective dehydration with no reaction, it results in the formation of lactones (Zhang et al. 2021).
- Dehydration conjointly typically results in the formation of anhydrides. For this reaction, liquid-based heterogeneous catalysts are mostly the most well liked (Zhang et al. 2021).
- Lactones are mainly used as solvents for various styles of reactions (Zhang et al. 2021).
- Polyglucaric esters and amides: Selective esterification of glucaric acid alongside the management of branching results in the formation of those compounds. Branching management is completed to regulate the relative molecular mass of the species being shaped and conjointly its properties. These compounds are

primarily employed in producing materials like Nylon and Kevlar as some carpet fibers (Zhang et al. 2021).

4.3.9 *Glutamic Acid*

This acid is a 5-carbon amino acid and is a building block of five-carbon polymers. This acid is also used in the manufacturing of many types of polyesters and polyamides. This acid does not undergo any chemical transformation but under go biotransformation and results in the fermented product (Garattini 2000). For the fermentation of glutamic acid, microbial biocatalysts are used along with other acid components. But here, better control of the reaction environment is required (Lin and Tao 2017).

4.3.9.1 Methodology for the Production of Glutamic Acid

The feedstock for glutamic acid production includes sugar compounds like glucose, fructose, and sucrose, also called refined sugars. Some substrates like starch hydrolysates, molasses, and methanol are also used. These feedstocks have a minimum yield of glutamic acid. The best yield of glutamic acid was obtained when wheat bran and extracts from rice bran were used for its production. Biomass by-products, such as wheat dried from distilleries, are also used. The starting step involves solubilizing the protein content available in the wheat. Although the complete solubilization of the protein content is the default, it is a crucial starting step. To increase the solubility of the protein, enzymes are used. These enzymes break down the protein into peptides and amino acids, thus increasing the solubility of the protein. Then the compound formed undergoes hydrolysis, resulting in the formation of glutamic acid. The hydrolysis steps are performed under mildly acidic conditions.

4.3.9.2 Chemicals Derived from Glutamic Acid

- 1,5-propanediol: This forms once glutamic acid undergoes selective chemical action followed by reduction and subtractive chemical action (Holladay et al. 2004).
- 1,5-propane diacid: This conjointly forms in a very similar approach as one, 5-propanediol, and needs gentle conditions like part pressures and low temperatures (Wang et al. 2017).
- Amino alcohol or 5-amino-1-butanol forms within the same approach because the previous two and the reaction conditions are similar (Balcı et al. 2021).
- The above three chemicals are used as monomers for various polyesters and polyamides (Ingole et al. 2016).
- Chemistry of glutamic acid

4.3.9.2.1 Ionization

Upon dissolving glutamic acid in water, the amino group (NH_2) present receives a proton (H^+) and depending on the acidity of the medium, the carboxylic group present receives a proton (H^+) (Bolan et al. 1991). When the amino group receives a proton in an acidic environment, the molecule produces $HOOCCH(NH^{+3})(CH_2)_2COOH$, a cation possessing only a single charge of the positive sign at pH levels between 2.5 and 4.1. The carboxylic acid present closest to the amine loses a proton, producing the zwitterion $OOCCH(NH^{+3})(CH_2)_2COOH$, which is neutral. Both forms appear equal at pH 2.10, showing a gradual shift. When the other carboxylic acid group's proton is removed, it produces an acid that is almost all glutamate and has similar properties. The anion $OOCCH(NH^{+3})(CH_2)_2COO$ possesses a single negative charge at extremely high pH levels. This form, devoid of protons in both the carboxylates, is shared within the physiological pH range (7.35–7.45). The amino loses its additional proton at a higher pH, and the doubly negative anion $OOCH(NH_2)(CH_2)2COO$ becomes the dominating species.

4.3.9.2.2 Optical Isomerism

The atom adjacent is connected to four distinct groups and is called chiral. There are three isomeric optical forms in which glutamic acid can exist (Nguyen et al. 2006). The three forms in which glutamic acid exists include the dextrorotatory L-form, d (−), and l(+). The one most generally occurring in nature is the l form of glutamic acid. The d form occurs in some particular contexts, including forms like the cell walls of bacteria; therefore, it is also produced in the liver of mammals.

4.3.10 Itaconic Acid

Chemical synthesis of acid is not a natural choice as it is costly. Thus, reduction within the synthesis steps is needed for its synthesis. Biotransformation of this acid takes place. In biotransformation, this acid undergoes fermentation and forms a fermentation product (Hugh and Leifson 1953). The mechanism used for its fermentation is aerobic plant life fermentation. In biotransformation, involving fermentation, raising the standard of the catalyst may be a significant challenge alongside reducing the formation of other acids and by-products. When acid undergoes chemical transformation, it is used as a polymer with styrene-butadiene polymers. When it undergoes biotransformation, it is used as a polymer in vinylbenzene hydrocarbon polymers (Willke and Vorlop 2001). The major challenge involved within the production (Willke and Vorlop 2001).

4.3.10.1 Methodology for the Production of Itaconic Acid

The production of itaconic acid from biomass takes through the microbial fermentation process. This fermentation reaction occurs in aerobic conditions, and the starting step involves transforming glucose to pyruvate through glycolysis. Then, through the metabolism of the pyruvate, acetyl-CoA is produced, and a molecule of carbon dioxide is produced. Then, the acetyl-CoA undergoes partial transformation and results in the formation of oxaloacetate, which takes place in the mitochondrion of the microbe. After the partial transformation step, citrate and cis-aconitate are produced in the citric acid, and this process is known as the tricarboxylic acid cycle or TCA cycle. In the next step, the mitochondrial tricarboxylate transporter carries the cis-aconitate in the cytosol of the microbe. In the final step, cis-aconitate helps to discharge carbon dioxide, forming the final product, itaconic acid. Industrially, itaconic acid is formed by the zymolysis of carbohydrates like glucose or molasses. In fermentation, molds such as Aspergillus itaconic or *Aspergillus terreus* are used (Willke and Vorlop 2001). The most elucidated pathway for the *Aspergillus terreus* is the itaconate pathway. Generally, in glycolysis, the tricarboxylic acid cycle is used to synthesize the itaconic acid, and this process also includes the decarbopalladation of cis-aconitate into itaconate via cis-aconitate-decarboxylase (Steiger et al. 2013; Kratky and Vinsova 2012). Macrophage lineage cells suffer a lot from creating itaconate; hence, they lose the power to perform phosphorylation at the mitochondrial substrate level (Pallotta 2020).

4.3.10.2 Reactions to Itaconic Acid

The heating of itaconic anhydride leads to the isomerization of the itaconic anhydride into citraconic acid anhydride, which is later hydrolyzed into citraconic acid (2-methyl maleic acid). The steps in converting acid to citraconic acid via aconitic and itaconic acids include the partial hydrogenation of acid over Raney nickel, which affords 2-methyl succinic acid. Itaconic acid is employed as a co-monomer in synthesizing acrylonitrile butadiene styrene and acrylate latexes with solicitation in the paper and architectural coating industry (Chiloeches et al. 2021).

4.3.10.3 Chemicals Derived from Itaconic Acid

- When acid undergoes selective reduction, it forms alkyl radical butanediol, butyrolactone, and tetrahydrofuran family of chemicals. The reaction takes place at cold and at gas pressure. These chemicals are used more for compounds like BDO, GBL, and THF (Plackett 2011).
- Amination of acid leads to the formation of pyrrolidinones (Bai et al. 2011).

- Polytechnic acid: once selective esterification and branching management is completed, the acid transforms into polytechnic acid. These chemicals can produce and synthesize recent polymers (Werpy and Petersen 2004).

4.3.11 Levulinic Acid

It is a vital chemical that is obtained from biomass. This acid has several industrial uses and is used together with the elements in petrol and conjointly as a biodiesel additive. Levulinic acid is one of the products that are made by the treatment of 6-carbon sugar carbohydrates, which is obtained from the starch and present in lignocellulosic compounds with acids. Chemically, levulinic acid is produced in one-step reaction. This one section includes the dehydration of the levulinic acid in the appearance of an acid catalyst, followed by the decomposition of cellulose and sugars. Selective dehydration with a connected reaction is needed. Levulinic acid is highly low cost to manufacture and is instantly out there from each five-carbon and six-carbon sugars, and due to this, it forms a crucial base-building compound for several industrial chemicals (Ahmed et al. 2021).

Biomass is treated with an acid during the formation of levulinic acid. The hydrated form of HMF is the intermediate which is the formation of this reaction. One of the critical side products formed in this reaction is formic acid, produced in an equimolar amount. Levulinic acid can also be produced from compounds like xylose, arabinose, etc., also called five-carbon carbohydrates, which are present in the hemicellulose of levulinic acid. This mechanism involves the reduction of furfuryl alcohol along with the acid treatment. On an industrial scale, levulinic acid is produced by the "Biofine Process." This process involves cracking the lignocellulosic feedstock in the presence of the dilute mineral acid. This reaction takes place at a moderate temperature. The fine technology uses a plug flow reactor operated at a high temperature and has a short residence time. This PFR is followed by CSTR, which runs at a moderate temperature and has an average residence time. This two-reactor system has the advantage of having a very high yield, conversion, and throughput. In the process, the cellulose in the biomass is broken down, resulting in levulinic acid. It also gives formic acid as a side product which is also an essential chemical. Also, in this process, the lignin, along with some degraded cellulose, is sent to the power station, where it is burnt to produce power and is also used for export purposes. Dutch chemist Gerardus Johannes Mulder first prepared levulinic acid by heating fructose in the presence of an acidic compound. Levulinic acid has reached billboard use in significant volume. The company Quaker Oats developed endless activities to assemble levulinic acid in 1953 (Zhang et al. 2022). Levulinic acid was recognized as a platform chemical with very high potential in 1956. In 2004, the US Department of Energy (U.S., DOE) identified levulinic acid as one of 12 potential base-building chemicals in the sphere of the biorefinery concept. Levulinic acid is produced by using chemicals such as hexoses (glucose, fructose) or starch in the presence of vitriol. Some partly insoluble by-products are also produced along with levulinic acid. These by-products are deeply colored physically, and their complete removal may be challenging for many technological processes (Zhang et al. 2022).

4.3.11.1 Applications of Levulinic Acid

Levulinic acid generally finds its application as a starting base chemical for many necessary pharmaceuticals, plasticizers, and other supplementary daily use products (Benisvy-Aharonovich et al. 2020). The building of aminolevulinic acid, a biodegradable herbicide used in South Asia, is the most critical application and use of levulinic acid. Levulinic acid is also used in various cosmetic products (Silva et al. 2022). Ethyl levulinate, a principal derivative of levulinic acid, is rigorously used and added to fragrances and perfumes. Levulinic acid may be a chemical building block or starting material for other important chemical compounds. These chemical compounds include chemicals like γ-valerolactone and 2-methyl-THF.

4.3.12 3-Hydroxybutyrolactone

3-Hydroxybutyrolactone is a cyclic compound with four carbon atoms or a C-4 compound formed from chemical transformations (Isikgor and Becer 2015). For the chemical transformation, dilute peroxide is additionally used. It can be used as N-intermediate for drug company compounds of upper values. 3-Hydroxybutyrolactone doesn't undergo any biotransformation (Kumbhar et al. 2016).

4.3.12.1 Chemicals Derived from 3-Hydroxybutyrolactone

- Furans and completely different analogs of pyrrolidines: 3-hydroxybutyrate undergoes selective reduction or hydroxyls in the presence of C=C to make furans and pyrrolidines (Pal et al. 2015). The reaction takes place at gas pressure and cold. In the presence of alcohol groups, selective reduction of aldehyde (RCHO) clusters conjointly happens. These chemicals are primarily used as solvents for synthesizing many completely different chemicals (Pal et al. 2015).
- Amin analogs to tetrahydrofuran: These chemicals are used for managing the rates of the reaction. Selective esterification is completed to regulate banking. These amines are used in the lycra fibers (Pal et al. 2015). 3-Hydroxybutyrolactone is employed for making new spinoff compounds. It is a specialty chemical employed for producing high-priced chemicals and conjointly as an artifact chemical for developing alternative chemicals (Pal et al. 2015).

4.3.13 *Glycerol*

Glycerol may be a versatile chemical employed in several standards of living products like soaps, face creams, and many makeup products (Demirbas 2008). Chemically, it is created from the transesterification of oils and the product-like products duct, oral products duct, several styles of medicine and prescribed drugs, foods, and beverages and conjointly for the synthesis of polyether, polyols, and polymer (Demirbas 2008). Biotransformation of alcohol involves protein transesterification. This esterification reaction occurs within the methanol-water resolution, and its prices are equivalent to the chemical esterification product (Demirbas 2008).

4.3.13.1 Methodology for the Production of Glycerol

Traditionally, glycerol is produced from fat separation or the transesterification of fats and oils. In a transesterification reaction, ion of alcohol reacts in the presence of a catalyst. The products formed including fatty acids, methyl esters, and glycerol are obtained as a by-product. The transesterification reaction can be carried out in both batch and continuous modes. In other processes, the separation of the feed stream occurs. Transesterification reactions can be acid-catalyzed, base-catalyzed, and enzyme-catalyzed. The saponification reaction also results in the formation of glycerol as a by-production. The saponification reaction involves the hydrolysis of fats and oils with an Alkali, forming two products. These are fatty acids (later used for soaps), and glycerol is formed as the by-product.

4.3.13.2 Chemicals Derived from Glycerol

- PLA congener and acid: Alcohols (ROH) regenerate to acids like RCOOH. Excess oxidants like peroxide are avoided, and dilute oxidants like gas from the air are employed. The reaction of aldehydes to acids and alcohols additionally takes place. PLA has superior chemical compound and is employed in several polyester fibers with new properties (Zheng et al. 2008).
- Propylene glycol: used as Nursing liquid agent, humectant, etc.
- 1,3-propanediol is employed in Sorona Fiber production (Kurian 2005).
- Branched polyols and polyesters: Selective esterification is finished, followed by branching management to manage the properties and the mass. These are used in synthesizing unsaturated polyurethanes and in resins that are used in insulations and alternative materials (Lang et al. 2020).

4.3.14 Sorbitol (Alcohol Sugar Formed from Glucose)

Hydrogenation of aldohexose results in the formation of sorbitol. The catalyst employed in the chemical action-reaction may be a Raney nickel catalyst and is created in a very batch process. Sorbitol is majorly employed in the food industries in situ of sugars obtained from sugarcane (Shivlata and Tulasi 2015). Sorbitol is employed as a compound for increasing the transition temperature of polymers. This chemical is additionally used in the assembly of PET bottles. Medication additionally finds several applications in Nursing (Hubbe et al. 2008). Hydrogenation of aldohexose results in the formation of sorbitol and provides a yield of 99%. Aldohexose does not endure any biotransformation.

4.3.14.1 Methodology for the Production of Sorbitol

There are two methods to produce sorbitol. One is the industrial method, and the other is the production of sorbitol through fermentation. The industrial process involves the catalytic hydrogenation of glucose. This process takes place batch-wise and results in the production in huge volumes. The catalyst used for the hydrogenation of sorbitol is nickel. The fermentation process is not used because it results in a meager sorbitol yield. The third method is the reduction method. The reduction method involves a reduction reaction involving glucose, and during this reaction, the aldehyde radical which is formed gets converted into a hydroxyl compound. This reaction is catalyzed by aldose reductase (Suzen and Buyukbingol 2003). The starting step of the polyol pathway involving glucose metabolism is the reduction of glucose, which has been witnessed in multiple diabetic problems. The above-described mechanism entails a tyrosine residue. This tyrosine residue reacts in the agile site of aldehyde reductase. After the reaction of the tyrosine residue, the hydrogen atom in the NADH gets detached and sent to the electrophilic aldehyde's carbon atom. After the hydrogen atom transfer, the electrons on the aldehyde carbon-oxygen double bond get carried to the oxygen atom. This oxygen atom draws off the proton on the tyrosine side chain, forming hydroxyl (Meunier et al. 2004). The aldehyde reductase tyrosine phenol group performs the primary role, and it functions as a general acid whose task is to supply proton to lessen the aldehyde oxygen in tin glucose. If the blood glucose level is high in patients with diabetes, then up to 1/3rd of their glucose could pass through the pathway that involves the reduction of the glucose (Smith and Singleton 2008). In this process, the NADH gets eventually consumed and damages the cells due to the consumption of the NADH. The alternative process for making sorbitol is through catalytic hydrogenation, which involves the enantiomer of glucose like d-glucose. From the d-glucose, d-sorbitol is synthesized. The yield of d-sorbitol in this reaction is close to 100% when the reaction of d-glucose occurs with the hydrogen in water (Awuchi and Echeta 2019).

4.3.14.2 Chemicals Derived from Sorbitol

Isosorbide and anhydrous sugars: Selective dehydration of sorbitol with non-facet reactions results in medication formation. The dehydration additionally results in the assembly of anhydrides and lactones. Heterogeneous catalysts are used for the reaction and have replaced the previously used liquid catalysts. Medication is majorly used in the assembly of PET bottles and additionally within the production of polymers like polyethene medication and terephthalates (Ochoa-Gómez and Roncal 2017; Brankova et al. 2003).

4.3.14.3 Uses and Applications of Sorbitol

4.3.14.3.1 Sweetener

Sorbitol is a surrogate of sugar, with the INS number and E no. 420, when it gets utilized in food. Sorbitol is less sweet than sucrose and is approximately 60% less sweet as sorbitol provides salutary energy: of 2.6 kcal (11 kJ)/g to contrast the typical 4 kcal (17 kJ) for carbohydrates. It also acts as a nutritive sweetener. Sorbitol is not used for energy by most bacteria. However, sorbitol is slowly fermented within the mouth of a bacterium that causes cavities called *Streptococcus mutans*. Many other sugar alcohols, including isomaltose and xylitol, are considered non-acidogenic. Sorbitol is found in many of the stone fruits and berries obtained from the Sorbus saplings.

4.3.14.3.2 Medical Applications

Generally, gastrointestinal distress is caused by foods that have sorbitol in them. When sorbitol is taken orally, then it acts as a laxative. Sorbitol draws water inside the massive intestine, which further helps stimulate bowel movements (Biswas et al. 2019). Sorbitol is found in some dried fruits. It is also found in fruits such as apples, plums, pears, cherries, dates, and peaches.

4.3.14.3.3 Miscellaneous Uses

When sorbitol is mixed with saltpeter, then this mixture is used as a solid rocket propellant. Sorbitol acts as an essential key chemical intermediate used to produce fuels derived from biomass resources (Yang et al. 2021). When the sorbitol is reduced completely, it opens up new pathways for synthesizing various compounds like alkanes. These compounds are used as biofuels. The aqueous phase catalytic reforming of sorbitol leads to hydrogen production, which is required for the successful completion of the reaction given below:

$$C_6H_{14}O_6 \rightarrow 13C_6H_{14} + 36CO_2 + 42H_2O \quad (4.3)$$

The above-given reaction is exothermic. From the stoichiometric analysis of the above reaction, it has been calculated that 1.5 moles of sorbitol produce approximately 1 mole of hexane. CO_2 is not produced in the above reaction when no hydrogen is provided in the feed of the reactor. Polyols derived from sorbitol are often found in the production of polyurethane foam which is later used in the construction industry (Awuchi and Echeta 2019).

4.3.14.4 Medical Importance of Sorbitol

The first enzyme-based compound produced in sorbitol-aldose reductase trace is aldose reductase. Aldose reductase is responsible for the conversion of glucose into sorbitol, and the trimming of galactose to a compound called galactitol (Awuchi and Echeta 2019). When a large amount of sorbitol gets trapped in the retinal cells, it results in extended hyperglycemia, which goes along with poorly controlled diabetes. Inhibitors based on the aldose reductase are currently being studied as a possible alternative idea to prevent complications like the ones mentioned above (Awuchi and Echeta 2019).

4.3.15 Xylitol/Arabinitol (Sugar Alcohols Made from Xylose and Arabinose)

Xylitol, a sweet-tasting compound, is used as a non-nutritive sweetener. Xylitol may be a carbon compound having five carbon atoms (C-5 compound) (de Arruda et al. 2019). It is used in Nursing and production of unsaturated polyester organic compound (UPR). Biotransformation results from the treatment path for the lignocellulosic process followed by severance from the opposite sugars (de Arruda et al. 2019).

4.3.15.1 Methodology for the Production of Xylitol

Xylitol is present in hemicellulose as the primary C-5-carbohydrate compound. When these C-5-carbohydrate compounds undergo hydrogenation, then it results in the formation of the isomers of xylitol. The starting base material for the production of xylitol is the corn cobs, and the waste residue left after the kernels of the corn have been extracted. The waste stream of the pulp and paper industry is also used as feedstock as it contains a high quantity of carbohydrates in them. One of the significant advantages of producing xylitol is that its products can be integrated into a pulp and paper plant as the process is very similar, as well as the residue or the black liquor, which contains the lignin that acts as the feedstock for the xylitol. The major challenge is producing clean xylitol.

4.3.15.2 Chemicals Derived from Xylitol

- Xylitol and Xylonic acids, Arabic acid, and Arabinonic acid: Selective reaction results in their formation, and these acids have new uses (Werpy and Petersen 2004).
- Polyols (propylene and olefin glycols) and carboxylic acid are used as nursing liquid agents (Kandlbauer et al. 2022).
- Xylitol, xylaric, xylionic polyesters and nylons are new polymers and have several vital applications in creating powerful parts. Selective esterification, additionally with controlled branching, is finished to provide these chemicals.

4.4 Summary

According to reports, India will become the third-largest energy consumer by 2030 and will have the most significant share of energy demand growth at 25% over the next two decades. Biomass is an essential component, the base ingredient for producing an alternative source of energy that is required to meet the significant energy demand. It also helps to reduce environmental concerns as it is a renewable source. Biomass is a renewable feedstock. Products of various chemicals from biomass are of terribly high importance chemicals fascinating in replacement of the presently accessible petrochemicals to create the biorefineries additional economical. Biorefineries need to be upgraded to compete with the presently well-established fuel-based refineries and may be upgraded into integrated biorefineries. Chemicals that are derived include 5-hydroxymethylfurfural (HMF-5), levulinic acid, furfurals, sugar alcohols, drinkable acids, carboxylic acid, and phenols. These chemicals are called "Platform Chemicals" because they are used for the assembly of a wide range of chemicals on an industrial scale. Some essential chemicals produced from the biomass are four carbon di-acids: malic acid, succinic acid, and fumaric acid, 2,5-furan di-carboxylic acid 3-hydroxypropionic acid, aspartic acid, glucaric acid, glutamic acid, itaconic acid, levulinic acid 3-hydroxybutyrolactone, glycerol, sorbitol, xylitol, etc. The significant acids derived from the four-carbon acids like succinic acid, fumaric acid, and malic acid include butanediol, pyrrolidinones, etc.

However, there are specific engineering challenges in synthesizing these base four carbon di-acids. These include problems like very little availability of research work. But specific engineering problems are associated with the proper green scale production of 2,5-Furan Di-Carboxylic Acid. Green production of FDCA requires using micro-organisms for a good yield, but the highest yield from the microorganisms is obtained after the start of the process. It poses a significant drawback as it is not feasible for large-scale production of this acid.

Furthermore, FDCA requires the use of many ingredients which must be nontoxic. Biomass is considered to be one of the clean sources of energy. But there are many engineering problems and issues regarding the projects and plants involving the production of chemicals from biomass. Some of the most common challenges include the project's high costs, the very low efficiency of the technologies, and the lack of technological advancement and research. Synthesis of chemicals from biomass has high risks involved, and at the same time, the rewards are also significantly less due to the ineffective technologies and lack of study and knowledge. But now, with the depletion of the traditionally available resources, governments worldwide are pushing such kinds of projects, and proper funding in many countries.

References

Abednatanzi S, Derakhshandeh PG, Depauw H, Coudert FX, Vrielinck H, Van Der Voort P, Leus K (2019) Mixed-metal metal-organic frameworks. Chem Soc Rev 48(9):2535–2565

Ahmed I, Zia MA, Afzal H, Ahmed S, Ahmad M, Akram Z, Iqbal H (2021) Socio-economic and environmental impacts of biomass valorization: a strategic drive for sustainable bioeconomy. Sustain For 13(8):4200

Anastassiadis S, Morgunov IG (2007) Gluconic acid production. Recent Pat Biotechnol 1(2): 167–180

Andanson JM, Baiker A (2010) Exploring catalytic solid/liquid interfaces by in situ attenuated total reflection infrared spectroscopy. Chem Soc Rev 39(12):4571–4584

Awuchi CG, Echeta KC (2019) Current developments in sugar alcohols: chemistry, nutrition, and health concerns of sorbitol, xylitol, glycerol, arabitol, inositol, maltitol, and lactitol. Int J Adv Acad Res 5:1–33

Bai H, Li C, Shi G (2011) Functional composite materials based on chemically converted graphene. Adv Mater 23(9):1089–1115

Balcı CM, Palabıyık D, Beşli S (2021) Amino alcohol based cyclotriphosphazene: effects of chain length and base factor on product diversity. Inorg Chim Acta 527:120577

Banerjee P, Mahanty A, Heriniaina RM, Li T, Tengjiao G, Mallick B, Wang B (2017) Comprehensive interpretation of nanostructured lipid carriers (NLC): a promising new generation lipid carrier for the delivery of cancer chemotherapeutics. Chemotherapeutics 6(5):225–262

Benisvy-Aharonovich E, Zandany A, Saady A, Kinel-Tahan Y, Yehoshua Y, Gedanken A (2020) An efficient method to produce 1,4-pentanediol from the biomass of the algae Chlorella ohadi with levulinic acid as intermediate. Bioresour Technol Rep 11:100514

Berger R (1984) Acute and subacute effects of an oil shale retort effluent on a freshwater cyprinid. In: Pimephales promelas (master's thesis), vol 47. San Jose State University, San Jose, pp 777–780

Bermejo-Deval R, Orazov M, Gounder R, Hwang SJ, Davis ME (2014) Active sites in Sn-Beta for glucose isomerization to fructose and epimerization to mannose. ACS Catal 4(7):2288–2297

Biswas A, Cheng HN, Kim S, Appell M, Boddu V, Alves CR, Furtado RF (2019) Preparation of sorbitol-based polyurethanes and their semi interpenetrating polymer networks. J Appl Polym Sci 136(22):47602

Bolan NS, Hedley MJ, White RE (1991) Processes of soil acidification during nitrogen cycling with emphasis on legume-based pastures. Plant Soil 134(1):53–63

Brankova T, Bekiari V, Lianos P (2003) Photoluminescence from sol-gel organic/inorganic hybrid gels obtained through carboxylic acid solvolysis. Chem Mater 15(9):1855–1859

Brown CJ, Toste FD, Bergman RG, Raymond KN (2015) Supramolecular catalysis in metal–ligand cluster hosts. Chem Rev 115(9):3012–3035

Cammas S, Guerin P, Girault JP, Holler E, Gache Y, Vert M (1993) Natural poly (L-malic acid): NMR shows a poly (3-hydroxy acid)-type structure. Macromolecules 26(17):4681–4684

Cecchi T, Carolis CD (2021) Bio cascading: platform molecules, value added chemicals, and bioactives. In: Biobased products from food sector waste. Springer, Cham, pp 169–229

Cheng KK, Zhao XB, Zeng J, Wu RC, Xu YZ, Liu DH, Zhang JA (2012) Downstream processing of biotechnological products produced succinic acid. Appl Microbiol Biotechnol 95(4):841–850

Chiloeches A, Funes A, Cuervo-Rodríguez R, López-Fabal F, Fernández-García M, Echeverría C, Muñoz-Bonilla A (2021) Biobased polymers derived from itaconic acid bearing clickable groups with potent antibacterial activity and negligible hemolytic activity. Polym Chem 12(21):3190–3200

Colantonio S, Cafiero L, De Angelis D, Ippolito NM, Tuffi R, Ciprioti SV (2020) Thermal and catalytic pyrolysis of a synthetic mixture representative of packaging plastics residue. Front Chem Sci Eng 14(2):288–303

Crosby J (1991) Synthesis of optically active compounds: a large scale perspective. Tetrahedron 47(27):4789–4846

de Arruda PV, Sene L, da Silva SS, de Almeida Felipe MDG (2019) Xylitol bioproduction: state-of-the-art, industrial paradigm shift, and opportunities for integrated biorefineries. Crit Rev Biotechnol 39(7):924–943

Delidovich I, Hausoul PJ, Deng L, Pfützenreuter R, Rose M, Palkovits R (2016) Alternative monomers based on lignocellulose and their use for polymer production. Chem Rev 116(3): 1540–1599

Demirbas A (2008) Liquefaction of biomass using glycerol. Energy Sour Part A 30(12):1120–1126

Dhakshinamoorthy A, Alvaro M, Garcia H (2011) Metal–organic frameworks as heterogeneous catalysts for oxidation reactions. Cat Sci Technol 1(6):856–867

Garattini S (2000) Glutamic acid, twenty years later. J Nutr 130(4):901S–909S

Giménez-Arnau A, Silvestre JF, Mercader P, De la Cuadra J, Ballester I, Gallardo F, Bruze M (2009) Shoe contact dermatitis from dimethyl fumarate: clinical manifestations, patch test results, chemical analysis, and source of exposure. Contact Dermatitis 61(5):249–260

Grochala W, Edwards PP (2004) Thermal decomposition of the non-interstitial hydrides for the storage and production of hydrogen. Chem Rev 104(3):1283–1316

Harmsen PF, Hackmann MM, Bos HL (2014) Green building blocks for bio-based plastics. Biofuels Bioprod Biorefin 8(3):306–324

He Z, Wang X (2012) Hydrodeoxygenation of model compounds and catalytic systems for pyrolysis bio-oils upgrading. Catal Sustainable Energy 1(2013):28–52

Holladay JE, Werpy TA, Muzatko DS (2004) Catalytic hydrogenation of glutamic acid. In: Proceedings of the twenty-fifth symposium on biotechnology for fuels and chemicals. Humana Press, Totowa, pp 857–869

Hronska H, Michalikova S, Rosenberg M (2017) Microbial production of speciality C4 dicarboxylic acids from maleic anhydride. J Food Nutr Res 56(3):219–231

Hubbe MA, Rojas OJ, Lucia LA, Sain M (2008) Cellulosic nanocomposites: a review. Bioresources 3(3):929–980

Hugh R, Leifson E (1953) The taxonomic significance of fermentative versus oxidative metabolism of carbohydrates by various gram negative bacteria. J Bacteriol 66(1):24–26

Iglesias-Juez A, Chiarello GL, Patience GS, Guerrero-Pérez MO (2022) Experimental methods in chemical engineering: X-ray absorption spectroscopy—XAS, XANES, EXAFS. Can J Chem Eng 100(1):3–22

Ingole PG, Baig MI, Choi WK, Lee HK (2016) Synthesis and characterization of polyamide/polyester thin-film nanocomposite membranes achieved by functionalized TiO2 nanoparticles for water vapor separation. J Mater Chem A 4(15):5592–5604

Isikgor FH, Becer CR (2015) Lignocellulosic biomass: a sustainable platform for the production of bio-based chemicals and polymers. Polym Chem 6(25):4497–4559

Islam MR, Beg MDH, Jamari SS (2014) Development of vegetable-oil-based polymers. J Appl Polym Sci 131:18

Joanna B, Michal B, Piotr D, Agnieszka W, Dorota K, Izabela W (2018) Sugar beet pulp as a source of valuable biotechnological products. In: Advances in biotechnology for the food industry. Academic, New York, pp 359–392

Kandlbauer A, Tondi G, Zaske OC, Goodman SH (2022) Unsaturated polyesters and vinyl esters. In: Handbook of thermoset plastics. William Andrew Publishing, Norwich, pp 97–158

Kolb HC, Finn MG, Sharpless KB (2001) Click chemistry: diverse chemical function from a few good reactions. Angew Chem Int Ed 40(11):2004–2021

Kratky M, Vinsova J (2012) Advances in mycobacterial isocitrate lyase targeting and inhibitors. Curr Med Chem 19(36):6126–6137

Krautkramer KA, Fan J, Bäckhed F (2021) Gut microbial metabolites as multi-kingdom intermediates. Nat Rev Microbiol 19(2):77–94

Kricheldorf HR (2006) Polypeptides and 100 years of chemistry of α-amino acid N-carboxyanhydrides. Angew Chem Int Ed 45(35):5752–5784

Kumbhar P, Sawant J, Ghosalkar A (2016) Catalysis for renewable chemicals. In: Industrial catalytic processes for fine and specialty chemicals. Elsevier, Amsterdam, pp 597–662

Kurian JV (2005) A new polymer platform for the future—Sorona® from corn derived 1,3-propanediol. J Polym Environ 13(2):159–167

Laffitte A, Neiers F, Briand L (2016) Characterization of taste compounds: chemical structures and sensory properties. Wiley-Blackwell, Oxford, pp 154–191

Lang K, Sánchez-Leija RJ, Gross RA, Linhardt RJ (2020) Review on the impact of polyols on the properties of bio-based polyesters. Polymers 12(12):2969

Lawson T, Simkin AJ, Kelly G, Granot D (2014) Mesophyll photosynthesis and guard cell metabolism impacts on stomatal behaviour. New Phytol 203(4):1064–1081

Liao W, Liu Y, Frear C, Chen S (2008) Co-production of fumaric acid and chitin from a nitrogen-rich lignocellulosic material–dairy manure–using a pelletized filamentous fungus Rhizopus oryzae ATCC 20344. Bioresour Technol 99(13):5859–5866

Lin B, Tao Y (2017) Whole-cell biocatalysts by design. Microb Cell Factories 16(1):1–12

Litter MI (1999) Heterogeneous photocatalysis: transition metal ions in photocatalytic systems. Appl Catal B Environ 23(2-3):89–114

Liu WJ, Li WW, Jiang H, Yu HQ (2017) Fates of chemical elements in biomass during its pyrolysis. Chem Rev 117(9):6367–6398

Liu X, An W, Turner CH, Resasco DE (2018) Hydrodeoxygenation of m-cresol over bimetallic NiFe alloys: kinetics and thermodynamics insight into reaction mechanism. J Catal 359:272–286

Lu L, Mikos AG (1996) The importance of new processing techniques in tissue engineering. Bulletin 21(11):28–32

Meissner M, Valesky EM, Kippenberger S, Kaufmann R (2012) Dimethyl fumarate–only an anti-psoriatic medication? JDDG 10(11):793–801

Meunier B, De Visser SP, Shaik S (2004) Mechanism of oxidation reactions catalyzed by cytochrome P450 enzymes. Chem Rev 104(9):3947–3980

Nguyen LA, He H, Pham-Huy C (2006) Chiral drugs: an overview. Int J Biomed Sci 2(2):85

Ochoa-Gómez JR, Roncal T (2017) Production of sorbitol from biomass. In: Production of platform chemicals from sustainable resources. Springer, Singapore, pp 265–309

Ockerman HW (1996) Chemistry of meat tissue. Dept. of Animal Science, Ohio State University. US (NAL TX556.M4O3 1996)

Ozcelik B, Brown KD, Blencowe A, Ladewig K, Stevens GW, Scheerlinck JPY, Qiao GG (2014) Biodegradable and biocompatible poly (ethylene glycol)-based hydrogel films for the regeneration of corneal endothelium. Adv Healthc Mater 3(9):1496–1507

Pal S, Singh G, Singh S, Tripathi JK, Ghosh JK, Sinha S, Chakraborty TK (2015) Tetrahydrofuran amino acid-containing gramicidin S analogues with improved biological profiles. Org Biomol Chem 13(24):6789–6802

Pallotta ML (2020) The nutritional status of individuals as an indicator of resilience against destabilisation in the Covid-19 pandemic crisis: a new era for developing sustainable and leading-edge food systems. EC Nutr 15:4–6

Papadopoulos L, Zamboulis A, Kasmi N, Wahbi M, Nannou C, Lambropoulou DA, Bikiaris DN (2021) Investigation of the catalytic activity and reaction kinetic modelling of two antimony catalysts in the synthesis of poly (ethylene furanoate). Green Chem 23(6):2507–2524

Patel S, Goyal A (2015) Applications of natural polymer gum Arabic: a review. Int J Food Prop 18(5):986–998

Plackett D (ed) (2011) Biopolymers: new materials for sustainable films and coatings. Wiley, Hoboken, pp 233–254

Pollard P, Wortham N, Tomlinson I (2003) The TCA cycle and tumorigenesis: the examples of fumarate hydratase and succinate dehydrogenase. Ann Med 35(8):634–635

Pujro R, García JR, Bertero M, Falco M, Sedran U (2021) Review on reaction pathways in the catalytic upgrading of biomass pyrolysis liquids. Energy Fuel 35(21):16943–16964

Richardson S, Gorton L (2003) Characterization of the substituent distribution in starch and cellulose derivatives. Anal Chim Acta 497(1-2):27–65

Rodriguez BA, Stowers CC, Pham V, Cox BM (2014) The production of propionic acid, propanol and propylene via sugar fermentation: an industrial perspective on the progress, technical challenges and future outlook. Green Chem 16(3):1066–1076

Romero EG, Muñoz GS, Alvarez PM, Ibáñez MC (1993) Determination of organic acids in grape musts, wines and vinegars by high-performance liquid chromatography. J Chromatogr A 655(1):111–117

Rotstein BH, Zaretsky S, Rai V, Yudin AK (2014) Small heterocycles in multicomponent reactions. Chem Rev 114(16):8323–8359

Safari R, Hoseinifar SH, Dadar M (2021) Enrichment of common carp (Cyprinus carpio) diet with Malic acid: effects on skin mucosal immunity, antioxidant defence and growth performance. Ann Anim Sci 21(2):561–573

Sajid M, Zhao X, Liu D (2018) Production of 2,5-furandicarboxylic acid (FDCA) from 5-hydroxymethylfurfural (HMF): recent progress focusing on the chemical-catalytic routes. Green Chem 20(24):5427–5453

Serrano-Ruiz JC, Luque R, Sepulveda-Escribano A (2011) Transformations of biomass-derived platform molecules: from high added-value chemicals to fuels via aqueous-phase processing. Chem Soc Rev 40(11):5266–5281

Shinde SB (2007). Studies in hydrogenation of carboxylic acids using supported metals catalysts. 2007-0103

Shivlata L, Tulasi S (2015) Thermophilic and alkaliphilic actinobacteria: biology and potential applications. Front Microbiol 6:1014

Silva JAC, Grilo LM, Vasconcelos MH, Lacerda TM (2022) Levulinic acid: perspectives of its biobased production and most promising derivatives. In: Production of Top 12 biochemicals selected by USDOE from renewable resources. Elsevier, Amsterdam, pp 387–414

Smith AG, Singleton JR (2008) Impaired glucose tolerance and neuropathy. Neurologist 14(1): 23–29

Steiger MG, Blumhoff ML, Mattanovich D, Sauer M (2013) Biochemistry of microbial itaconic acid production. Front Microbiol 4:23

Suzen S, Buyukbingol E (2003) Recent studies of aldose reductase enzyme inhibition for diabetic complications. Curr Med Chem 10(15):1329–1352

Varadarajan S, Miller DJ (1999) Catalytic upgrading of fermentation-derived organic acids. Biotechnol Prog 15(5):845–854

Wang L, Eguchi Y, Chen EYX (2017) Streamlined synthesis of biomonomers for bio sourced materials: bis furan diacids, diols, and diamines via common bis furan bromide intermediates. Ind Eng Chem Res 56(40):11380–11387

Werpy T, Petersen G (2004) Top value-added chemicals from Biomass: volume I–results of screening for potential candidates from sugars and synthesis gas (No. DOE/GO-102004-1992). National Renewable Energy Lab, Golden

Willke T, Vorlop KD (2001) Biotechnological production of itaconic acid. Appl Microbiol Biotechnol 56(3):289–295

Wirth TE, Gray CB, Podesta JD (2003) The future of energy policy. Foreign Aff 2003:132–155

Yang ST, Zhang K, Zhang B, Huang H (2011) 3.16-Fumaric acid. In: Murray MY (ed) Comprehensive biotechnology, 2nd edn. Academic, Burlington, pp 163–177

Yang Y, Ren Z, Zhou S, Wei M (2021) Perspectives on multifunctional catalysts derived from layered double hydroxides toward upgrading reactions of biomass resources. ACS Catal 11(11): 6440–6454

Zhang Z, Deng K (2015) Recent advances in the catalytic synthesis of 2,5-furandicarboxylic acid and its derivatives. ACS Catal 5(11):6529–6544

Zhang Q, Wan Z, Iris KM, Tsang DC (2021) Sustainable production of high-value gluconic acid and glucaric acid through oxidation of biomass-derived glucose: a critical review. J Clean Prod 312:127745

Zhang M, Wang N, Liu J, Wang C, Xu Y, Ma L (2022) A review on biomass-derived levulinic acid for application in drug synthesis. Crit Rev Biotechnol 42(2):220–253

Zheng Y, Chen X, Shen Y (2008) Commodity chemicals derived from glycerol, an important biorefinery feedstock. Chem Rev 110(3):1807

Zhou H, Xu H, Wang X, Liu Y (2019) Convergent production of 2,5-furandicarboxylic acid from Biomass and CO2. Green Chem 21(11):2923–2927

Chapter 5
Forestry Biomass as Carbon Neutral Source for the Production of Biofuels and Aromatics

Uplabdhi Tyagi, Neeru Anand, Arinjay Kumar Jain, and Deepak Garg

Abstract The production of a wide range of value-added products is heavily dependent on conventional fuels. With an annual growth rate of around 7%, the waste to energy sector in India could reach 14 billion USD by 2025. Despite continuous increase in energy demands, several types of underutilized wastes are generated from energy crops, anthropogenic, forestry, and municipal, agricultural, and industrial activities in huge quantities to maintain the country's growth. It is estimated that nearly 560 billion tons of biomass is available on Earth, while the total primary production is ~100 billion tons/year. There is a surplus of agricultural and forest land in India, which produces about 500 million metric tons of biomass every year. In the last 50 years, agricultural production has increased by more than three times due to the expansion of agricultural soil. As agricultural products become more popular, this demand is expected to increase. Agricultural wastes include crop waste, livestock waste, food processing wastes, and animal wastes. Also, in agriculture farm, 5.3 kg of manure is generated per 100 kg of live weight (wet weight) each day. Biofuels and aromatics can be produced from lignocellulosic biomass, which is an economical, renewable, and abundant alternative to fossil fuels. As part of a bio-based economy and a biorefinery, there is a significant opportunity for the development of biodegradable building blocks (monosaccharides, oligosaccharides, biofuels, and polymers) and materials (fiber products, cellulose nanofibers, starch derivatives, and furfurals). Bioenergy needs to be modernized to fit into a sustainable, environmentally friendly, economically viable, and socially responsible development pathway. Various conversion technologies and pathways are capable of solubilizing diverse chemical constituents and are used for the production of a wide range of bio-derived intermediates and end products. The efficiency of conversion technologies differs greatly and depends on the type of biomass used as raw material, which may contain different fractions and compositions of cellulose, hemicellulose, and lignin. With the rising demand, high cost, and emerging

U. Tyagi · N. Anand (✉) · A. K. Jain · D. Garg
University School of Chemical Technology, Guru Gobind Singh Indraprastha University, New Delhi, India
e-mail: neeruanand@ipu.ac.in; deepak@ipu.ac.in

D. B. Pal, A. K. Tiwari (eds.), *Sustainable Valorization of Agriculture & Food Waste Biomass*, Clean Energy Production Technologies,
https://doi.org/10.1007/978-981-99-0526-3_5

environmental concerns of fossil fuels, the monetization of biomass waste can provide significant benefits. This chapter explores the potential of waste biomass as a promising source for commercially viable industrial chemicals such as monosaccharides, biofuels, and oligosaccharides.

Keywords Biomass · Cellulose · Hemicellulose · Biofuels · Carbon neutral source and aromatics

5.1 Introduction

Global energy demand is intensifying daily, leading to an energy catastrophe and environmental pollution. The world's energy demand is expected to increase by 48% over the next 20 years due to the rapid growth of the world's population. Fossil resources are utilized not only for the production of energy but also for the manufacturing of value-added products. Currently, 80% of energy requirement is met by fossil fuels, while 97% production of aromatics relies on crude oil. Literature suggests that three major basic aromatics, including benzene, toluene, and xylene, were industrially produced in huge quantities, i.e., 95 million tons/year in 2012. About 40 million tons of benzene and xylene are consumed annually, while toluene is used as a chemical raw material in lesser amounts. The current global crude oil consumption is 85 million barrels per day (Mbpd), out of which 600 to 1200 million metric tons are naphtha (15–30% of crude oil), which contains linear and cyclic paraffinic hydrocarbons and is potentially consumed for aromatics production (Prakash et al. 2021). Unfortunately, the progressive consumption of fossil fuels and their growing concerns about exhaustion, climate change, energy security, and environmental sustainability is a serious threat to the nations. Hence, there is a global need for a shift to renewable and sustainable resources for the production of fuels, chemicals, and materials.

Lignocellulosic biomass is a renewable organic material which is derived from plants and animals. It is composed of cellulose, hemicellulose, and lignin, each of which has its unique and complex structure. Literature suggests that forestry biomass covers 4.03 billion hectares worldwide, equivalent to about 30% of the world's total land area. The photosynthesis process on Earth produces about 170 billion tons/year of lignocellulose by immobilizing carbon from atmospheric carbon dioxide. In addition, the agricultural waste produced alone in Europe stood up to ~250 million tons/year, and according to the World Bank, more than 2.24 billion tons of municipal waste is generated annually (Thangavelu et al. 2022). Table 5.1 summarizes the supply of bioenergy derived from biomass in the year 2000 to 2019.

Lignin is an alkyl-aromatic polymer found in the cell walls and the second most readily available biopolymer in biomass. It consists of a highly irregular structure due to the presence of three phenolic building blocks, p-coumaryl (H-unit), synapyl (S-unit), and coniferyl (G-unit), thus forming a 3-D spatial structure due to the linkage of ether and C–C linkages. Depending upon the origin of lignin, it consists of varying proportions of these monomers. Hardwood mainly consists of S and G

Table 5.1 Biomass energy supply from 2000 to 2019

Supply of bioenergy supply from biomass (EJ)						
Year	Total amount	Total municipal solid waste	Total industrial waste	Total solid biomass	Total biogas	Total liquid biofuels
2000	41.6	0.74	0.50	39.5	0.29	0.52
2019	56.9	1.45	1.14	48.5	1.43	4.30

units in equal amounts, but on the other hand, softwood contains high amounts of G units. The world produces about 50 million tons of lignin each year. Out of the total available lignin, lignin derived from the sulfite pulp industry contributes to only 1–2% for the production of useful aromatics, while the remaining 98–99% of lignin is incinerated for the production of steam and energy (Fírvida et al. 2021). Hence, lignin has a high potential for conversion to various essential aromatic compounds abbreviating the dependency on nonrenewable resources. Due to its complex structure, low reactivity, and low solubility, lignin cannot be utilized or transformed into valuable products on a large scale in biorefineries. Therefore, an effective extraction followed by pretreatment is highly critical to overcoming the recalcitrant properties of lignocellulosic biomass for downstream biological transformations. Numerous researches have been demonstrated for the valorization of lignin for the effective production of aromatics. Saraeian et al. (2020) reported lignin valorization for the production of alkenes and aromatics through pyrolysis and vapor phase hydrodeoxygenation. Results revealed that hydrodeoxygenation process using MoO_3 as a catalyst was highly effective and leads to the generation of mono-aromatics (17–29 C%) and alkenes and alkanes. In addition, the char formation was also significant (50%), while partial depolymerization leads to increase in the aromatic yield (53–55 C%). Singh and Dhepe (2019) reported the conversion of lignin using catalytic ionic liquids with varying combinations of various cations and anions under mild operating conditions (120 °C, 1 h, ambient pressure). Results revealed that imidazolium cation and HSO_4 as an anion were found to be promising as compared to other ionic liquids and showed highest activity, i.e., 78% yield, respectively. Khan et al. (2022) critically reviewed recent progressions in the field of lignin valorization techniques for the sustainable production of aromatics. This article also addresses the potential benchmarks to fossil refinery products and explores the breakthrough of lignin pretreatment, extraction and isolation, thermocatalytic conversion, and high-end aromatic products.

5.2 Properties, Potential, and Current Application of Biomass

Lignocellulosic materials are primarily composed of cellulose, hemicellulose, and lignin polymers. These polymers are arranged randomly and form a heterometric structure, and the characteristics of all these polymers vary differently depending on the type of biomass, composition, and source of the biomass. Each type of lignocellulosic biomass is optimized for energy conversion based on the relative availability of holocellulose and lignin. A wide variety of biomass including woody, herbaceous, and aquatic plants, grasses, agricultural residues, manures, pesticides, and municipal waste contain different fractions of holocellulose, lignin, and extractives. Generally, plant biomass contains ~45–55% cellulose (except for cotton and hemp bast fiber, which contain 80% cellulose), 20–40% hemicellulose, and 25–35% lignin. As a result of the inherent properties of biomass sources, biomass can be recalcitrant to bioprocessing. Several properties including lignin content, cellulose crystallinity, and accessibility to cellulase determine the overall solubility and digestibility of the biomass (Yu et al. 2021). The relationship between the heterometric structure and the fraction of carbohydrates of any biomass type reflects the complexity of any biomass type. Factors that cause recalcitrance in biomass include cellulose sheathing by hemicellulose, cellulose crystallinity and degree of polymerization, protection of cellulose and hemicellulose by structured lignin, accessible surface area, and fiber strength. Depending on the biomass feedstock, this variability contributes to differences in digestibility/hydrolysis. The effective removal of lignin selectively enhances the digestibility of biomass and improves the rate of hydrolysis or microbial digestibility. Amorphous cellulose has also been shown to be less susceptible to cellulase attack than highly crystalline cellulose. The highly crystalline biopolymers result in low product yield and require high energy input and operating costs. This can be improved by targeting the upstream activity such as size reduction by any mechanical means and physicochemical pretreatments such as acid/alkaline hydrolysis.

5.2.1 Efficacy of Pretreatment

Pretreatment of lignocellulosic biomass requires several key features. It should offer the following advantages: cost-effective both in terms of capital and operation, effective on a different type of biomass, compatible with varied loadings of biomass, high recovery of lignocellulosic components in a usable form, low formation of inhibitory compounds, requiring low preparation/handling or preconditioning steps, and economical solvents and catalysts. All these features are considered in such a way that the efficacy of pretreatment balances the cost of downstream processing and manufacturing cost (biomass and reagent) and is comparable with operational cost. However, it is difficult to compare and determine the efficacy of pretreatment technologies due to variations in their upstream processing cost, chemical recycling, capital investment, downstream processing cost, and waste treatment systems (Domínguez et al. 2020). The use of mass and energy balance analysis can, however, validate the effectiveness of a pretreatment technique for any given biomass type when integrated into an industrial system or biorefinery. Many researchers have investigated the efficiency of several pretreatment methods including oxidative, acid/alkaline, reductive, and ozonolysis, and further technological progression is still underway (Thangavelu et al. 2022; Domínguez et al. 2020). For industrial processes, determining the best pretreatment method requires a thorough economic analysis of a particular feedstock, particularly when colocated with existing plants that have inexpensive power, steam, chemical reagents, or default treatment.

5.2.2 Assessment/Evaluation of Pretreatment

The ideal and effective pretreatment process generates highly amorphous substrate which can be readily solubilized without the formation of sugar, degradation products (furans), and inhibitors. To evaluate the efficacy of biomass pretreatment, it is necessary to consider a parameter called the "severity factor." Temperature, acidity, and pretreatment duration combine to determine the effectiveness of the severity factor (Khan et al. 2022). Pretreatment assessment can be performed using the following ways such as analyzing/quantifying the sugars contents, insoluble fractions after pretreatment, fermentation of hydrolyzed, and enzymatic hydrolysis of residue. For instance, ammonia fiber explosion pretreated (AFEX)-derived water-insoluble solids can be effectively used as a nutritional and potential supplement in cattle feeds.

The chemical composition of biomass waste varies greatly and has shown potential for the development of economically viable biorefineries. Residues from agriculture and forestry contain high levels of cellulose and hemicellulose, which are used to produce fermentable sugars, aromatics, and biofuels (ethanol, methanol, and butanol). Food processing waste, such as carrot waste and apple pomace, also contains a high amount of holocellulose. Therefore, such wastes can be converted into sugar without any complex pretreatment. Spent coffee grounds comprises of ~30.1% hemicellulose and ~19.5 mannan, which is the main polysaccharide in the coffee grounds (Saraeian et al. 2020). This composition makes spent coffee grounds highly potential source for the production of oligosaccharides and its derivatives. Several biomass waste sources have the potential to produce xylooligosaccharides, including rice straw, corn cobs, wheat straw, corn stover, sugarcane bagasse, and switchgrass which are rich in xylan. Pectic oligosaccharides are considered an emerging class of prebiotics and can be produced from pectin-containing agricultural residues. This includes citrus residues (~35%), apple/orange pulp (~21%), sugar beet/beetroot pulp (~17.9%), olive/carrot pomace (~35%), potato/pineapple residue (~18%), soy hull/groundnut (~17%), and onion/garlic skin (~28–35%) (Yu et al. 2021; Khan et al. 2022). Fruit and vegetable waste-derived phenolic acids, flavonoids, carotenoids, and their derivatives are major bioactive compounds which possess anti-inflammatory, antioxidant, antithrombotic, antiallergenic, antimicrobial, and arthrogenic properties and can be effectively utilized in nutraceutical, pharmaceutical, and cosmetics industries. In recent years, new research has focused on producing economical nanocellulose from a variety of fiber-based waste materials. Lignin being the most abundant natural polymer can be directly isolated from wood wastes and agricultural residues. In addition to classical chemical applications, lignin can also be used in innovative future platforms. The effective lignin valorization is 10 times better in terms of high-end products compared to directly burning it for the generation of steam and electricity. As a result, converting waste into energy, nanocomposites, platform chemicals, aromatics, or polymers will not only be economically efficient but will also benefit the environment. Following are the different types of waste generated that creates severe problems for human health and the environment.

5.2.2.1 Wastewater and Industrial Wastes

Such wastes include discharge from industries such as vegetable packaging, animal manure, milk processing units, and black liquor obtained from the paper and pulp industry and breweries. Dumping of sewage and other wastes causes severe water and soil pollution. The transformation of wastes on land, and the release of the organic matter directly into the groundwater or surface water leads to severe health problems and also increases the fish mortality rate.

5.2.2.2 Food Industry Wastes

These wastes include a wide variety of restaurants, hotels, and kitchen wastes such as vegetable flay, fruit and vegetable residues, stale food (uneaten bread and rice), and confectionaries (nonstandard food, filter sludge, fruits and vegetable scraps, pulp, solids, and fibers obtained from extraction and isolation of sugars and starch). These wastes are typically disposed of in landfill by the process of incineration due to their promising characteristics and potential raw material for the generation of biogas by anaerobic digestion. In addition, liquid wastes are also generated from the cleaning of poultry and fish, fruit and vegetables, wine-making processes, and meat washing processes which contain a significant amount of sugar and starch (Singh and Dhepe 2019). Thus, the production of biogas and further fermentation to bio-alcohol is the most feasible and sustainable alternative to cover such a huge quantity of waste.

5.2.2.3 Animal Wastes

The main components of animal manure are ash, organic material, and moisture. The animal manure degradation can be performed under aerobic or anaerobic conditions.

Anaerobic conditions produce extra CH_4, whereas aerobic conditions produce carbon dioxide and stabilized organic materials (SOM). India has a great potential for CH_4 production due to the significant production of animal manure. This enables frequent and more advance transformation of animal manure into useful products.

5.2.2.4 Municipal Solid Waste

Each year, millions of tons of household waste are collected, and most of it is disposed of in open fields. MSW in India is primarily composed of paper and plastic, which account for 80% of the total. Several methods are available for the conversion of municipal solid waste into energy, including anaerobic digestion and direct combustion. The landfills produce methane and carbon dioxide in a 1:1 ratio through natural decomposition. To produce energy, these gases are collected from the collected materials and then swabbed before being fed into gas turbines or internal combustion (IC) engines (Sethupathy et al. 2022). Biogas can be generated from the municipal solid waste by fermenting organic fractions anaerobically in high-rate biomass digesters for steam and electricity generation.

5.2.2.5 Sewage Waste

Similar to other animal wastes, sewage is also a potential source of bioenergy and a promising feedstock for the production of value-added chemicals. To produce biogas, the anaerobic digestion process can be used to extract energy from the sewage (Pang et al. 2021).

In India, there is a large quantity of biomass waste. Several biological and thermochemical processes can be used to convert these wastes into biofuels. Biomass wastes derived from agriculture include rice straw, maize straw, wheat straw, paper waste, corn straw, sugarcane bagasse, and rice husk, which generate 731, 256, 354, 323, 204, 181, and 115 million tons, respectively, each year. The USA and Canada alone produce approximately 72.5 Mt of forestry residues every year from harvesting and processing. There is also a substantial amount of biomass waste generated by the food industry every year, including rapeseed meal (37 Mt), citrus waste (16.9 Mt), banana waste (11.5 Mt), grape pomace (5.8–9.3 Mt), and apple pomace (4.5–6.8 Mt) (Singh and Dhepe 2019). On the other hand, ~30 Mt of waste is produced annually only from the olive oil industry in areas of the Mediterranean, which is a serious environmental concern. Also, ~7.4 Mt of spent coffee grounds are produced by the coffee agro-industry, as well as large quantities of coffee pulp, silver skin, and cherry husks are produced in a significant amount, which are damaging the ecosystem due to their degradable characteristics (Pang et al. 2021; Ye et al. 2021). These huge quantities of biomass waste are usually disposed of in open land or groundwater; hence, an effective management system that can utilize and manage such a huge quantity of waste is essential. Valorization of wastes not only resolves disposal problems but also produces a wide range of value-added products effectively and sustainably.

Biomass waste conversion to energy is being practiced by some agencies and industries in India and has reported huge benefits. The share of biomass in global energy consumption will rise from 15% in 2001 to 30% in 2040 (Radhakrishnan et al. 2021). The Indian scenario demonstrates the enormous potential for converting biomass waste to energy. By implementing these technologies, industries can reduce their energy costs in-house, resulting in increased profits. These technologies are also being adopted by various industries with the help of government and nongovernment agencies. This includes several government agencies like MNRE, DBT, CSIR, and DST and other academic institutions like the Indian Institute of Science (IISc), National Institute of Technology (NITs), and Indian Institutes of Technology (IITs), as well as other nongovernment organizations. Biomass reserves in India are estimated to be about 500 metric tons per year and have the potential to generate 17,500 MW of power. While ~5500 MW of electricity can be produced using the available 130–180 Mt of biomass. This profuse biomass is generated from several existing industries, while most of the quantity is generated from the sugar industry. According to existing combustion technology available for biomass transformation, ~6.2 EJ of direct heat is generated from industrial and residential sectors, while ~3.8 to 5.9 EJ of heat is generated from combined heat and power (CHP) plants. It is estimated that ~73,000 MW of energy will be generated from biomass as well as bagasse cogeneration by 2032 (Yang et al. 2021). Also, ~94,125 MW of power is generated using renewable energy resources, i.e., wind power contributes ~52.20%, small hydropower contributes ~20.98%, cogeneration bagasse contributes ~5.31%, biomass power contributes 18.63%, and waste to energy contributes 2.88%, respectively. In various states of India, biomass power generation projects are being installed for supplying energy requirements through biomass, while platform chemical production is still underway. Table 5.2 summarizes the estimated economic benefits of different biomass technologies for the production of electricity and biofuels. The developing states for power projects and chemical manufacturing using biomass are Maharashtra, Karnataka Chhattisgarh, Tamil Nadu, Uttar Pradesh, and Andhra Pradesh. Several states including Uttar Pradesh, Maharashtra, Andhra Pradesh, Tamil Nadu, and Karnataka have taken the position of leadership toward bagasse cogeneration and chemical manufacturing. There have been 130 biomass power projects installed by MNRE, which aggregate 999.0 MW, as well as 158 bagasse cogeneration projects in sugar mills that combine to generate 1666.0 MW of power (Bilal et al. 2021). A total of 35 biomass power projects contributing to 385 MW of energy are under construction at various pilot stages. Over 70 cogeneration projects are currently being implemented with a capacity of 895 MW.

Table 5.2 Estimation of the economic benefits of different biomass technologies for the production of electricity and biofuels

Biofuels			
Process	Plant size (million liters/ year)	US dollar/l	Conversion efficiency (liters of ethanol or biodiesel/ ton) (%)
Enzymatic hydrolysis	235	0.39–0.68	31–35
Fischer-Tropsch	48–365	0.37–0.97	15
Gasification	121–148	0.44–0.49	19–25
Transesterification	35–289	0.65–0.98	96–99
Thermochemical process	250	0.38–0.90	25–34
Fast pyrolysis	135–219	0.15–0.18	22–35

5.3 Biomass Waste and Its Environmental Impact

Annually, a huge amount of biomass waste is generated around the world. Traditionally, biomass was either burnt or transformed into organic fertilizers naturally under favorable process conditions. In recent years, biomass waste has become increasingly problematic due to its serious environmental concerns. In undeveloped countries, despite the adverse effect on the environment including the release of toxic gases such as nitrogen dioxide, carbon monoxide, hydrogen sulfide, and nitrous oxide, the burning of agricultural and forestry waste is a common practice. This results in the formation of toxic gases such as ozone, sulfur dioxide and nitric acid are formed, which contribute to acid deposition and pose a serious threat to human and animal health. Also, as a result of insufficient skills, climate change, poor infrastructure, and natural calamities, fruit and vegetable wastes are primarily generated during the production and storage stages. The majority of these wastes are dumped in landfills which emit methane and carbon dioxide, contaminate surface water, and groundwater, emit odors, and contaminate the soil. Methane emitted from landfill contributes significantly to greenhouse gas emissions (GHGs). Agricultural, landfill, wastewater, and fossil fuel production and transportation account for ~65% of methane emissions globally (Acciardo et al. 2022). Further, landfill leachate comprises high levels of NH_3-N, organic and inorganic compounds, heavy metals, and other hazardous compounds which influence the growth of plants. The release of leachate from irrigation activities affects soil characteristics, including salinity and biotoxicity. Consequently, the accumulation, degradation, and inefficient treatment of biomass materials can cause serious environmental impacts. For instance, Usmani et al. (2021) studied the bioprocessing of waste biomass for the sustainable development of value-added products and minimizes their associated environmental impact. This study explores the aspects of advancement in the bioeconomy and lifecycle assessment and determines the avenues of process optimization. Similarly, Ahmed et al. (2021) demonstrated the socioeconomic and environmental aspects of biomass valorization and provided insight details of physiochemical characteristics of waste generated by domestic and industrial sectors.

5.4 Current Status of Biomass Valorization

In modern biorefineries/bioenergy toward a climate-neutral future, biomass valorization plays a key role in the production of various value-added biochemicals and biofuels. Table 5.3 summarizes the production of bioproducts in various operational plants along with their plant capacities.

The conversion process is another important part of waste biomass transformation. It is not recommended or even possible to utilize waste biomass in its pristine state for several reasons. Depolymerization and subsequent conversion are the most commonly employed processes. However, the main challenge in converting lignocellulosic biomass is recalcitrance, which occurs due to the natural resistance of plants. To overcome this problem, several pretreatment options including physical, thermal, chemical, biological, or combinations of these are available as depicted in Fig. 5.1. There have been several pretreatment processes developed for increasing the accessibility of cellulose and hemicellulose that effectively breaks the complex linkages between these polymers. The operating conditions and efficacy of different pretreatment techniques have been extensively studied by researchers, so they are not included in this chapter. The screening of an effective and suitable pretreatment technique should be given great attention since it is the most expensive and essential step of biomass isolation and extraction. In selecting a suitable pretreatment method, the following parameters should be taken into account: (1) type of biomass, (2) conversion/transformation technology, (3) inhibitory compounds, (4) desired end product, (5) concurrent and consecutive step(s), and (6) feasibility including social, economic, environmental, and political aspects (Cassoni et al. 2022). Despite this, it is hard to discover a completely trouble-free pretreatment technique. As a result, a relatively effective biomass pretreatment technique may demonstrate the following characteristics: (1) scale-up capability, (2) environment friendly, (3) minimum processing steps, (4) economical, (5) non/less toxic, (6) low energy consumption, (7) avoid/minimize the formation of inhibitory compounds, and (8) high downstream process efficiency. In addition, the type of feedstock, pretreatment technique, and the final product could influence the choice of waste biomass conversion technology as shown in Fig. 5.2.

This section discusses some latest technologies for managing and valorization waste biomass. Recently, some of the promising technologies including ionizing and nonionizing radiation, ultrasound radiation, high-pressure steam explosion, pulsed electrical field, and solar-driven conversion are becoming sustainable and green pretreatment options for the upgradation of biomass valorization at large scale. These technologies still have a lot of challenges to overcome in terms of capital

Table 5.3 Summary of current operational plants with their plant capacity

Company name	Operational year and origin	Raw material	Products	Annual production
Reverdia VOF	2012; Italy, Europe	Starch	Levulinic acid and succinic acid	15 kt
BioAmber Inc. and Mitsui & Co.	2014; Canada, North America	Glucose, fructose	Levulinic acid and succinic acid	35–55 kt
Diamond Green Diesel	2013; United States	Inedible corn oil, animal fats, and used cooking oil	Diesel	280–680 MG
DuPont Tate & Lyle Bio Products Company, LLC	2004; United States	Starch, glucose	1,3-propanediol	ND
Myriant Technologies LLC	2013; United States	Sorghum and sugars derived from lignocellulosic	Levulinic acid and succinic acid	15.5 kt
BioAmber (DNP/ard)	2010; France, Europe	Glucose derived from wheat	Levulinic acid and succinic acid	4.5 kt
Novamont S.p.A.	2016; Italy, Europe 2018; France, Europe	Sugars Other plant origins	BDO Bio-polyester	35.5 kt 105 kt
Succinity	2014; Spain, Europe	Different renewable materials	Levulinic acid and succinic acid	10 kt
BioAmber Inc. and Mitsui & Co.	In process, North America	ND	Levulinic acid and succinic acid	70–200 kt
Nature Works™ LLC	ND	ND	Lactic acid	140 kt
PTT MCC BioChem	2010; Thailand 2015; Thailand	Lignocellulosic biomass Lignocellulosic biomass	Succinic acid Polybutylene succinate	36 kt 20 kt
Huangshi Xinghua Biochemicals Co. Ltd.	China	Sweet potato, fruit waste, and corn	Citric acid and other citrate salt	65 kt
Avantium	2018; Netherlands, Europe	Wood chips, bagasse, and grasses	Reducing sugars and lignin	135 kt
COFCO Corporation	2011; China	Corn stalk and wheat straw	Ethanol	3.9 MG
CalBio Co. Ltd.	2019; United States	Dairy waste	Electricity	320 MW

(continued)

Table 5.3 (continued)

Company name	Operational year and origin	Raw material	Products	Annual production
Mascoma Co. Ltd.	2014; United States	Waste derived from wood	Ethanol	25–85 MG
VärmlandsMetanol AB (publ)	2006; Sweden	Residue from forestry waste	Methanol	135,200 m^3
Enerkem	2015; Canada, North America	Sorted municipal solid waste	Methanol and ethanol	15 MG
Abengoa Bioenergia SA	2011; United States 2010; Netherlands, Europe	Corn residual waste Corn or wheat residues	Ethanol, methanol Ethanol, methanol	27 MG 130 MG
Abengoa	2014; United States 2014; Belgium, Europe	Wood chips residue Agro-residues	Electricity Thermal power fuel	220 110 MW
BlueFire Renewables	2013; United States	Sorted municipal solid waste, agricultural residues, and wood waste	Ethanol and methanol	22 MG
Flexible Solutions	2009, Canada, North America	Reducing sugar	Aspartic acid and lactic acid	5–15 kt

BDO 1,4-butanediol, kt 1000 t and MG million gallons

cost and other constraints. One of the disadvantages of commercially upgraded pretreatment techniques is their high energy consumption, which increases production costs. For the transformation of waste biomass, the adoption of economical energy sources and viable energy-saving processes has been and will continue to be the core research areas. Hydrothermal carbonization has become increasingly popular due to its ability to valorize food waste at relatively low temperatures (170–370 °C) using wet feedstocks. In contrast, most common traditional techniques such as incineration, gasification, and pyrolysis require higher temperatures (450–600 °C and 350–1100 °C, respectively) and low moisture substrates, thus adding time and energy to the process (Chakravarty et al. 2022; Bilal et al. 2021). In the past, syngas was generated by electrolyzing water with alkali at low temperatures and pressures, while modern technology uses a completely off-grid solar-driven route for the production of cellulosic-derived ethanol. In this regard, the cold hydrolysis process was found to be significant for the production of ethanol from various biomass feedstocks. In this method, hydrolysis of starch is performed below gelatinization temperature which reduces energy requirements, processing and pretreatment steps, and operation costs and maximizes the process productivity. It is considered one of

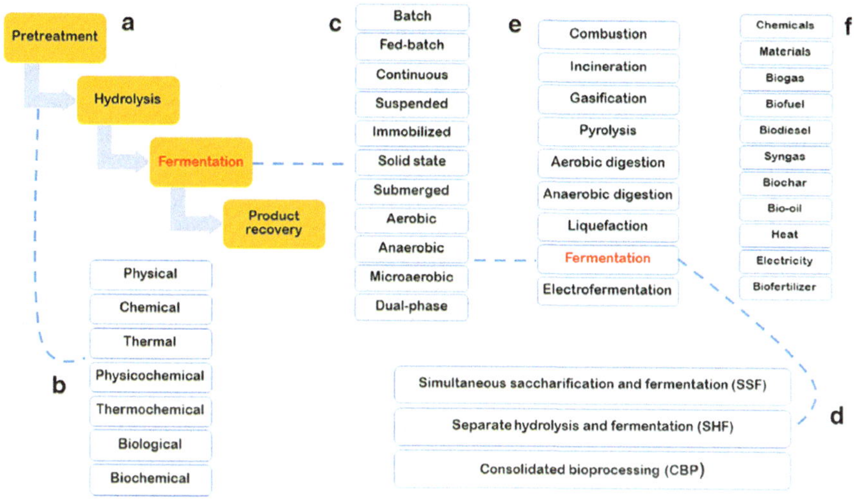

Fig. 5.1 Schematic representation of waste lignocellulosic biomass to desired products: (**a**) Overview of the scheme/process, (**b**) methods/modes of pretreatment, (**c**) types of fermentation, (**d**, **e**) conversion methods of waste biomass, (**f**) desired valuable products

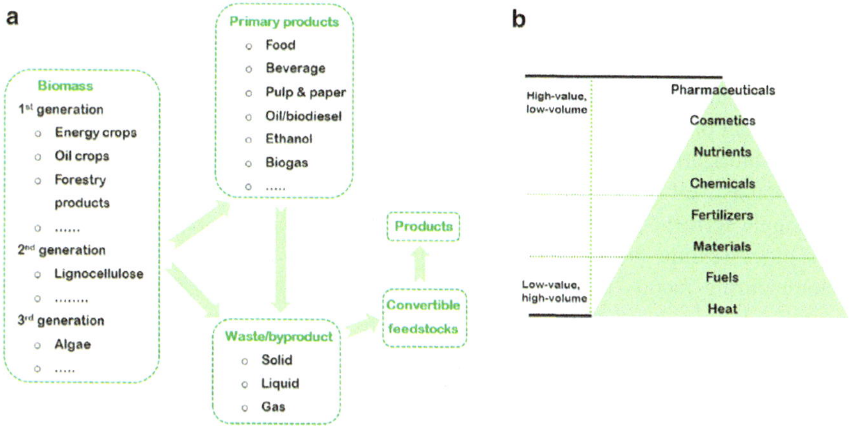

Fig. 5.2 (**a**) Integration of biorefinery processes (**b**) Low-value with high-volume and high-value with low-volume products

the key innovations in the starch-to-ethanol industry. Besides the effective production of starch and ethanol, this process also offers the advantage of producing a wide variety of other value-added products. Furthermore, biological pretreatment exhibits interesting characteristics such as nontoxicity, minimal energy consumption, low formation of inhibitory compounds, and environmental friendliness. The disadvantage of these methods is that they require longer retention times, so for large-scale applications, these methods need to be combined with chemical or physical pretreatment. Valorization of biomass wastes via biological methods includes pretreatment by bacteria, fungi, and enzymes. There is a substantial cost associated with the use of commercially available enzymes. Bioethanol production has been reported to be the most expensive due to enzyme purchases (up to 40% of the total costs). The development of more effective pretreatment methods that do not need enzymatic hydrolysis, or the use of more cost-effective and sustainable strategies to produce enzymes, are alternatives to enzymatic hydrolysis. Some studies reported that fruit and vegetable waste hydrolysates were fermented with enzymes derived from the solid-state fermentation of fungi to produce succinic acid, while some studies reported that submerged fermentation of filamentous fungi directly degraded agricultural residues without additional pretreatment. Also, recent studies investigated the possibility of developing natural degradation systems for lignocellulosic biomass utilizing microbes in animals' digestive systems. Animals usually digest plant materials with the help of microbes present within their guts; hence, to develop an efficient biorefinery system, it is necessary to understand the underlying mechanism. In this regard, the utilization of ruminant animals has facilitated the isolation of a significant number of microorganisms. The other captivating technique is consolidated bioprocessing, where lignocellulosic biomass is transformed into desired products in a single step without enzymes. During this process, the given biomass is supposed to be metabolized by microbes that can produce essential products with high yield and selectivity. Unfortunately, there are no naturally occurring microbes with such characteristics; hence, genetic engineering can play a vital role. R&D advancements in the field of metabolic engineering, catalysis, hydrolysis, ozonolysis, and oxidative processes with modern tools and strategies lead to the development of tunable and tailored microbes, solvents, catalysts, and other reagents for the valorization of biomass. The saccharification of biomass followed by fermentation as well as the synergistic effect of pretreatment followed by fermentation in a single fermenter reduces process time and costs. According to some studies, such processes are capable of improving product quality, facilitating multiproduct manufacturing facility design, and providing high chemical recovery rates. Commonly, cellulose, hemicellulose, and lignin are separated from lignocellulosic biomass and then depolymerized, converted, and finally regenerated as the desired product. However, these processes have several disadvantages, including loss of biomass functionality, complex separation, low regeneration, formation of side products, and high energy requirements (Nguyen et al. 2021). In recent years, the complete conversion of waste biomass into desired products in a single pot has received considerable interest. A multistep process to utilize biomass simultaneously entails the aforementioned disadvantages. It was, therefore, proposed to eliminate complex pretreatments and tedious separations or extraction by directly catalyzing lignocellulosic biomass. Many researchers have demonstrated the transformation of biomass to produce platform chemicals and aromatics and also provide insights of thermochemical, physiochemical, biological, and integrated processes. In addition, several researches have demonstrated the effective utilization, recycling of residue from pyrolysis, and other thermochemical processes (Okolie et al. 2022; Roy et al. 2021). The authors critically evaluated the use of nanotechnology in lignocellulosic biomass valorization including pretreatment and hydrolysis. Different nanoparticles have been utilized in these processes including enzyme immobilization as well as end-product utilization (Sarwer et al. 2022; Flores et al. 2021; Devi et al. 2022).

5.5 Current Status of Different Bioproducts Obtained Biomass Valorization

5.5.1 Monosaccharides

It is necessary to deconstruct biomass into its constituent sugars to produce biofuels, aromatics, platform chemicals, or biopolymers. Table 5.4 summarizes the important conversion technologies and their associated products. Fermentation of sugars can also be used as a source of carbon for the conversion of sugars into building blocks. The United States Department of Energy has recognized 12 different sugar-derived building blocks, including 1,4-diacids (succinic, fumaric, and malic), 3-hydroxy propionic acid, 2,5-furan dicarboxylic acid, glutamic acid, aspartic acid, xylitol/ arabinitol, glucaric acid, 3-hydroxybutyrolactone, itaconic acid, levulinic acid, glycerol, and sorbitol which can be transformed into emerging classes of platform chemicals and effectively replaces the commonly used petroleum-derived products. Plant cell walls contain polymers such as cellulose, hemicellulose, and lignin that make up heterometric biomass matrix. A combination of acid hydrolysis followed by enzyme hydrolysis produces monomeric sugars from cellulose and hemicellulose

Table 5.4 Main conversion technologies and their corresponding products

Process/ technology		Feedstock	Usable end product
Thermochemical	Combustion	Agriculture waste, municipal waste, and animal wastes	Heat, energy, and electricity
	Pyrolysis	Agricultural waste, forestry waste, and wood waste	Pyrolyzed oil, solid char, and producer gas
	Gasification	Forestry waste, agricultural waste, and wood waste	Producer gas, solid char, and liquid fuels
	Liquefaction	Agricultural waste, algal biomass, and wood waste	Fertilizers, pesticides, syngas, and liquid fuels
Biochemical	Anaerobic digestion	Animal wastes, municipal waste, and sewage sludge	Liquid fuels, electricity, heat, and biogas
	Fermentation	Agricultural waste, starch, perennial grasses, and sugars	Liquid fuels and heat
Physiochemical	Esterification or transesterification	Vegetable oils, animal fats, and waste oils	Liquid fuels, electricity, and glycerol

(Radhakrishnan et al. 2021). Enzymatic hydrolysis is a favorable process as it offers the following advantages: high conversion yield, mild process conditions, minimum inhibitory byproducts, high selectivity, and low corrosion (Nguyen et al. 2021; Usmani et al. 2021). An effective pretreatment step reduces the recalcitrance or heterogeneity of biomass thereby increasing hydrolysis yield. There has been considerable interest in pretreatment-assisted enzymatic hydrolysis of lignocellulosic sugars in the laboratory and on a pilot scale. In this regard, the American Process company has developed a sustainable and cost-effective process that utilizes a combination of sulfur dioxide and ethanol to depolymerize and isolate biomass components from its matrix. As a result of autohydrolysis, cellulose can be easily transformed into glucose, while hemicellulose produces sugar by enzymatic hydrolysis. Comet Biorefining constructed a sugar plant in 2018 which has a plant capacity of 60 million pounds of sugar annually from corn stover using enzymatic hydrolysis. Similarly, Virdia Ltd. developed a cold acid solvent extraction technology for the transformation of wood chips and other cellulosic materials into industrial sugar and resulting in a maximum yield of ~95–97%. In the past few years, Renmatix has developed a technology that can produce 1 Mt of cellulosic sugar annually from local agricultural residues, municipal wastes, grasses, and woody plants (Ahmed et al. 2021). Hemicellulose and cellulose are frequently fractionated using supercritical and subcritical water, resulting in the formation of sugars as a downstream product. While in the case of sugarcane bagasse, hot-compressed water treatment at an optimum condition (180 °C for 30 min at 1 MPa) can easily extract a significant amount of xylose (85%), while solid fraction (rich in cellulose) can be utilized in other processes (Roy et al. 2021; Acciardo et al. 2022). AMG Energy Group discovered that agricultural and yard waste can be converted into commercially viable cellulosic sugars by means of mechanical, chemical, and dry processes. In contrast to current acid- or enzymatic-based processes, this could offer a more efficient alternative route. Since the pretreatment process releases a significant amount of inhibitors, therefore, refining sugar products is an important step in biomass-to-sugar technology. This includes furan derivatives (furfural, polychlorinated biphenyl, 5-hydroxymethylfurfural, and dioxins), phenolic compounds (vanillin, hydroxycinnamic acids, phenols, gallic acid, and p-hydroxybenzoic acid), and carboxylic acids (acetic, propanoic acid, formic, butanoic acid, and levulinic acid). In addition to being sustainable and flexible, membrane technology is becoming one of the emerging technologies that consume less energy. Its unique capability of separating and purifying intermediates or products has drawn considerable interest. It has been demonstrated that nanofiltration and reverse osmosis are effective in separating C5 and C6 sugars from acetic acid, furfural, p-hydroxybenzoic acid, 5-hydroxymethyl furfural, polychlorinated biphenyl, vanillin, and gallic acid and results in high sugar rejection (Okolie et al. 2022; Sarwer et al. 2022). Commercially available membrane processes can effectively concentrate the sugar streams which results in a higher concentration of downstream products and reduces the energy consumption required for the recovery of desired products. In addition, osmosis has shown great results in enhancing rice straw sugars. Researchers have demonstrated that reverse osmosis and nanofiltration can be effectively used to concentrate sugars and subsequently facilitate the removal of acetic acid and furfurals from the lignocellulosic hydrolysate. Wheat straw hydrolysate is capable of concentrating glucose and recycled cellulase through a combination of ultra and nanofiltration. To improve the economic viability of sugar production and its further processing, it is important to recover hydrolytic enzymes. Toray Industries and Mitsui Sugar have developed a pilot-scale plant that produces ~1550 tons of high-quality cellulosic sugar annually from sugarcane bagasse using membrane technology (Devi et al. 2022; Flores et al. 2021).

5.5.2 *Oligosaccharides*

Production of oligosaccharides and its derivatives from lignocellulosic residues has shown tremendous growth in the past few years as these are considered as potential prebiotics. A prebiotic oligosaccharide is a short-chain carbohydrate that cannot be digested by human digestive enzymes. Due to their high stability at low pH and high temperature, these oligosaccharides reach the lower gut intact and are fermented by probiotic bacteria. In addition to non-carcinogenic properties, they stimulate bacterial growth and fermentation and possess antioxidant, immunomodulatory, antiallergenic, selective cytotoxic, and antimicrobial activity as well as improve mineral absorption. Currently commercialized prebiotic oligosaccharides include lactosucrose, fructooligosaccharides, soybean oligosaccharides, galactooligosaccharides, xylooligosaccharides, and isomalto-oligosaccharides. Hydrolysis of hemicellulose results in the production of several oligosaccharides including xylooligosaccharides, manno-oligosaccharide, and arabino oligosaccharide (Khan et al. 2022). Manno-oligosaccharide provides significant health-promoting effects in humans, but the research on their production from mannan-rich agricultural wastes is still underway. Several methods can be used to prepare oligosaccharides, including direct hydrolysis, acid or enzymatic hydrolysis, or pretreatment with thermal or chemical agents. Acid hydrolysis, however, can also produce undesirable byproducts, thus decreasing xylooligosaccharide yields and resulting in more complex and expensive purification steps. In the nutraceutical and pharmaceutical industries, xylan conversion to xylooligosaccharide is preferred because it does not produce toxic or unwanted products. On the other hand, biomass waste containing pectin can be depolymerized to produce pectic oligosaccharide (POS), which is an emerging prebiotic that prevents and treats numerous chronic diseases.

5.5.3 Biofuels

In addition to fossil fuels, biofuels can also be manufactured from a variety of renewable biological resources such as plants and plant-derived materials. As an environmentally benign and cost-effective replacement for petroleum, biofuels are often recommended. Table 5.5 summarizes the production of liquid biofuels and biogas. Bioethanol, one of the downstream products of sugar manufacturing, is an important and common liquid fuel. Besides serving as an alternative energy source, bioethanol can be used to produce ethylene, ethylene glycol, polyethene, polyethene terephthalate, and their derivatives. It is expected that global ethanol production will more than double by 2024, i.e., from 114 billion liters to ~134.5 billion liters (Thangavelu et al. 2022). Production processes for bioethanol include consolidated bioprocessing, simultaneous saccharification, and cofermentation and separate hydrolysis and fermentation and simultaneous saccharification and fermentation. While butanol is another bio-based renewable and green fuel which is produced via alcoholic fermentation. Typically, butanol is produced through the combustion of fossil fuels, but it can also be produced through the fermentation of biomass. Due to its longer hydrocarbon chain, butanol blends better with gasoline than ethanol because of its lower polarity. Similar to gasoline, butanol has a high energy density. Despite its limitations, it is not considered a viable biofuel like ethanol due to production difficulties. Biobutanol is produced from different kinds of feedstocks via acetone-butanol-ethanol (ABE) fermentation process. In addition to low butanol yield, low butanol titer, high substrate cost (grains, molasses), end-product inhibition, and poor product recovery are some of the major limitations of ABE fermentation. Methanol is another alcoholic fuel that is easily distributable, has low volatility, and has a high-octane rating (Prakash et al. 2021). The chemical and physical properties of methanol are similar to those of ethanol. The most common method for generating methanol is through the catalytic conversion of syngas from fossil sources, but it can also be generated from lignocellulosic biomass materials. Besides its use as a fuel for vehicles, methanol may also be used as a feedstock for methyl tertiary-butyl ether, which is an additive for gasoline.

Table 5.5 Liquid biofuels and biogas production

Year	Liquid biofuels (kt)	Liquid biofuels (billion liters)	Biogas supply (EJ)	Biogas supply (in Bm^3)
2000	15,992	19.2	0.29	12.4
2005	30,851	37.1	0.54	23.4
2010	86,420	104	0.89	38.7
2015	109,152	131	1.34	58.0
2016	112,229	135	1.35	58.7
2017	115,702	139	1.38	59.8
2018	123,262	148	1.41	61.4
2019	132,313	159	1.43	62.3

5.5.4 *Bioactive Compounds*

The solid waste generated from fruit and vegetable processing contains diverse molecules that may have biological activities. The dry peels of citrus contain 3.8% D-limonene and flavonoids such as hesperidin, luteolin, naringin, tangeretin, and eriocitrin, which are used in the food, pharmaceutical, and cosmetic industries. In addition to its high polyphenolic content, grape pomace possesses great scavenging activity against free radicals and exhibits anti-inflammatory and antiproliferative properties. These potential characteristics and properties make them highly suitable for cancer therapy. There is a significant amount of tannin (17–29%) and other polyphenolic compounds (3–7%) in grape skins, such as catechins, proanthocyanidins, anthocyanins, gallic acid, quercetin, ellagic acid, and resveratrol. Approximately 60% of the polyphenols in grapes are found in grape seeds, with a significant concentration of epicatechins, flavan-3-ols, and catechins (Sethupathy et al. 2022). On the other hand, apple pomace is rich in polyphenols and flavonoids contents and leads to the production of several bioactive compounds including phloretin glycosides, procyanidins, hydroxycinnamates, quercetin glycosides, and catechins. Olive pomace which is a byproduct of the olive processing industry consists of 98% phenolic compounds including cinnamic acid derivatives (verbascoside, caffeic acid, and p-hydroxycinnamic acid), secoiridoids (oleuropein aglycone and glycoside), tyrosol, and flavonoids (apigenin, luteolin, and rutin). While coffee byproducts consist of 1.5% total polyphenols and exhibit excellent anti-allergenic, antioxidant, and anti-inflammatory activities due to the presence of chlorogenic acids (Yang et al. 2021).

5.5.5 *Lignin Byproducts*

The lignin polymer consists primarily of substituted aromatic monolignols, i.e., coniferyl alcohol and sinapyl alcohol, while grasses contain β-coumaryl alcohol. The heterogeneity and recalcitrance of lignin make its use in biorefineries both challenging and advantageous. Depending on the type, origin of biomass, and technology used, isolated lignin shows great variations in terms of purity, quality, structure, and molecular weight. Table 5.6 summarizes the production of several bioproducts from technical lignin. A large amount of lignin is generated in the lignosulfonate industry by the sulfite process which is commercialized by Borregaard LignoTech with a maximum capacity of 1 metric ton annually. High-quality technical lignin from a kraft pulp mill was extracted and successfully upgraded by LignoBoost Technology, with a maximum capacity of 27,000 tons of kraft lignin, and can be further utilized for various fuel applications such as dispersant, activated carbon, antioxidant, binder, and carbon fibers (Chakravarty et al. 2022). Renmatix produces lignin as a coproduct from supercritical technology which exhibits high reactivity and low production cost. This offers a benefit during

Table 5.6 Bioproducts derived from technical lignin

Sector/area	Biofuel	Lignin type
Biofuels	• Pyrolysis and liquefaction derived bio-oil • Biodiesel obtained from the Fischer-Tropsch process • Syngas produced from gasification	Pyrolytic, lignosulfonates, kraft, and organosolv
Chemicals	Flocculants, phenolic compounds and aromatics, dispersants, adhesives, and emulsifiers	Lignosulfonates, soda, kraft, steam explosion, and organosolv
Materials/polymers	Bioplastics, nano-carbon fibers, additives for cement/paints, biocomposites, activated carbons, bio-adsorbents, and binders	Lignosulfonates, soda, kraft, and organosolv
Environmental	Pesticides, bio-fertilizers, herbicides, soil and dust agents, and water treatment agents	Kraft and organosolv
Pharmaceuticals	Antioxidants, prebiotics, and cosmetics	Kraft

commercialization as the produced lignin can be used for a wide range of applications including a replacement for wood adhesives. Lignin is also used to produce a polymer called Renol which can be blended with existing thermoplastics. Licella Holdings Ltd. and Canfor Pulp Ltd. developed a biorefinery to transform woody waste using supercritical H_2O for the production of bio-crude oil. In recent studies, lignin-modifying enzymes such as peroxidase, lignin peroxidase, dye-decoloring peroxidase, manganese-dependent peroxidase, and laccase have been extensively utilized in the paper and pulp industry, food industry, dye, and cosmetic industry (Cassoni et al. 2022). Several enzymes degrade lignin, including glyoxal oxidase, hydrolases, pyranose 2-oxidase, oxidoreductases, aryl alcohol oxidase, ligases, and glucose dehydrogenase, while future lignin valorization may be achieved by these enzymes.

5.6 Indian Biomass Energy Conversion Policy and Initiatives by the Government

In recent years, the population and economic growth of India have increased energy consumption. This demand is likely to increase significantly with the rapid urbanization of India and the improvement of living standards for millions of households. The Indian government is, therefore, making numerous policies and plans in the energy and chemical sectors. As sustainable development and green energy is now the world's foremost goal, renewable resources are also being considered for the generation of power and the production of value-added products. In this field, the Ministry of New and Renewable Energy of India (MNRE) has developed many projects and policies and provided subsidies and incentives to encourage the adoption of these methodologies. As a part of the 12th five-year plan, the Indian government is allocating a substantial amount of Rs 46.00 crores for biomass gasifier schemes, including promotional activities, tenders, and other administrative tasks. In addition, the following components have been included during the implementation of programs during the 12th five-year plan:

- Implementation of an off-grid or distributed power scheme using biomass gasifiers for rural areas to fulfil electricity demand.
- Grid-connected power programs based on biomass gasifiers are supported at the megawatt (MW) level with 100% engines based on producer gas.
- A biomass-based boiler turbine generator (BTG) with a maximum capacity of 2 MW would be supported.
- Several promotional activities, publicity, seminars, and training are also included in the programs.

To promote the growth of bio-based platform chemicals and bio-power, the Indian government provides a variety of subsidies. They are working on attracting various investors and sponsors in the field of bio-based platform chemicals and bio-power markets using off-grid and on-grid schemes. Private and government sectors are also receiving various types of subsidies from MNRE. Furthermore, the Department of Biotechnology (DBT) has mandated the development of cost-effective biofuels by enhancing the quality of feedstock, development in production technologies, and development of high-throughput enzymes/microbes/algae/fungi for high yield of platform chemicals and aromatics. In 2018 under Swachh Bharat Mission, the prime focus was on generating energy from different wastes including municipal solid waste (MSW) and municipal liquid waste (MLW). The associated technologies and the obtained products developed at the following 4 DBT (Department of Biotechnology)-Bioenergy Centers are taken forward to be scaled up and have already been evaluated at a pilot scale. The majority of the research outcomes were published in more than 35 publications and protected by more than 5 process patents. In addition, India's first continuous steam explosion pilot plant and extractor system have been installed and commissioned by the DBT-IOC Centre for Advanced Bioenergy. To develop sustainable and green technology for the production of 2G ethanol which is agnostic to various feedstocks (cotton, mustard stalk, etc.), a 300 kg/day pretreatment pilot facility is being constructed. In third-party premises, the Centre has successfully upgraded its enzyme technology to a 5 KL fermenter with an activity level of 8 FPU/ml. At M/s Praj Industries, Pune, the entire enzyme broth material was evaluated at a 1 MT/day pilot plant. The hydrolysis efficiency of the enzyme was comparable with that of the commercial enzyme. In the past few months, DBT-ICT Centre has achieved two major milestones. In partnership with Perkin-Elmer and India/Germany/USA, the Centre has installed a robotic platform explorer TM G3 Workstation for high-throughput microbial and enzyme screening and molecular biology routines. As a result of the facility, advances in enzyme and microbial innovations have been accelerated along with synthetic biology involving multiple microbial chassis for various desired products. The DBT has supported research projects on generating energy from waste as part of the National Swachh Bharat Mission. The development of eight novel and viable waste-to-energy technologies has been initiated for cleaner, pollution-free environments and the generation of energy from municipal solid waste (MSW).

Apart from improving the access to electricity, CNG, and LPG in India over the past few decades, biomass is also growing continuously as a traditional fuel and now dominates the fuel mix of rural and urban households. According to studies, 64% of rural households in India use firewood to cook, while another 26% use crop wastes and animal wastes, while almost 30% of urban households use traditional energy for cooking and heating.

In comparative statistics of household energy usage in India and China, it was reported that solid fuels (biomass and coal) accounted for 80% of the total household energy usage, while in India, firewood is widely used as cooking fuel. The rural sectors of India consume about 235 MT of firewood annually for cooking. Depending on income level, firewood accounts for 65–80% of energy consumption. In addition to residential uses, biomass is used in traditional and rural enterprises including hotels, rice par-boiling, bakeries, restaurants, potteries, brickmaking, and charcoal production.

5.7 Prospects of Biomass Valorization

In comparison to existing conversion processes, thermochemical conversion of waste biomass is less expensive, more productive, and faster and has established infrastructure. Even though these processes produce substantial amounts of biofuels, aromatics, and other platform chemicals, some major technical details are still missing. This includes suitable preconditioning of feedstocks, unappealing economics, catalysts screening, and biomass engineering (particle size, pressure, porosity, temperature, surface area, residence time, and biomass loading). These challenges must be resolved before implementing them on a commercial scale. The current thermochemical technologies including gasification, pyrolysis, and liquefaction are quite noneconomic and complex and thus need to be simplified at a commercial scale. To resolve the techno-economic issues associated with the production of biomass-derived products, a major thrust needs to be made in the direction of research and development. Pretreatment of biomass remains another bottleneck during the bioprocessing of waste lignocellulosic for the production of biofuels and other bioproducts. Even though some pretreatment methods may appear to be advantageous, one method will not be the best option for all types of biomass. For optimal substrate quality, co-digestion of nonfood cellulosic biomass feedstocks with different pretreatment technologies should be investigated. Hence, the extent of pretreatment for any given feedstock is, therefore, governed by biorefining interests. The key to industrializing the processing of lignocellulosic on an industrial scale may be to discover or create markets for the byproducts of biomass pretreatment technologies. It is possible to enhance biomass digestion by using multiple or combinatorial pretreatments. This could be performed and regulated under various operating conditions that can maximize product selectivity and product recovery and regeneration while minimizing the generation of inhibitory compounds.

5.8 Conclusion

This chapter highlights a robust analysis of different biomass resources and revealed huge potential toward biomass conversion to energy and wide range of platform chemicals. Surplus biomass availability of ~230 million metric tons per annum from agriculture sector revealed the great potential of about 28 GW. More than 70% of the country's population depends on biomass for its energy needs, and biomass still supplies 32% of the country's total primary energy. Several sources including agricultural, industrial, and food wastes are available in large volumes and exhibit the tendency to replace conventional sources. Globally, 4.40 EJ of renewable energy was used in the transport sector in 2019, with liquid and gaseous biofuels accounting for 91% of that energy. Moreover, the effective disintegration of biomass into its constituent sugars can produce biofuels, aromatics, platform chemicals, or biopolymers. These bio-based products can be successfully utilized in various sectors including cosmetic, chemical, pharmaceutical, and food industries. However, biomass waste has become increasingly problematic in recent years due to its potential to cause significant environmental impacts. This includes release of toxic gases such as nitrous oxide, carbon monoxide, and nitrogen dioxide. In addition to these gases, ozone and nitric acid are formed, which contribute to acid deposition, which is hazardous to human and ecological health. Hence, biorefining interests represents the extent of waste biomass utilization for any biomass. Combinatorial or multiple pretreatments operated under suitable conditions can easily enhance biomass digestibility and maximize the selectivity of products, improves product recovery, and minimizes the generation of inhibitory products.

References

Acciardo E, Tabasso S, Cravotto G, Bensaid S (2022) Process intensification strategies for lignin valorization. Chem Eng Process-Process Intensif 171:108732

Ahmed I, Zia MA, Afzal H, Ahmed S, Ahmad M, Akram Z, Iqbal HM (2021) Socio-economic and environmental impacts of biomass valorisation: a strategic drive for sustainable bioeconomy. Sustain For 13(8):4200

Bilal M, Qamar SA, Yadav V, Cheng H, Khan M, Adil SF, Iqbal HM (2021) Exploring the potential of ligninolytic armory for lignin valorization–A way forward for sustainable and cleaner production. J Clean Prod 326:129420

Cassoni AC, Costa P, Vasconcelos MW, Pintado M (2022) Systematic review on lignin valorization in the agro-food system: From sources to applications. J Environ Manag 317:115258

Chakravarty I, Gahane D, Mandavgane S (2022) Recent advances in lignin valorization. Adv Biofuel Technol 2022:365–388

Devi A, Bajar S, Kour H, Kothari R, Pant D, Singh A (2022) Lignocellulosic biomass valorization for bioethanol production: a circular bioeconomy approach. Bioenergy Res 2022:1–22

Domínguez E, del Río PG, Romaní A, Garrote G, Gullón P, de Vega A (2020) Formosolv pretreatment to fractionate paulownia wood following a biorefinery approach: isolation and characterization of the lignin fraction. Agronomy 10(8):1205

Fírvida I, del Río PG, Gullón P, Gullón B, Garrote G, Romaní A (2021) Alternative lime pretreatment of corn stover for second-generation bioethanol production. Agronomy 11(1):155

Flores EM, Cravotto G, Bizzi CA, Santos D, Iop GD (2021) Ultrasound-assisted biomass valorization to industrial interesting products: state-of-the-art, perspectives and challenges. Ultrason Sonochem 72:105455

Khan RJ, Lau CY, Guan J, Lam CH, Zhao J, Ji Y, Leu SY (2022) Recent advances of lignin valorization techniques toward sustainable aromatics and potential benchmarks to fossil refinery products. Bioresour Technol 346:126419

Nguyen LT, Phan DP, Sarwar A, Tran MH, Lee OK, Lee EY (2021) Valorization of industrial lignin to value-added chemicals by chemical depolymerization and biological conversion. Ind Crop Prod 161:113219

Okolie JA, Epelle EI, Tabat ME, Orivri U, Amenaghawon AN, Okoye PU, Gunes B (2022) Waste biomass valorization for the production of biofuels and value-added products: a comprehensive review of thermochemical, biological and integrated processes. Process Saf Environ Prot 159: 323–344

Pang T, Wang G, Sun H, Sui W, Si C (2021) Lignin fractionation: effective strategy to reduce molecule weight dependent heterogeneity for upgraded lignin valorization. Ind Crop Prod 165: 113442

Prakash DG, Gopinath KP, Vinatha V, Shreya S, Sivaramakrishnan R, Chi NTL (2021) Enhanced production of hydrocarbons from lignin isolated from sugarcane bagasse using formic acid induced supercritical ethanol liquefaction followed by hydrodeoxygenation. Chemosphere 285: 131491

Radhakrishnan R, Patra P, Das M, Ghosh A (2021) Recent advancements in the ionic liquid mediated lignin valorization for the production of renewable materials and value-added chemicals. Renew Sust Energ Rev 149:111368

Roy S, Dikshit PK, Sherpa KC, Singh A, Jacob S, Rajak RC (2021) Recent nanobiotechnological advancements in lignocellulosic biomass valorization: a review. J Environ Manag 297:113422

Saraeian A, Aui A, Gao Y, Wright MM, Foston M, Shanks BH (2020) Evaluating lignin valorization via pyrolysis and vapor-phase hydrodeoxygenation for production of aromatics and alkenes. Green Chem 22(8):2513–2525

Sarwer A, Hamed SM, Osman AI, Jamil F, Al-Muhtaseb AAH, Alhajeri NS, Rooney DW (2022) Algal biomass valorization for biofuel production and carbon sequestration: a review. Environ Chem Lett 2022:1–55

Sethupathy S, Morales GM, Gao L, Wang H, Yang B, Jiang J, Zhu D (2022) Lignin valorization: status, challenges and opportunities. Bioresour Technol 2022:126696

Singh SK, Dhepe PL (2019) Lignin conversion using catalytic ionic liquids: understanding the role of cations, anions, and hammett acidity functions. Ind Eng Chem Res 58(47):21273–21284

Thangavelu K, Desikan R, Uthandi S (2022) Optimization of combined lime and hydrodynamic cavitation for pretreatment of corncob biomass using response surface methodology for lignin removal. Biomass Convers Biorefinery 2022:1–13

Usmani Z, Sharma M, Awasthi AK, Sivakumar N, Lukk T, Pecoraro L, Gupta VK (2021) Bioprocessing of waste biomass for sustainable product development and minimizing environmental impact. Bioresour Technol 322:124548

Yang C, Chen H, Peng T, Liang B, Zhang Y, Zhao W (2021) Lignin valorization toward value-added chemicals and fuels via electrocatalysis: a perspective. Chin J Catal 42(11):1831–1842

Ye K, Liu Y, Wu S, Zhuang J (2021) A review for lignin valorization: challenges and perspectives in catalytic hydrogenolysis. Ind Crop Prod 172:114008

Yu B, Fan G, Zhao S, Lu Y, He Q, Cheng Q, Song G (2021) Simultaneous isolation of cellulose and lignin from wheat straw and catalytic conversion to valuable chemical products. Appl Biol Chem 64(1):1–13

Chapter 6
Biomass (Algae) Valorization as an Energy Perspective: Review of Process Options and Utilization

Aman Kumar, Amit Kumar Tiwari, Sumit Kumar Jana, and Dan Bahadur Pal

Abstract Recent geopolitical shocks have highlighted the need for countries to shift their dependency from the fossil fuels to renewable and domestically sustainable energy as the primary source of energy. Algae are the original green technology that is renewable, sustainable, and environmentally friendly and has great potential due to its abundant, high carbohydrate and absence of lignin properties. Apart from making biodiesel, algae can be utilized to create a variety of other products like nutraceuticals, pharmaceuticals, and cosmetics. In this review, I have summarized the recent technological advances in the production of algal biomass as a feedstock and its conversion to value-added products. Economics of the production and its recovery as well as the main challenges and of refineries using algae as biomass is also discussed.

Keywords Renewable · Bio-diesel · Algae · Biomass · Valorization · Energy perspective

6.1 Introduction

Biomass is any organic material that is used for the production of fuel and electricity. Biomass as a source of fuel has been used by humans since ancient times in the form of wood. It was the largest source of energy consumption until the discovery of fossil fuels such as coal and oil and nuclear fuels such as Uranium. The Industrial Revolution which began 200 years ago catapulted us into a new era of fossil fuels (Wrigley 2013). But this short-lived era has once again forced us to go back to biomass and use its potential to lead us into a new era of biofuels. Biomass can be

---

A. Kumar · A. K. Tiwari · S. K. Jana
Department of Chemical Engineering, Birla Institute of Technology, Ranchi, Jharkhand, India

D. B. Pal (✉)
Department of Chemical Engineering, Harcourt Butler Technical University, Kanpur, India

D. B. Pal, A. K. Tiwari (eds.), *Sustainable Valorization of Agriculture & Food Waste Biomass*, Clean Energy Production Technologies,
https://doi.org/10.1007/978-981-99-0526-3_6

converted to energy using different methods like burning directly to produce heat or using thermochemical and biological processes to get various solid, liquid, and gaseous fuels. Currently, the industrial and transportation sector use the largest amounts of biomass, about 78%, followed by the residential, commercial, and electrical power sectors (Poudyal et al. 2019).

Among the sustainable biomass, food crops such as soy, maize and corn that were used as the raw materials to produce biofuels like biodiesel and bioethanol are regarded as the first-generation of bio-energy (Lee and Lavoie 2013). Second-generation bio-energy is defined as energy produced from a large array of different feedstock that ranges from lingo-cellulose-based feedstocks to local solid wastes. The fundamental disadvantage of first-generation bio-energy is that they are made from food-grade biomass. When there isn't enough food to feed everyone, this becomes an issue. Second-generation biofuels are made from nonedible biomass; however, even they still compete for land used for food production. This is why third-generation biofuels has had an exponential interest. Third-generation bio-energy is the energy produced from photosynthetic organisms like marine biomass especially algae. Microalgae have high rate of production and can be grown on ground that isn't suitable for agriculture, and thus, we do not have to worry about farmable land scarcity. They are photosynthetic organisms; hence, their use helps to reduce greenhouse gas emissions (Morales et al. 2021). As a result, microalgae have gotten a lot of interest for their potential as biodiesel or other lipid-based biofuel producers.

Algal biomass can be classified into the unicellular microalgae and the multicellular macroalgae. Red, green, and brown macroalgae are the most common types, with color derived from natural pigments and chlorophyll. Aside from lipids and proteins, macroalgae are mostly composed of polysaccharides, which are classified as energy storage polysaccharides or structural polysaccharides (Kraan 2012) based on their biological roles.

Microalgal biodiesel, in stark contrast to the other oil-producing crops, has the potential to totally replace petroleum-derived transportation fuels without compromising food and other agricultural goods. Microalgal biodiesel is also a better alternative to bioethanol from sugarcane, which is now the most extensively used transport biofuel.

6.2 Algae

The term "algae" refers to a diverse group of organisms that can produce oxygen through photosynthesis which is the process of converting light energy from the sun to generate carbohydrates. It can be single-celled, multicelled, microscopic, and macroscopic as well as filamentous (Kaštovský et al. 2019).

6.2.1 *Macroalgae*

Thousands of species of macroscopic, multicellular, marine algae are referred to as seaweed, or macroalgae. Chlorophyta (green), Phaeophyta (brown), and Rhodophyta (red) macroalgae are also a part of macroalgae. Seaweeds such as kelps offer key nursery habitats for fisheries and other marine animals, thereby protecting food sources; other species, such as planktonic algae, play an important part in carbon capture, generating at least half of the world's oxygen (Russell 2013).

6.2.2 *Microalgae*

Microalgae, often called microphytes, are microscopic algae that are invisible to the naked eye. They are phytoplankton that may be found in both freshwater and marine environments, living in both the water column (Schlesinger and Bernhardt 2020) and the sediment. Depending on the species, their sizes can range from a few micrometers (m) to a few hundred micrometers. They are single-celled creatures that may live alone, in groups, or in chains. Microalgae do not have roots, stalks, or leaves, unlike bigger plants. They've evolved to be thriving in a world ruled by viscous forces. Around half of the oxygen produced on the planet are produced by tiny invisible organisms that mainly live in the water. These are called microalgae. It is the most valuable living organisms on the planet that use a simple photosynthesis, converting carbon dioxide to oxygen, thereby sustaining life on the planet. It all started three billion years ago when in the lack of oxygen, prokaryotic life first appeared on the Earth as a by-product of photosynthesis. This was the cyanobacteria also known as the blue green algae which formed and began releasing oxygen into the atmosphere. However, surface rocks exposed to metals like iron oxidized and devoured oxygen, and because the huge ocean absorbed oxygen (Canfield 2004), it took roughly a billion years for oxygen levels to grow appreciably. However, two and a half billion years ago, the amount of oxygen in the atmosphere began to grow as the exposed minerals were completely oxidized and absorption by the oxygen-rich top layers of the ocean decreased.

6.2.3 *Factors Affecting Algae Growth*

(a) **Oxygen:** To create 1 kg of microalgal biomass, the photosynthetic process produces roughly 2 kg of oxygen stoichiometrically. As a result, when there is a lot of microalgal development, a lot of oxygen is produced. When there is a lot of oxygen in the air, the production of microalgae decreases considerably due to photorespiration and photoinhibition effects (Kazbar et al. 2019).

(b) **Light:** Light availability is the most essential component in photosynthetic microorganism growth and production (Xu et al. 2016). Because light is the primary source of energy for photosynthetic organisms, it must be maximized. Excessive light, on the other hand, can harm the photosynthetic system, especially when combined with a low temperature or high oxygen content. As a result, the cultivation system's light supply must be maximized by proper architecture (orientation and design) (Tredici and Zittelli 1998).

(c) **Contaminants:** Because the open algae growth lakes are exposed to the outdoors, bacteria, viruses, fungus, and other microalgae species, as well as predators, can readily contaminate them (Di Caprio 2020). Shifting the lakes within greenhouses with controlled climatic conditions is a solution; however, this is economically unsustainable for large-scale biofuel production.

Nevertheless, it is well established that very few pollutants can persist in harsh environments. By growing some extremely resistant algae cultures at high pH or high salinity, contamination can be prevented (Qiu et al. 2017).

6.2.4 Microalgae as a Source of Fuel

New species of microalgae are being discovered all the time, and there are already more than several million species discovered. But out of all these, very few are usable to produce clean and sustainable biofuels. A very few of them are able to survive captivity and grow in a large scale. Table 6.1 highlights some of the commercially viable algae for biofuels and other value-added products.

The following are the primary benefits of employing microalgae in a number of industrial applications (Khan et al. 2018):

- Algal-producing facilities can be colocated on terrain that is otherwise unproductive or not for agriculture.
- They grow very quickly and have a higher solar energy conversion efficiency compared to most terrestrial plants.
- They can exploit waste CO_2 sources, potentially reducing greenhouse gas emissions into the atmosphere.
- They can be gathered in batches or continually for virtually the whole year.

Table 6.1 Important algae species suitable for commercial production (Darzins et al. 2010)

Algae class	Main species
Chlorophyceae	*Neochloris oleoabundans*
Prasinophyceae	*Tetraselmis* spp.
Cyanobacteria (blue green algae)	*Arthrospira (Spirulina) platensis*
Bacillariophyceae (diatoms)	*Cyclotella cryptica; Chaetoceros* sp.; *Skeletonema* sp.
Eustigmatophyceae	*Nannochloropsis* spp.
Haptophyceae	*Chrysotila carterae, Isochrysis galbana*

- They can produce a large range of feedstocks from which safe, biodegradable biofuels and lucrative coproducts can be made.
- They may exploit salt and waste sources of water that traditional agriculture is unable to employ.

6.3 Production of Microalgae

Despite the high market for the algae-based products, microalgae production is still limited to about 10 to 20 thousand tons of dry matter per year, and the cost of production is still too high for applications like energy production, feed production, or treatment of wastewater. The problem of low-cost biomass harvesting is one of the key challenges limiting large-scale microalgae production. There are no well-defined and proven industrial-scale technologies for extracting and isolating oils and lipids from algal biomass, which is essential for biofuel production. Extraction techniques now available are mostly suited to analytical and laboratory-scale processes, as well as the recovery and removal of high-value products. Extraction procedures must be efficient and successful in order to manufacture algal biofuels as a competitive product. Figure 6.1 showed the popular microalgae to biofuel production pathways.

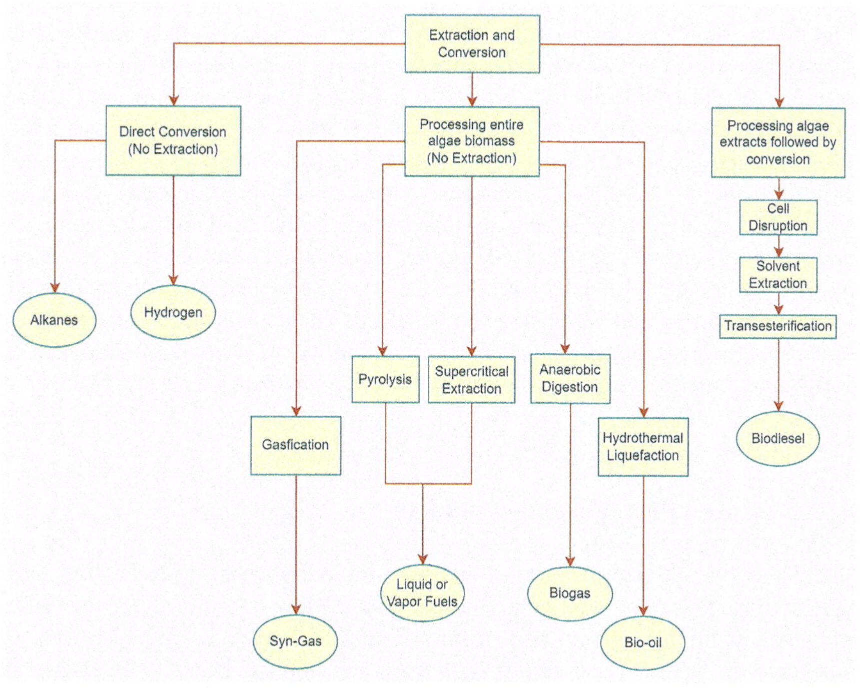

Fig. 6.1 Popular microalgae to biofuel production pathways

6.3.1 Methods of Cultivation

Plants, mosses, macroalgae, microalgae, cyanobacteria, and purple bacteria are examples of organisms that employ photosynthesis to build biomass from light and carbon dioxide. A photo-bioreactor (PBR) is a reactor, used for carrying out reactions in a biologically sustained environment that cultivates phototrophic bacteria using light (Masojídek and Torzillo 2014). Specific parameters are carefully managed for individual species inside the artificial environment of a photo-bioreactor. As a result, a photo-bioreactor may achieve substantially greater growth rates and purity levels than natural or equivalent environments. In a photo-bioreactor, phototropic biomass may theoretically be produced from nutrient-rich sewage water and flue gas carbon dioxide. Open culture systems, such as open ponds, tanks, and raceway ponds, and controlled closed cultivation systems, which use various types of bioreactors, are the two most popular ways of microalgae production.

6.3.1.1 Open Pond

For the growth of microalgae, the open pond farming technique employs shallow ponds ranging in size from 1 acre to big ponds exposed to natural radiation from the sun. Because of its cheaper operating and investment costs, the open pond cultivation method is widely used in the industry. Some examples of open pond systems include raceway ponds (these ponds are called racing ponds because their forms are identical to and comparable to those of a racetrack), stirred and unstirred surface ponds. Microalgae can also be farmed in a large open environment such as rivers, lakes, seas, ponds, sewage waters, etc., in the open pond setup. This setup is the most easy-to-maintain, well-known, inexpensive, and contemporary method of growing algae. The most notable benefit of this system is that it does not require any additional equipment for a light source or temperature control. Open ponds can also be constructed artificially with concrete or polymers like PVC, and baffles and paddles for stirring and ventilation can be added. The main issues of an open pond system are the management of operating conditions and undesirable contaminations, which may be addressed by using highly selective microalgae species (Costa et al. 2019).

6.3.1.2 Closed PBR (Photo-Bioreactors)

PBRs may be built and tuned to fit the strain of your choosing. Closed cultivation systems, also known as closed photo-bioreactors (PBRs), are more effective in terms of overall quality since they can be operated under more precise circumstances and so overcome the drawbacks of open culture systems (Wang et al. 2012). Unlike the pond system, they are not open to the atmosphere and are always covered using transparent materials like sheets or tubes. This closed system takes up minimal space while enhancing light availability and significantly reducing contamination concerns. However, overheating, algae development at the bottom surfaces, biofouling, cleaning challenges, a large buildup of dissolved oxygen resulting in growth limits, and, most critically, very high capital expenses for developing and managing PBRs are some of the drawbacks.

It is of three main types:

(a) ***Tubular***: These photo-bioreactors are made up of long transparent glass tubes or transparent polymers that are placed horizontally, vertically, or in helix-style or tilted to enhance sunlight capture. A mechanical pump or an airlift device circulates the microalgae culture within the loop (Stephenson et al. 2010). The most significant disadvantage of the tubular photo-bioreactor design is its poor mass transmission across the system, since the lengthy tube utilized in the bioreactor design may result in disparities in substrate and product concentrations along the tubes.

(b) **Vertical:** A transparent vertical cylindrical tube and a sparger that pumps in air bubbles to homogenize the culture and allow carbon dioxide and oxygen transfer between the air and the microalgae culture make up a vertical column bioreactor. Due to the capacity of the sparser utilized in this reactor to create bubbles which are less in size, giving a higher total surface area for more effective transfer of mass, this PBR delivers the largest mass transfer(gas-liquid) efficiency in comparison to other systems. Furthermore, the design's simplicity allows for a reduced energy requirement and a simpler operating method. The cylindrical-shaped container, on the other hand, does not supply the microalgae with enough light to efficiently undertake photosynthesis (Fernandez Sevilla et al. 2004).

(c) **Flat:** Out of all photo-bioreactor designs, this one has the highest total surface area for light and the least amount of oxygen buildup, resulting in the best photosynthetic efficiency. The rectangular-shaped compartment of the flat-plate photo-bioreactor, on the other hand, is composed of transparent material like polyvinyl chloride (PVC). A looping airlift system is used to stir the culture inside the reactor. However, the photo-bioreactors aeration design causes stress damage to the microalgae cells (Huang et al. 2017).

6.3.1.3 Hybrid

While the majority of organizations have opted to focus only on open pond systems or closed PBRs, others are attempting to combine the benefits of each system by employing "hybrid" procedures that mix two or more of the technologies. Small PBRs can be used with big ponds, larger PBRs can be used in conjunction with ponds, or ponds and fermenters can be utilized in a sequential order. The goal of hybrid systems is to optimize the benefits of each process individually (Yun et al. 2018).

6.3.2 Harvest Technologies

Harvesting algae is the process of concentrating diluted algal suspension into a thick algae mud with the goal of producing slurry that contains at least 2%–7% algal suspension on a dry matter basis. Due to this microalgal biomass, collection necessitates the use of effective solid-liquid separation technology. A broad range of methods are available because solid-liquid separation is a crucial unit operation in many industrial processes. These can be classified into two categories: methods that separate particles from the liquid phase using gravity or buoyancy (sedimentation, flotation, and centrifugation) compared to methods that remove particles from the liquid manually using a filter or screen. Algal biomass harvest is, in general, a difficult phase in the algal biomass conversion cycle. Because the algal cells are tiny (3–30 μm in diameter) and the cell concentration is low (1 gramme per liter in an open pond system and 5 gramme per liter in a photo-bioreactor culture), a considerable amount of solution must be treated. As a result, harvesting algal biomass is an extremely expensive procedure; it's estimated that recovering biomass from the culture suspension costs 20%–30% of the overall cost of generating biomass (Fasaei et al. 2018).

Sedimentation, filtering, and centrifugation are all methods for harvesting microalgae. The size and features of an algae strain influence the harvesting procedure chosen. Open pond culture methods use sedimentation and flotation harvesting processes, whereas photo-bioreactors use filtration and centrifugation. The harvest technique chosen must be capable of handling a big amount of algae culture slurry (Singh and Patidar 2018).

6.3.2.1 Gravity-Based Separation

Gravity-based separation technologies are commonly used in industries as a method to separate two components, such as a suspension or a dry granular combination, where gravity separation is feasible (it is necessary that the two components of the mixture differ in specific weights). Gravity settling, flotation, and centrifugation discussed here have this commonality in the sense that gravity is the main force in all of the three approaches.

6.3.2.1.1 Gravity Settling

Allowing algal cultures to settle on their own without any interference is one of the most basic strategies for collecting them. This has been done for a variety of algal strains, and it has a wide range of applications in wastewater treatment plants where high bacterial loads and nutrient levels encourage clumping and settling. Due to poor compaction and sluggish settling velocities, this technique usually results in a moist, thick sludge. The settling characteristic of suspended materials is influenced by the density and radius of algal cells, as well as the generated sedimentation velocity. Lamella separators and sedimentation tanks can help improve microalgal harvesting through sedimentation. Flocculation is widely used to improve gravity sedimentation efficiency. The density of microalgal particles has a significant impact on the effectiveness of solids removal by gravity settling. Microalgae with naturally high sedimentation rates can benefit from a gravity sedimentation system. This is done in thickeners or clarifiers as part of typical water-treatment plant activities. The initial investment and ongoing operating costs are both minimal. A flocculation agent can be employed if the strain has poor settling qualities (Kamyab et al. 2019).

6.3.2.1.2 Flotation

Flotation may be thought of as upside down gravity settling, and it is favorable since microalgae prefer to float rather than settle. High overflow rates, short detention times, a compact footprint, and a thicker concentrate are all advantages of flotation versus sedimentation. Flotation produces sludge with a solid content of 2%–7%. The economic viability of flotation has been reported to range from high capital and running expenses to cheap initial and ongoing costs, minimal energy input, and simple operation. Despite these variations, flotation has been identified as the most cost-effective approach for harvesting microalgae in bulk (Laamanen et al. 2016). But a limitation is that there are difficulties in the process if a mock tail of algae species is present. Flotation types:

(a) **Flotation with dispersed air:** Air is added to the system by a mechanical agitator or air injection through a porous media in dispersed air flotation. Because of the comparatively high bubble size in the range of 700–1500 m, this is less energy expensive than dissolved air flotation.
(b) **Flotation with dissolved air:** A part of the water that has been separated is recycled and pressured in this method. This recycling ratio, which is normally 5%–15% and pressured between 400 and 650 kPa, controls the number of bubbles in the operation (Ratnayaka et al. 2009).
(c) **Electrocoagulation-Flotation:** Electrocoagulation-flotation (ECF) is based on the release of cations, but it does so through electrochemistry, making it simpler to manage and prevent overdose. During ECF, the anode oxidation releases aluminum and iron ions into solution, forming metal hydroxides that function as coagulation or flocculation agents. ECF creates microbubbles that cause the algae flocks to float to the top for simple skimming without adding any anion other than hydroxyl ion to the solution (Mollah et al. 2001).

6.3.2.1.3 Centrifugation

Centrifugation is a technique for separating molecules with various densities in solution by spinning them at high speeds around an axis (in a centrifuge rotor).

The denser particles will migrate outward in this situation, while the less dense particles would move inward. A rise in the concentration of biomass as well as an increase in the efficiency of harvesting may be attained in a small interval of time with this strategy. The total effectiveness of the centrifugation process is mostly determined by the settling properties of the cells, the depth of settling (which may be reduced by centrifuge design), and the slurry holding time in the centrifuge that is regulated by the flow rate (Najjar and Abu-Shamleh 2020). About 80%–90% of biomass can be recovered successfully in 2–5 min using centrifugation. Another significant benefit of this procedure is that it does not necessitate the use of chemical additives. As a result, the algae biomass can be effectively stored for a large interval of time while maintaining its quality. Although centrifuges are often utilized in the food and pharmaceutical sectors, a lot of newly constructed centrifuges (Heasman et al. 2000) have lately been adopted in biodiesel production.

6.3.2.2 Flocculation

Flocculation is a chemical process in which colloidal particles come out of suspension and settle as flock or flake, either naturally or as a result of the addition of a clarifying agent. Prior to flocculation, colloids are just suspended in a liquid in the form of a stable dispersion and are not really dissolved in solution, as is the case with precipitation. The scattered microalgal cells combine and create bigger particles with a faster sedimentation rate during flocculation (Gerde et al. 2014). Flocculation can be triggered in a variety of ways. At the industrial scale, induced chemical flocculation utilizing Zn^{2+}, Al^{3+}, Fe^{3+}, or other chemical flocculants is used, particularly in wastewater treatment facilities (Buelna et al. 1990). Although this approach is simple and effective, it is not suitable for low-cost and long-term microalgae harvesting in large-scale microalgae production facilities because excess cationic flocculants must be removed from the medium before it can be reused, resulting in additional operational expenses. Another approach for induced flocculation of microalgae that has been proposed is biologically induced flocculation with bacteria, which has been used effectively in wastewater treatment. But it requires additional substrate and an additional energy source for bacterial development, resulting in unwanted bacterial contamination of the algal production facility. Flocculation can also be caused by applying severe pH, nutrient depletion, temperature variations, and changes in the quantity of dissolved O_2 in the culture.

6.3.2.3 Filtration

Filtration is a physical separation procedure that uses a filter material with a sophisticated system through which only the fluid can pass to separate solid particles and fluid from a mixture. Oversize solid particles cannot pass through the filter media, and the fluid that flows through is known as the filtrate (Aramaki et al. 2021). Filtration is regarded as one of the most relevant ways for separating microalgae from their growth media, taking into account several parameters (such as efficiency, energy demand, prices, algae species, biomass quality, biomass quantity, toxicity, and processing time). The membrane filtering technique is chemical-free, nontoxic, has no phase shifts, and may reach high efficiency of separation (up to 100%). It also allows for continuous/discontinuous microalgae separation and broth recyclability (Bilad et al. 2014). Microalgae harvesting utilizing membrane-based technology, on the other hand, necessitates a rather high energy input, which, together with fouling, is the key obstacle for extensive industrial uses. It also has high flux requirements and high membrane filtration operating and maintenance costs. Membrane fouling from cake formation, blocking, and adsorption of gel-forming foulants like intracellular and extracellular organic matter (e.g., polysaccharides, proteins, lipids, etc.) reduces flux through the membrane, increases costs, and decreases efficient use in the long run. As a result, the main aims of membrane filtration microalgae harvesting systems are to decrease expenses by reducing fouling and boosting flux through it and increasing efficiency of energy.

6.3.3 Extraction and Conversion

Despite its auspicious qualities, no one has yet been able to commercially produce algae on a large scale for the purpose of creating fuels. The main reason for this sluggish growth is that it is not cost-effective. Growing and converting algae into fuel costs more than the fuel itself is worth (Hannon et al. 2010). If algae-based biofuel is to be commercially feasible, it must also be energy viable. In principle, this implies that the fuel must have more energy than it took to make it, with the additional energy coming from the sun through photosynthesis. In algae production, energetic viability is a key concern. During each stage of the production process, including growing the algal culture and harvesting it and extracting biofuels, the energy required multiplies. To create the feedstocks for biofuel production, the gathered biomass (containing roughly 5%–20% of solids, depending on the harvesting method utilized) must be processed further. Biogas may also be produced by anaerobic fermentation, which uses methanogens (microorganisms that produce methane) to process the biodegradable content of the feedstock (Ward et al. 2014). Microalgae may be used to create biodiesel because of their high lipid content, but their propensity to collect starch and cellulose also makes them appropriate for bioethanol and biogas generation. Bioethanol is mostly generated by anaerobic fermentation of sugary and starchy sources using yeast as the microbial culture. Biodiesel is synthesized by transesterifying oil with methanol to form fatty acid methyl esters (FAME) (Bhatia 2014). For the conversion of algal biomass to fuels, there are several different approaches. The following three categories can be used to classify these pathways:

- Direct conversion of useable fuel molecules like hydrogen and ethanol from algae bypassing the extraction step.

- Processing entire algae biomass to obtain fuel.
- Processing algae extracts like lipids and carbohydrates to obtain fuel.

6.3.3.1 Direct Conversion

Direct biofuel synthesis from algae biomass via heterotrophic fermentation and growth provides certain economic benefits since it eliminates numerous steps involved, like extraction of oil, and their related expenses in the entire fuel production process. Heterotrophic development also enables the maintenance of tightly regulated conditions, which might be geared first toward biomass production and later toward oil production. A system like this may produce a lot of biomasses (hundreds of grams per liter) and a lot of lipids out of it (>50%). The system can easily be upscale, and there's a lot of room to employ different algae cultures thereby lowering manufacturing costs. Through heterotrophic fermenting of starch, algae like *Chlorella vulgaris* and *Chlamydomonas perigranulata* may produce ethanol and other alcohols (Hon-Nami 2006).

This can be performed by producing and storing starch inside the algae through photosynthesis or by explicitly feeding sugars to the algae and then anaerobic fermenting of these carbon pools to make ethanol under nocturnal circumstances. If these alcohols can be isolated directly from the algae culture, the approach might be far less expensive and time-consuming than competing algal biofuel technologies. The procedure would obviate the need to separate the biomass from the water as well as extract and treat the oils. Some microalgae exude oil or alcohol directly into the growing media, eliminating the requirement for extraction and conversion. The green algae *Botryococcus braunii* is a microalga that releases long-chain hydrocarbons. These hydrocarbons have the benefit of being easily turned into biofuels (Lozoya-Gloria et al. 2019). Although *Botryococcus braunii* produces stunning blooms in nature, it is difficult to regulate in open ponds, owing to competition with fast-growing microalgae. The majority of investigations on the effects of various parameters on biomass and hydrocarbon production were conducted in the laboratory utilizing cylindrical batch airlift systems. Experimental cultures in tubular photo-bioreactors submerged in water for cooling have been carried out up to a capacity of 200 L under natural light, but they have not been tested in large-scale production setups. Apart from alcohol, algae may also create hydrogen without the requirement for any sort of extraction (Lam et al. 2019). There are four options for doing so:

(a) **Direct photolysis:** When water is exposed to light, it dissociates into hydrogen and oxygen. Because they have chlorophyll and photosynthetic mechanisms, green microalgae can use light to perform photosynthesis. The enzyme hydrogenase is particularly sensitive to oxygen; thus, when a specific oxygen concentration is present, hydrogenase activity is reduced and the enzyme stops generating hydrogen (Yu and Takahashi 2007).

(b) **Indirect photolysis:** The first stage consists of breaking up of the water molecules in the presence of sunlight, which results in the formation of protons and oxygen. Second, carbon dioxide is fixed, and storage carbohydrate is created, followed by hydrogenase producing hydrogen gas. Blue-green algae produce hydrogen with this method (Dalena et al. 2017).

(c) **Fermentation in the dark:** Production of hydrogen in the absence of sunshine, water, and oxygen in a dark atmosphere. Microorganisms carry out fermentation thereby hydrolyzing complex polymers (organic) to monomers, which are then transformed by the necessary hydrogen-producing bacteria into a mixture of organic acids and alcohols.

(d) **Photo-fermentation:** It is the utilization of sunlight as an energy source to ferment organic substrates into hydrogen and carbon dioxide (Alice and Luisa 2020).

6.3.3.2 Processing Entire Algae Biomass

In addition to producing biofuels directly from algae, the entire algae may be converted into fuels rather than removing oils first and then processing. These techniques benefit from lower costs (since no extraction) and the added benefit of being able to handle a wide variety of algae, but some dewatering is still necessary. Pyrolysis, gasification, anaerobic digestion, and hydrothermal liquefaction are the four popular methods used.

6.3.3.2.1 Pyrolysis

Pyrolysis is the heating-induced chemical breakdown of a condensed material. It doesn't entail any interactions with oxygen or other reagents, yet it does happen often in their presence. Depending on the reaction conditions, the thermochemical treatment of algae or other biomass can produce a wide range of products. Short residence durations, quick heating rates, and moderate temperatures improve liquid product yield. Pyrolysis has one significant benefit over other conversion methods: It is incredibly quick, with reaction times ranging from seconds to minutes (Huber and Dumesic 2006). This enables the transformation of goods into a distinct, and often superior, character than the original biomass. Microalgae pyrolysis produces three types of products: condensed liquid (bio-oil), gaseous products, and biochar (Yang et al. 2019). Because pyrolytic liquids often contain 30%–50% water, they generate two layers of products at the same time: the water phase and the oily phase, which are referred to as aqueous products and bio-oil, respectively. The product yields for water soluble, bio-oil, gases, and biochar vary from (in weight %) 15% to 30%, 18% to 57.9%, 10% to 60%, and 15% to 43%, respectively. Microalgal species, growth circumstances, and reaction conditions all contribute to the vast variety of product distributions. Microwave pyrolysis is a mild, medium-speed pyrolytic technique that falls between slow and fast pyrolysis in terms of heating rate. Microwave-assisted pyrolysis research (Chen et al. 2012) on *Chlorella vulgaris* used activated carbon, calcium oxide, silicon carbide, and microalgal biochar as catalysts. Without the use of catalysts, a maximum liquid yield of 34.9% was achieved. When pyrolyzing microalgae-based biomass with one of the extra materials, it was discovered that the new material boosted gas output but decreased liquid yield, despite the overall conversion of microalgae biomass increasing. Meanwhile, pyrolysis with a low power input (740 W) yielded 93 weigh percent biochar, showing that microwave-assisted pyrolysis might be a viable biochar manufacturing method.

6.3.3.2.2 Gasification

Gasification is the incomplete burning of biomass that produces combustible gases such as hydrogen (H_2), carbon monoxide (CO), and trace amounts of methane (CH_4). This combination is also known as producer gas. Gasification technology has been put to a lot of work in order to get it prepared for the market and expand it into a robust, effective, and economical technology. The diversity of the gases generated is the primary cause of interest in biomass gasification. Gasification, in comparison to the well-established and developed combustion technology, faces hurdles such as tar and other trace contaminants, which present operational issues for downstream gas usage. Such gas usage may be found in gas engines or turbines, as well as when catalysts are utilized to reform flammable gas. One of the most frequent technologies for on-site CO_2 collecting is chemical looping gasification (CLG) (Zeng et al. 2017).

This method's gasification process is broken into multiple sub-reactions that are completed in two reactors. The first one is a fuel reactor, which is used for biomass gasification, and the other is an air reactor, which is used for char cremation and oxygen carrier (OC) regeneration. By avoiding direct contact of fuel with air, the quality of syngas generated by this process improves significantly. CLG is the most popular approach for boosting hydrogen (H_2) and carbon monoxide (CO) at the moment since it is cost-effective and produces acceptable results when compared to conventional techniques. *Scenedesmus almeriens* algae was gasified at temperatures below 700 °C with a nickel-based catalyst (Díaz Rey et al. 2015), yielding hydrogen-rich syngas with a calorific value of roughly 25 Mega joules per normal cubic meter. The hydrogen output was higher than the 10 mol/kg algae during hydrothermal gasification of *Saccharina latissima, Chlorella vulgaris, and Spirulina platensis* (Onwudili et al. 2013). Macroalgae such as *Fucus serratus* has also been steam gasified to produce over 45 mol hydrogen per gram algae biomass (Duman et al. 2014).

6.3.3.2.3 Anaerobic Digestion (AD)

Microorganisms decompose organic matter in the absence of oxygen through a series of processes known as anaerobic digestion (Ward et al. 2014). The method is used to manage trash and create energy in both industrial and home settings. Anaerobic digestion is utilized in a lot of commercial and home fermentation processes to make food and drink items. Anaerobic digestion happens naturally sometimes in soils and sediments from lakes and ocean basins, where it has been termed as anaerobic activity. The input components are first hydrolyzed by bacteria, which starts the digestive process. Organic polymers that are insoluble, such as carbohydrates, are broken down into soluble derivatives that may be used by other bacteria. The sugars and amino acids are subsequently broken down into carbon dioxide, hydrogen, ammonia, and organic acids by acid-producing bacteria. Bacteria transform these organic acids into acetic acid, along with extra ammonia, hydrogen, and carbon dioxide, among other chemicals, during acetogenesis. Finally, methanogens decompose these compounds into methane and CO_2. The populations of methanogens are critical in the treatment of anaerobic wastewater.

Microalgae may be used as a suitable renewable substrate for AD to create biogas while also producing fertilizer and biofuels. The cost-effectiveness and sustainability of the integration and combining phase AD for biogas generation may be improved. The AD of microalgae and their leftovers should be maximized to play a major part in the potential of renewable energy by a deeper understanding of the varied microalgae species and their distinct potentials. For untreated *Microcystis* sp., the lowest gas generation from freshwater microalgae biomass was 70 mL/g volatile solids.

Verstraete and co-authors have been showed the maximum methane production with a gas output of 600 mL/g volatile solids for a mixed unknown freshwater microalgae population. Algal methane production capacity is influenced by the relative amounts of carbs, proteins, and lipids. Lipids, rather than glucose or protein, produce more biogas per gram of organic matter. At the same time, algal composition varies greatly depending on species, season, and environmental circumstances, as well as thallus portion (holdfast, stipe, etc.) in the case of seaweed, posing a problem for biomethane generating repeatability. Furthermore, the bacteria involved in AD's methane generation are sensitive to the chemical makeup of the feedstock (Chen 1987). It has been hypothesized that macroalgae's low lipid content makes them "particularly appropriate" for AD biogas generation. Unlike seaweed, microalgae may be high in lipids, with levels ranging from 20% to 50%, and in some cases exceeding 70% when "stressed" during development. Triglycerides and long-chain fatty acids (LCFAs) have a high methane potential, but when present in large quantities, they can cause process obstructions, resistant floating fatty crusts, and LCFAs sticking to the surface of acetoclastic and methanogenic bacteria, impeding their growth (Anderson et al. 2003). Despite the fact that macroalgae generate 100 times more wet biomass than microalgae, much of the research on algal biofuels has concentrated on microalgae rather than macroalgae. This is primarily due to the possibility of selecting or developing microalgal strains with a high triacyl-glycerides (TAG) concentration suited for transesterification.

6.3.3.2.4 Hydrothermal Liquefaction

In the presence of a catalyst and hydrogen, hydrothermal liquefaction creates bio-oil from a relatively wet biomass at a relatively low temperature (300 °C–400 °C) and high pressure (40–200 bar). There is also the production of gas and char. A representative product may comprise 11%–74% bio-oil, 8%–20% gas, and 0.2%–0.5% char (by weight) (Elliott et al. 2015). The bio-oil produced includes less oxygen than pyrolysis-derived bio-oil. With or without the aid of a catalyst, hydrothermal liquefaction (HTL) transforms wet biomass into liquid biocrude oil. Compared to pyrolysis and gasification, hydrothermal liquefaction is a much more costly procedure. Furthermore, if the water content of the biomass surpasses 90%, the energy balance of hydrothermal liquefaction is likely to be negative. Bio-oil can be recovered in the range of 23% to 49% of the original dry mass, depending on the microalga. This bio-oil might contain up to 75% of the original biomass's energy.

When comparing the two methods for treating microalgae (pyrolysis and hydrothermal liquefaction), certain major difficulties for pyrolysis emerge. The drying step, which is necessary before entering a pyrolysis reactor, is always a vital phase, and it is especially important for microalgae feed owing to the reduced concentration of algae in the growth medium, which normally ranges from 10%–20% even after harvesting. Pyrolysis oil has certain advantages, such as a reduced viscosity that is equivalent to vegetable oil, but the HTL looks to be a more intriguing technology in terms of yield, nitrogen and oxygen content, and hence heating value.

The feedstock's uniqueness poses significant challenges for the reactor (Tews et al. 2014). The methods developed so far are based on lignocellulosic materials' expertise, but microalgae have higher inorganic and ash content, a unique state of aggregation dependent on harvesting and pretreatment, and so on. Because of the more critical pressure and temperature operation conditions of HTL, the use of salty water for algae culture raises the problem of corrosion and solid deposition. When it comes to determining the technical readiness level of the processes, pyrolysis looks to be the most advanced, with prototype scale plants currently operational. Microalgae HTL is progressing from applied research to small-scale development.

6.3.3.2.5 Supercritical Extraction

The method of isolating one component (the extracting) from the other (the matrix) using supercritical fluids as the extracting solvent is known as supercritical fluid extraction (SFE). The SFE method is a fluid extraction technique that uses fluids at temperatures above their critical point of temperature. The density of the supercritical fluid is similar to that of liquids, but its viscosity is equivalent to that of a gas. Its diffusivity, on the other hand, is between gases and liquids. These features allow supercritical fluids to permeate solid matrices deeper and quicker. Carbon dioxide is the most often used supercritical fluid (SC-CO_2). Its critical temperature and pressure are both low (304.2 K and 7.38 MPa, respectively), which makes it easier to operate. Furthermore, its lack of toxicity, flammability, low cost, and high purity availability make it a competitive fluid for industrial applications, among others (Nikolai et al. 2019).

Supercritical processing is a new entry to the arsenal of procedures capable of extracting and converting oils into biofuels at the same time. Supercritical fluids are selective, resulting in high purity and concentrations of product. Supercritical fluid extraction of microalgae biomass is significantly more efficient than typical solvent separation procedures, and it has also been shown to be exceptionally effective in extracting other components from algae. Microalgae oil extracts can also benefit from this supercritical transesterification method. Furthermore, neither the extract nor the wasted biomass contains any organic solvent residues. Extraction is most effective at low working temperatures, such as less than 50 °C, assuring optimal product stability and quality. Furthermore, supercritical fluids may be utilized on complete algae without the need for dewatering, boosting the process' efficiency. An example in literature (Mendes et al. 2003) is the supercritical extraction of entire, crushed, and slightly crushed states of *Arthrospira maxima, Chlorella Vulgaris, Dunaliella salina, and Botryococcus braunii* that were supercritical extracted at a temperature of 40–60 °C and pressures of 125, 200, and 300 bar, respectively.

6.3.3.3 Processing Extracts Followed by Conversion

The high lipid content of microalgae (approximately 51% of ash-free dry weight) obtained in experimental and cultivation field studies has made it possible to extract lipids, in the form of oil, for fuel generation. In terms of the amount of water removed for efficient extraction of the oil from the algae biomass, extraction processes can be classified as wet or dry. The oil is commonly turned to biodiesel by transesterification after extraction. The typical way of biofuel generation from algae is the processing of extracts generated from algal sources by extraction and then conversion. As a result of the evident and crucial relationship between the kind of extraction method utilized and the product composition, a fundamental and comprehensive understanding of the many types of inputs to conversion technologies is required. Lipid-based algae extracts, such as triacyl-glycerides, are the most prevalent form of algae extract under consideration (Zhou et al. 2022), as they may be turned into biodiesel. The chemical, biological, and catalytic methods that can be used to transform algae extracts are considered in this segment.

Removal of oil by solvent treatment, cell disruption, or a combination of the two is the popular extraction methods used. Cell disruption methods liberate lipids trapped in the cells by destroying the cell wall, whereas solvent extraction techniques extract oil from the cells of the microalgae by diffusion.

6.3.3.3.1 Cell Disruption

The majority of microalgae's important components are stored inside the cell, behind a strong and impenetrable cell wall. As a result, energy- or solvent-intensive processes are required to break down this physical barrier and extract the necessary substances (lipids, proteins, etc.) effectively. To make cell components available without losses, a gentle cell disruption approach is required (Gomes et al. 2020). Some algae, such as *Dunaliella*, lack a cell wall and are, thus, vulnerable to cell disruption. *Scenedesmus* is made up of three layers: an inner cellulosic layer that separates individual cells, a thin middle algaenan-based layer, and an outside pectic layer that connects the cells. *Haematococcus* is a motile cell with a thick gelatinous extracellular matrix made up of interconnecting fibers, granular, and crystalline materials. Several ways may be used to break down such molecules, which are part of the cell wall and have varying degrees of stiffness and provide protection from environmental stimuli. Physical, chemical, and enzymatic procedures are used to break cell walls.

Mechanical methods use physical force to break down cell wall components. The most extensively utilized mechanical procedures for microalgal cell disruption are the bead mill, high-pressure homogenization, and ultrasonication. One of the major successful approaches is bead milling, which employs kinetic energy to cause tiny beads of steel, plastic, ceramic, or glass to crash with one other and with the algal cells. Shaking containers are less effective than agitated beads. Different press mechanisms such as screws, pistons, and expeller are also employed. The expeller press compresses microalgae cells and squeezes substance out of them using mechanical force. Despite the many benefits of bead milling, such as its suitability for large-scale manufacturing, this process still consumes a lot of energy (Alhattab et al. 2019).

The high-speed homogenizer uses a combination of hydrodynamic cavitations caused by high-revolutions per minute swirling and physical compression to shatter cell walls. One of the first strategies for disrupting microalgae was high-pressure homogenization. Under high pressure (140–399 MPa), microalgal concentrate is forced through a tiny orifice (81–200 m) through a valve, and the suspension is then discharged into a reduced pressure compartment. When a cell moves from valve to chamber, high-pressure contact of an accelerating cellular spray on the stable valve surface causes cell wall breakage, as does a pressure drop-induced shear stress. The different valve-seat layouts devised allow for maximum cell disruption effectiveness while minimizing valve seat damage caused by cavitations.

The dispersion of electromagnetic waves in the irradiated material causes microwave heating. The dissipating power is influenced by the dielectric characteristics and the average electric field. Microwaves induce water and other polar molecules in wet biomass to vibrate, causing temperature rises in the intracellular liquids, causing the water to evaporate and put pressure on the cell walls, resulting in cell rupture. Microwave-assisted extraction (MAE), a technique that combines microwave with solvent extraction, has been described (Kumar 2019) as having lower operational costs and extraction time than traditional procedures, as well as greater lipid extraction than other extraction techniques. Some few nonmechanical methods of cell disruption are discussed below. Chemical approaches, which frequently rely on particular interactions with cell wall elements, can be more selective than mechanical ones. Furthermore, energy consumption is lower, cell disruption efficiency is better, and upscaling is more straightforward. The cost of chemicals and the quality of the goods, on the other hand, may limit the advantages.

Enzymatic hydrolysis is a low-energy, high-selective approach for cell destruction that functions under moderate circumstances. Commercial enzymes such cellulases, proteases, lysozyme, and glucanases are widely used and often immobilized to extend their lifetime and stability while limiting catalytic activity loss. The long process time, poor production capacity, and economic feasibility are the primary disadvantages of utilizing enzymes over mechanical or chemical approaches. Oxidizing chemicals such as hydrogen peroxide and other inorganic peroxides react with cell wall components, causing cell wall deterioration and disruption (Grabski 2009). This pretreatment can improve the efficiency rate of extraction; however, reaction periods should be maintained short to avoid chemical oxidation of the cell components. Ozonolysis has proven to be appealing and promising, with advantages, over other methods, such as low production of inhibition causing compounds, negligible effects on carbohydrates, no chemical demands, lack of liquid phase, mild conditions, direct ozone production, and the creation of readily biodegradable subproducts. However, high operational expenses, flammability, toxicity, reactivity, and corrosivity are some disadvantages.

6.3.3.3.2 Solvent Extraction

To preferentially extract lipids from a complex mixture of organic compounds, many organic solvents or combinations of organic solvents have been suggested. The Folch technique (Folch et al. 1957) extracts lipids from endogenous cells using a mixture of chloroform and methanol in a volume ratio of 2:1. The homogenized cells were kept in equilibrium with a quarter volume of saline solution and well mixed. The lipids settled in the top phase of the resultant mixture, which was left to split into two layers. This approach is one of the first in lipid extraction, and it served as the foundation for the development of improved extraction processes in the future. The following approach, with minor modifications, is still employed as it has the ability to analyze a large number of samples quickly and easily. Another widely used extraction technique is the Bligh and Dyer method (Annika et al. 2016). The basic principles for this is same as that of the Folch method using lesser amounts of chloroform and methanol, and also the solvent/solvent and solvent/tissue ratios are different. In this, the proteins are deposited at the interface of two liquid phases combining lipid extraction with partitioning. The lipids are extracted from a homogenized cell culture using a 1:2 (v/v) chloroform/methanol solution. The lipids from the chloroform phase are subsequently extracted and treated using a variety of methods.

In the presence of insoluble contaminants, Soxhlet extraction has historically been utilized for a solid material with limited solubility in a solvent. The Soxhlet extractor's main chamber is filled with a porous thimble containing a solid sample. The extraction cycle is often repeated multiple times by refluxing the solvent through the thimble using a condenser and a syphon side arm. Soxhlet extraction is a tried-and-true technology that allows for unassisted extraction. However, it necessitates a lengthy extraction period and a big volume of solvent (Aravind et al. 2021). Because the solvent has restricted access to the lipids in the algal cells, Soxhlet often does not work as well as the Bligh and Dyer technique. Accelerated fluid extraction (AFE) is comparable to Soxhlet extraction, with the exception that the solvents are utilized near their supercritical region, where they have superior extraction efficiency (Naqvi et al. 2021). The high temperature in that physical region allows for high solubility and high diffusion rate of lipids in the solvent, while the high pressure keeps the solvent underneath the boiling point, allowing for high solvent penetration in the sample. When compared to Soxhlet extraction, which takes several hours and uses 8 to 26 times the solvent, AFE allows for excellent extraction efficiency with a small solvent volume (16–40 mL) and a quick extraction time (14–20 min). After the oil has been extracted from the algae, it can be transesterified to make biodiesel or hydrotreated to make hydrocarbons.

6.3.3.3.3 Transesterification

Transesterification is a three-step process in which a triglyceride is sequentially transformed into diglycerides, then monoglycerides. When these monoglycerides react with methanol, they produce FAME (biodiesel) and crude glycerol. Transesterification is a reversible reaction that is carried out by mixing the reactants. Triglycerides made up of three alkyl groups of fatty acid chains are used in the transesterification process (Baskar et al. 2019). The triglycerides combine with the alcohol to generate three FAME (fatty acid methyl esters) molecules and one glycerol molecule through a displacement process. However, the presence of a catalyst, a strong acid (sulfuric acid) or strong base (sodium hydroxide), speeds up the conversion, and a little amount of extra alcohol is required to tip the balance toward the production of fatty acid alkyl esters and glycerol (Wahlen et al. 2011). Transesterification of microalgal oils for biodiesel synthesis has been done using both homogeneous and heterogeneous catalysis. Homogeneous alkaline catalysis is one of the most extensively used methods for biodiesel synthesis because it catalyzes the process at low temperatures and pressures while producing a high conversion yield in a short space of time.

Oil, an acyl acceptor, typically an alcohol, and an enzyme called lipase serve as the raw ingredients for enzymatic transesterification (Tran et al. 2012). This approach has the following advantages: high substrate specificity, extensive substrate variety, full catalysis of free fatty acids, good product quality, mild reaction temperatures, low alcohol to oil ratio, no saponification, and no effluent formation. However, due to its high cost, sluggish reaction rates, enzyme inhibition, and loss of function, enzyme transesterification is not frequently employed. As a result, additional advancements to lower the price, speed up the process, or lessen enzyme inhibition are needed.

6.3.4 *Applications of Algae*

Microalgae if extracted to its full potential can be a valuable source of carbon compounds for biofuels, medicines, pharmaceuticals, and cosmetics. They may also be used to clean wastewater and reduce carbon dioxide levels in the atmosphere. Some of the most common applications are discussed below.

6.3.4.1 Animal Feed

Microalgae as feed components have a lot of promise as a replacement for basic food crops like maize and soybean, reducing rivalry with the human food chain and helping to agriculture sustainability. The kind of microalgae and their nutritional contents, such as protein, carbs, lipids, vitamins, and antioxidants, have a significant impact on the use of microalgae as animal feed. The animal's adaption to the substance is another aspect that influences the outcome (Macke et al. 2017). There are various advantages of using microalgae biomass as feed, particularly for animal physiology, such as improved immune response, disease resistance, and antibacterial and antiviral activity. Microalgal biomass has been found to be useful as a feed additive in several nutrition and ecological research (Becker 2007). Pet animals such as cats, dogs, horses, birds, and cattle are just a few instances of how *Arthrospira* is used. *Chlorella, Isochrysis, Pavlova, Phaeodactylum,* and *Chaetoceros* are the most significant microalgae species for aquaculture farms, particularly for feeding fish larvae. Microalgae can be used as a partial replacement for traditional proteins in chicken feed and to improve the yellow color of broiler skin and egg yolk.

6.3.4.2 Human Food

Algae have been eaten in many civilizations for thousands of years. The earliest evidence of algae usage as a dietary source comes from Chile, where archaeological records date back (Buchholz et al. 2012) to 14,000 years ago. However, none of these algae have been developed as a major crop, owing to a variety of production and societal acceptance challenges, as algae are very different from typical crops. Protein and lipids make up a large portion of algae's dry biomass. *Arthrospira platensis* has been reported (Ljubic et al. 2018) to have up to 70% of its biomass as protein, whereas *Auxenochlorella protothecoides* has been found to have up to 70% of its dry biomass as lipids. Other nutrients, such as vitamins and minerals, are essential in the human diet in addition to protein and fats. The majorities of them are not created by animals, but rather by plants or other creatures, and are subsequently consumed by humans and animals. Algae are high in vitamins and minerals, much like regular vegetable diets. *Dunaliella tertiolecta*, a green alga that has been demonstrated to be a good source of vitamin E, vitamin A, vitamin B9, and vitamin B1 is one example. Pigments like B-carotene and astaxanthin are also commercially produced from algae (Ana et al. 2011). B-carotene (natural) is utilized as a food color agent and is converted into the vital vitamin A in the human body. Beta-glucans (Kwangwook et al. 2019) are a kind of polysaccharide made up of D-glucoses linked together by links that are commonly found in the cell walls of plants, bacteria, and fungi. When consumed by humans, glucans act as soluble fiber, lowering LDL cholesterol and lowering the risk of heart disease.

6.3.4.3 Fertilizers

The most pressing issue in agriculture today is the lack of inexpensive chemical fertilizers. The limiting element in food production has been identified as nitrogen fixation. The capacity of cyanobacteria to fix nitrogen in the soil is dependent on the microalgae's ability to thrive in soil.

Oscillatoria, Nostoc, Anabaena, Mastigocladus, Tolypothrix, and *Cylindrospermum* are microalgae (Ammar et al. 2022) that may fix nitrogen aerobically in heterocyst's (thick-walled modified cell, which is considered as site of nitrogen fixation by the nitrogenase enzyme). *Oscillatoria* and *Phormidium*, nonheterocyst-forming microalgae, can fix nitrogen in the absence of oxygen as well as in the presence of nitrogen and carbon dioxide.

6.3.4.4 Polymers

Algal biomass is used to make biopolymers (Azeem et al. 2017) that have significant benefits and allow higher level molecular interactions. Biopolymers may be employed in a range of applications due to their favorable features, including as medication delivery, tissue engineering, and 3D printing in the biomedical sector. Biopolymers may be made from algal biomass in three different ways. Biopolymer products are made by fermenting algal biomass with microorganisms. Natural biopolymers is produced by cell factories inside algae biomass; composite algal biopolymers are created by combining algae biomass and additives. Algal biomass is converted into bio-products including biopolymer using algae generating enzymes during the fermentation process. When *cyanobacteria* on microalgae like *Chlorella* and *Nostoc* (Mastropetros et al. 2022) are subjected to appropriate growth conditions, they produce polyhydroxybutyrate (PHB) and its copolymer polyhydroxybutyrate-co-valerate (PHBHV).

6.3.4.5 Biofuels

The residual algal biomass is generally anaerobically digested (Jankowska et al. 2017) to create biogas after lipid (oil) extraction. Biogas may be utilized directly for heating or power production if it is improved, that is, the methane concentration of the gas stream is increased. To capture the bulk of the remaining energy and nitrogen in the algal biomass after oil and/or carbs extraction, converting wet algal biomass to biomethane using a liquid fertilizer also called as anaerobic digestate is a good technique. Biogas from microalgae was found to be 7%–13% (Mussgnug et al. 2010) more enriched in terms of methane concentration than biogas from maize; however, productivity was still hindered by the comparatively high nitrogen content of biomass. A few microalgae contain carbohydrates that can be exploited as a source of ethanol. It should be noted that the method of producing alcohols from microalgae is fairly straightforward since microalgae cells do not contain lignin or hence no pretreatment (enzymatic or chemical) is necessary to extract sugars. However, a simple and affordable physical treatment method such as bead method and mechanical compression is still necessary to disrupt cell. Alcohol fuels made from algae still face significant problems, such as the algal biomass' low fermentable carbohydrate content (de Farias Silva and Bertucco 2016).

The conversion of lipids from algae by indirect transesterification in two phases is typical for biodiesel synthesis from algae. The dewatering of algae and drying of algal biomass are usually the initial steps, followed by the extraction of lipids, which are subsequently transesterified for biodiesel production. Biodiesel has the benefit of emitting 78% less carbon dioxide, 98% less sulfur, and 50% less particulate matter when burned (Helena et al. 2011). As discussed above, hydrogen can be produced from microalgae using direct and indirect photolysis, fermentation in the dark and photo-fermentation. Hydrogen is predicted to become more essential as a clean fuel in the future due to growing energy costs and the issues of a world that is becoming increasingly carbon-constrained. Despite the fact that biohydrogen generation using microalgae is currently not commercially viable, it deserves to be discussed because it is one of the most environmentally friendly ways to generate energy (Show et al. 2019).

6.4 Summary

This study examined literature to offer an analysis of the various algae biomass-to-fuel conversion processes. The majority of approaches are classified into three main processes: (1) direct conversion, (2) processing entire algae biomass, and (3) processing of algae extracts (extraction followed by conversion). *Chlorella vulgaris* is one of the first microalgae to be isolated and grown in pure culture and therefore is one of the most researched and most used algal biomasses for the production of biofuel. For the purpose of commercially producing biofuels, centrifugation appeared to be the most favored way of harvesting the biomass from their cultures. To increase recovery, a flocculation phase is used prior to centrifugation. When the microalgae being recovered are delicate, centrifugation is not a preferable option, and microfiltration is used. For extracting components inside the cell, bead milling is used. Microalgae have a high capacity for converting carbon dioxide from the atmosphere into beneficial products for industries like energy, pharmaceuticals, food, fertilizers, polymers, etc. However, there are significant limits and hurdles that must be overcome in order to move the technology from the prototype to the commercial stage. Enhancing microalgal cultivation, harvest, and dewatering systems, as well as the extraction and transesterification process improvement, must be done in order for it to be a truly renewable, sustainable energy and a worthy competitor to other biomass and renewable fuels.

References

Alhattab M, Kermanshahi-Pour A, Brooks MSL (2019) Microalgae disruption techniques for product recovery: influence of cell wall composition. J Appl Phycol 31:61–88

Alice F, Luisa G (2020) Chapter 28—Microalgal biorefineries. In: Jacob-Lopes E, Maroneze MM, Queiroz MI, Zepka LQ (eds) Handbook of microalgae-based processes and products. Academic Press, San Diego, CA, pp 771–798

Ammar EE, Aioub AAA, Elesawy AE, Karkour AM, Mouhamed MS, Amer AA, El-Shershaby NA (2022) Algae as bio-fertilizers: between current situation and future prospective. Saudi J Biol Sci 29:3083–3096

Ana G, Amaro HM, Malcata FX (2011) Microalgae as sources of carotenoids. Mar Drugs 9(4): 625–644

Anderson K, Sallis P, Uyanik S (2003) 24. Anaerobic treatment processes. In: Mara D, Horan N (eds) Handbook of water and wastewater microbiology. Academic Press, New York, pp 391–426

Annika S, Eggers LF, Schwudke D (2016) Liquid extraction: bligh and dyer. In: Wenk M (ed) Encyclopedia of lipidomics. Springer, Dordrecht

Aramaki T, Watanabe MM, Nakajima M et al (2021) Dewatering of microalgae suspensions by cake filtration with filter cloths. J Appl Phycol 33:1977–1985

Aravind S, Barik D, Ragupathi P, Vignesh G (2021) Investigation on algae oil extraction from algae Spirogyra by Soxhlet extraction method. Mater Today Proc 43:308–313

Azeem M, Batool F, Iqbal N, Ikram-ul-Haq (2017) Chapter 1—Algal-based bio-polymers. In: Zia KM, Zuber M, Ali M (eds) Algae based polymers, blends, and composites. Elsevier, Amsterdam, pp 1–31

Baskar G, Kalavathy R, Aiswarya I, Selvakumari A (2019) 7—Advances in bio-oil extraction from nonedible oil seeds and algal biomass. In: Azad K (ed) Advances in eco-fuels for a sustainable environment. Woodhead Publishing Series in Energy, Cambridge, pp 187–210

Becker EW (2007) Micro-algae as a source of protein. Biotechnol Adv 25:207–210

Bhatia SC (2014) 22—Bio-diesel. In: Advanced renewable energy systems. Woodhead Publishing, New Delhi, pp 573–626

Bilad MR, Arafat HA, Vankelecom IFJ (2014) Membrane technology in microalgae cultivation and harvesting: a review. Biotechnol Adv 32:1283–1300

Buchholz CM, Krause G, Buck BH (2012) Seaweed and man. In: Seaweed biology: novel insights into ecophysiology, ecology and utilization, pp 471–493

Buelna G, Bhattarai KK, de la Noue J, Taiganides EP (1990) Evaluation of various flocculants for the recovery of algal biomass growth on pig-waste. Biol Wastes 31:211–222

Canfield DE (2004) The early history of atmospheric oxygen: homage to Robert M. Garrels. Annu Rev Earth Planet Sci 33:1–36

Chen PH (1987) Factors influencing methane fermentation of micro-alga. Dissertation, University of California, Berkley

Chen C, Hu Z, Ma X (2012) A study on experimental characteristic of microwave-assisted pyrolysis of microalgae. Bioresour Technol 107:487–493

Costa JAV, Freitas BCB, Santos TD, Mitchell BG, Morais MG (2019) Chapter 9. Open pond systems for microalgal culture. In: Pandey A, Chang J-S, Soccol CR, Lee D-J, Chisti Y (eds)

Biomass, bio-fuels, biochemicals, bio-fuels from algae, 2nd edn. Elsevier, Amsterdam, pp 199–223

Dalena F, Senatore A, Tursi A, Basile A (2017) 17—Bio-energy production from second- and third-generation feedstocks. In: Dalena F, Basile A, Rossi C (eds) Bio-energy systems for the future. Woodhead Publishing, Sawston, pp 559–599

Darzins A, Pienkos P, Edye L (2010) Current status and potential for algal bio-fuels production. Bio-energy report

de Farias Silva CE, Bertucco A (2016) Bio-ethanol from microalgae and cyanobacteria: a review and technological outlook. Process Biochem 51:1833–1842

Di Caprio F (2020) Methods to quantify biological contaminants in microalgae cultures. Algal Res 49:101943

Díaz Rey MR, Cortés Reyes M, Herrera C, Larrubia MA, Amadeo NE et al (2015. 257; Part 2; 11–2015) Hydrogen-rich gas production from algae-biomass by low temperature catalytic gasification. In: Catalysis today. Elsevier Science, Amsterdam, pp 177–184

Duman G, Uddin MA, Yanik J (2014) Hydrogen production from algal biomass via steam gasification. Bioresour Technol 166:24–30

Elliott DC, Biller P, Ross AB, Schmidt AJ, Jones SB (2015) Hydrothermal liquefaction of biomass: developments from batch to continuous process. Bioresour Technol 178:147–156

Fasaei F, Bitter JH, Slegers PM, van Boxtel AJB (2018) Techno-economic evaluation of microalgae harvesting and dewatering systems. Algal Res 31:347–362

Fernandez Sevilla JM, Ceron Garcia MC, Sanchez Miron A, Belarbi EH, Garcia Camacho F, Molina Grima E (2004) Pilot plant-scale outdoor mixotrophic cultures of Phaeodactylum tricornutum using glycerol in vertical bubble column and airlift photo-bioreactors: studies in fed-batch mode. Biotechnol Prog 20:728–736

Folch J, Lees M, Sloane Stanley GH (1957) A simple method for the isolation and purification of total lipids from animal tissues. J Biol Chem 226:497–509

Gerde JA, Yao L, Lio JY, Wen Z, Wang T (2014) Microalgae flocculation: Impact of flocculant type, algae species and cell concentration. Algal Res 3:30–35

Gomes TA, Zanette CM, Spier MR (2020) An overview of cell disruption methods for intracellular biomolecules recovery. Prep Biochem Biotechnol 50(7):635–654

Grabski AC (2009) Chapter 18. Advances in preparation of biological extracts for protein purification. In: Burgess RR, Deutscher MP (eds) Methods in enzymology, vol 463. Academic Press, San Diego, CA, pp 285–303

Hannon M, Gimpel J, Tran M, Rasala B, Mayfield S (2010) Bio-fuels from algae: challenges and potential. Biofuels 1(5):763–784. https://doi.org/10.4155/bfs.10.44

Heasman M, Diemar J, O'connor W, Sushames T, Foulkes L (2000) Development of extended shelf-life microalgae concentrate diets harvested by centrifugation for bivalve molluscs—a summary. Aquacult Res 31:637–659

Helena M, Amaro A, Guedes C, Malcata FX (2011) Advances and perspectives in using microalgae to produce bio-diesel. Appl Energy 88:3402–3410

Hon-Nami K (2006) A unique feature of hydrogen recovery in endogenous starch-to-alcohol fermentation of the marine microalga, Chlamydomonas perigranulata. Appl Biochem Biotechnol 131(1–3):808–828

Huang Q, Jiang F, Wang L, Yang C (2017) Design of photobioreactors for mass cultivation of photosynthetic organisms. Engineering 3(3):318–329

Huber GW, Dumesic JA (2006) An overview of aqueous-phase catalytic processes for production of hydrogen and alkanes in a biorefinery. Catal Today 111(132):119

Jankowska E, Sahu AK, Oleskowicz-Popiel P (2017) Biogas from microalgae: review on microalgae's cultivation, harvesting and pretreatment for anaerobic digestion. Renew Sustain Energy Rev 75:692–709

Kamyab H, Chelliapan S, Kumar A, Rezania S, Talaiekhozani A, Khademi T, Sharma S (2019) Microalgal biotechnology application towards environmental sustainability. In: Application of

microalgae in wastewater treatment: volume 2: biorefinery approaches of wastewater treatment, pp 445–465
Kaštovský J, Fučíková K, Veselá J, Carías CB, Vegas-Veilarrúbia T (2019) Chapter 5—Algae. In: ValentíRull TV-V, Huber O, Señaris C (eds) Biodiversity of Pantepui. Academic Press, New York, pp 95–120
Kazbar A, Cogne G, Urbain B, Marec H, Gouic B, Tallec J, Takache H, Ismail A, Pruvost J (2019) Effect of dissolved oxygen concentration on microalgal culture in photo-bioreactors. Algal Res 39:101432. https://doi.org/10.1016/j.algal.101432
Khan MI, Shin JH, Kim JD (2018) The promising future of microalgae: current status, challenges, and optimization of a sustainable and renewable industry for bio-fuels, feed, and other products. Microb Cell Fact 17:36
Kraan S (2012) Algal polysaccharides, novel applications and outlook. In: Chang C-F (ed) Carbohydrates—comprehensive studies on glycobiology and glycotechnology. IntechOpen, London. https://doi.org/10.5772/51572
Kumar A (2019) The scenario of pharmaceuticals and development of microwave assisted extraction techniques
Kwangwook K, Ehrlich A, Perng V, Chase JA, Raybould H, Li X, Atwill ER, Whelan R, Sokale A, Liu Y (2019) Algae-derived β-glucan enhanced gut health and immune responses of weaned pigs experimentally infected with a pathogenic E. coli. Anim Feed Sci Technol 248:114–125
Laamanen CA, Ross GM, Scott JA (2016) Flotation harvesting of microalgae. Renew Sustain Energy Rev 58:75–86
Lam MK, Loy ACM, Yusup S, Lee KT (2019) Chapter 9. Biohydrogen production from algae. In: Pandey A, Mohan SV, Chang J-S, Hallenbeck PC, Larroche C (eds) Biomass, bio-fuels, biochemicals, biohydrogen, 2nd edn. Elsevier, Amsterdam, pp 219–245
Lee RA, Lavoie JM (2013) From first-to third-generation biofuels: challenges of producing a commodity from a biomass of increasing complexity. Anim Front 3(2):6–11
Lozoya-Gloria E, Cruz X, Ozawa-Uyeda T (2019) The colonial microalgae Botryococcus braunii as biorefinery. In: Vítová M (ed) Microalgae—from physiology to application. IntechOpen, London. https://doi.org/10.5772/intechopen.88206
Ljubic A, Safafar H, Holdt SL, Jacobsen C (2018) Biomass composition of Arthrospira platensis during cultivation on industrial process water and harvesting. J Appl Phycol 30:943–954
Macke E, Tasiemski A, Massol F, Callens M, Decaestecker E (2017) Life history and eco-evolutionary dynamics in light of the gut microbiota. Oikos 126(4):508–531
Masojídek J, Torzillo G (2014) Mass cultivation of freshwater microalgae☆. In: Reference module in earth systems and environmental sciences. Elsevier, Amsterdam
Mastropetros SG, Pispas K, Zagklis D, Ali SS, Kornaros M (2022) Biopolymers production from microalgae and cyanobacteria cultivated in wastewater: recent advances. Biotechnol Adv:107999
Mendes RL, Nobre BP, Cardoso MT, Pereira AP, Palavra AF (2003) Supercritical carbon dioxide extraction of compounds with pharmaceutical importance from microalgae. Inorg Chim Acta 356:328–334
Mollah MYA, Schennach R, Parga JR, Cocke DL (2001) Electrocoagulation (EC)—science and applications. J Hazard Mater 84(1):29–41
Morales M, Aflalo C, Bernard O (2021) Microalgal lipids: a review of lipids potential and quantification for 95 phytoplankton species. Biomass Bioenergy 150:106108
Mussgnug JH, Klassen V, Schlüter A, Kruse O (2010) Microalgae as substrates for fermentative biogas production in a combined biorefinery concept. J Biotechnol 150(1):51–56
Najjar Y, Abu-Shamleh A (2020) Harvesting of microalgae by centrifugation for bio-diesel production: a review. Algal Res 51:102046
Naqvi SAR, Akbar N, Shah SMA, Ali S, Abbas A (2021) Green approaches for the extraction of bioactive from natural sources for pharmaceutical applications. In: Green sustainable process for chemical and environmental engineering and science. Elsevier, pp 269–291

Nikolai P, Rabiyat B, Aslan A, Ilmutdin A (2019) Supercritical CO_2: properties and technological applications—a review. J Therm Sci 28(3):394–430

Onwudili J, Lea-Langton AR, Ross AB, Williams PT (2013) Catalytic hydrothermal gasification of algae for hydrogen production: composition of reaction products and potential for nutrient recycling. Bioresour Technol 127:72–80

Poudyal R, Loskot P, Nepal R, Parajuli R, Khadka SK (2019) Mitigating the current energy crisis in Nepal with renewable energy sources. Renew Sust Energ Rev 116:109388

Qiu R, Gao S, Lopez PA, Ogden KL (2017) Effects of pH on cell growth, lipid production and CO2 addition of microalgae Chlorella sorokiniana. Algal Res 28:192–199

Ratnayaka DD, Brandt MJ, Michael Johnson K (2009) Chapter 7. Storage, clarification and chemical treatment, water supply, 6th edn. Butterworth-Heinemann, pp 267–314

Russell C (2013) Algae: the world's most important "plants" an introduction. Mitig Adapt Strat Glob Chang 18:5–12

Schlesinger WH, Bernhardt ES (2020) Chapter 8. Inland waters. In: Schlesinger WH, Bernhardt ES (eds) Biogeochemistry, 4th edn. Academic Press, New York, pp 293–360

Show K-Y, Yan Y-G, Lee D-J (2019) Chapter 13. Biohydrogen production from algae: perspectives, challenges, and prospects. In: Pandey A, Chang J-S, Soccol CR, Lee D-J, Chisti Y (eds) Biomass, bio-fuels, biochemicals, bio-fuels from algae, 2nd edn. Elsevier, Amsterdam, pp 325–343

Singh G, Patidar SK (2018) Microalgae harvesting techniques: a review. J Environ Manage 217: 499–508

Stephenson AL, Kazamia E, Dennis JS, Howe CJ, Scott SA, Smith AG (2010) Life-cycle assessment of potential algal bio-diesel production in the United Kingdom: a comparison of raceways and air-lift tubular bioreactors. Energy Fuel 24(7):4062–4077

Tews IJ, Zhu Y, Drennan C, Elliott DC, Snowden-Swan LJ, Onarheim K, Beckman D et al (2014) Biomass direct liquefaction options. Technoeconomic and life cycle assessment (No. PNNL-23579). Pacific Northwest National Lab (PNNL), Richland, WA

Tran DT, Yeh KL, Chen CL, Chang JS (2012) Enzymatic transesterification of microalgal oil from Chlorella vulgaris ESP-31 for bio-diesel synthesis using immobilized Burkholderia lipase. Bioresour Technol 108:119–127

Tredici MR, Zittelli GC (1998) Efficiency of sunlight utilization: tubular versus flat photobioreactors. Biotechnol Bioeng 57:187–197

Wahlen BD, Willis RM, Seefeldt LC (2011) Biodiesel production by simultaneous extraction and conversion of total lipids from microalgae, cyanobacteria, and wild mixed-cultures. Bioresour Technol 102(3):2724–2730

Wang B, Lan CQ, Horsman M (2012) Closed photo-bioreactors for production of microalgal biomasses. Biotechnol Adv 30:904–912

Ward AJ, Lewis DM, Green FB (2014) Anaerobic digestion of algae biomass: a review. Algal Res 5:204–214

Wrigley EA (2013) Energy and the English industrial revolution. Philos Trans R Soc A Math Phys Eng Sci 371(1986):20110568

Xu Y, Ibrahim IM, Harvey PJ (2016) The influence of photoperiod and light intensity on the growth and photosynthesis of Dunaliella salina (chlorophyta) CCAP 19/30. Plant Physiol Biochem 106: 305–315

Yang C, Li R, Zhang B, Qiu Q, Wang B, Yang H, Ding Y, Wang C (2019) Pyrolysis of microalgae: a critical review. Fuel Process Technol 186:53–72

Yu J, Takahashi P (2007) Biophotolysis-based hydrogen production by cyanobacteria and green microalgae. In: Communicating current research and educational topics and trends in applied microbiology, vol 1, pp 79–89

Yun J-H, Cho D-H, Lee S, Heo J, Tran Q-G, Chang YK, Kim H-S (2018) Hybrid operation of photo-bioreactor and wastewater-fed open raceway ponds enhances the dominance of target algal species and algal biomass production. Algal Res 29:319–329

Zeng J, Xiao R, Zhang H, Wang Y, Zeng D, Ma Z (2017) Chemical looping pyrolysis-gasification of biomass for high H2/CO syngas production. Fuel Process Technol 168:116–122

Zhou J, Wang M, Saraiva JA, Martins AP, Pinto CA, Prieto MA, Simal-Gandara J, Cao H, Xiao J, Barba FJ (2022) Extraction of lipids from microalgae using classical and innovative approaches. Food Chem 384:132236

Chapter 7
Bio-Hydrogen Production Using Agricultural Biowaste Materials

Tefera Kassahun Zerfu, Fiston Iradukunda, Mulualem Admas Alemu, Makusalani Ole Kawanara, Ila Jogesh Ramala Sarkar, and Sanjay Kumar

Abstract Hydrogen is measured as one of the capable substitute fuel for fossil fuels. Creation of clean energy sources and consumption of dissipated materials make natural hydrogen production a unique and promising candidate to meet the escalating energy requirements as an alternative of fossil fuel. Bio-hydrogen production from agricultural waste is very advantageous since agricultural wastes are abundant, economical, renewable, and eco-friendly. Water splitting, steam reforming of hydrocarbons, and autothermal techniques are the best-applicable methods for hydrogen gas production, but not economical owing to high energy needs. In comparison to chemical methods of hydrogen gas, bio-production has considerable merits like bio-photolysis of water by algae and dark fermentation and photo-fermentation of untreated materials through bacteria. Diverse agriculture desecrated materials are presented to produce bio-hydrogen to fulfill its demand. New technique for bio-hydrogen production is the dark fermentation and photo-fermentation scheme. This chapter will summarize the manufacturing of bio-hydrogen using agriculture waste materials with current developments and qualified advantages.

Keywords Bio-hydrogen · Agriculture waste · Dark and photo-fermentations · Waste bioprocessing

7.1 Introduction

The scarcity of energy resources and fossil fuel degradations are scary issues that we are facing in over the decades since the industrial revolution. The challenges of development of new technology and an increasing world population are major concerns to worldwide leaders. According to *2021 at a Glance—Statistical Review of World Energy* 2022, the total energy consumption was over 583.9 joules. This indicates that the projection of energy consumption is increasing exponentially as

T. K. Zerfu · F. Iradukunda · M. A. Alemu · M. O. Kawanara · I. J. R. Sarkar · S. Kumar (✉)
Department of Chemical Engineering, Faculty of Technology, Marwadi University, Rajkot, India

D. B. Pal, A. K. Tiwari (eds.), *Sustainable Valorization of Agriculture & Food Waste Biomass*, Clean Energy Production Technologies,
https://doi.org/10.1007/978-981-99-0526-3_7

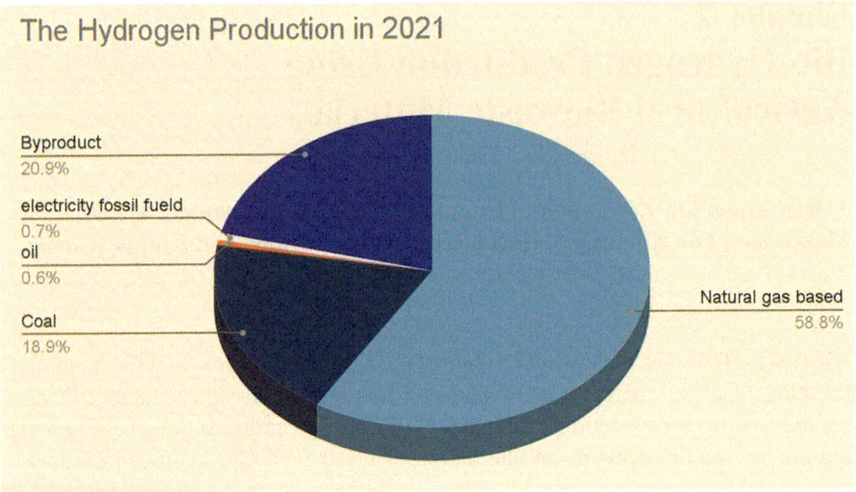

Fig. 7.1 The hydrogen production report in 2021. (*2021 at a Glance—Statistical Review of World Energy* 2022)

time changes. We all are aware that the prominent energy is being consumed at a severe level of fossil fuels, almost 80% of the total energy used annually (Ni et al. 2006). Burning fossil fuels increases the carbon dioxide concentration and methane gas as greenhouse gases (GHGs) are released into the atmosphere, leading to climate change and global warming (Lay et al. 2005). In recent years, researchers have started probing to see if they can bring novel technology that can produce green energy to save the source of nonrenewable energy and mitigate those issues resulting from using non-eco-friendly energies.

To reduce environmental degradation, global warming, the energy price crisis, and human and animal (aquatic and nonaquatic) health problems, we need to replace fossil fuels with renewable energy (Mafizur and Eswaran 2020). The report (Fig. 7.1; *2021 at a Glance—Statistical Review of World Energy* 2022) states that renewable energy has grown up by more than 8 EJ. These statistics demonstrate the progress made in the transition from nonrenewable energy to renewable energy. Lack of political support, low efficiency, and the high cost of renewable energy are some of the factors that impede a 100% transition to renewable energy and the sustainable development of energy (Nicoletti et al. 2015). Renewable energy resources include hydropower, geothermal energy, wind energy, photovoltaic energy, and biomass (Akkerman et al. 2002). In this chapter, we focus on biomass conversion, specifically bio-hydrogen production from waste of agricultural materials. By definition, "energy comes from the sun and heat." Biomass energy was the best option because the raw materials are available abundantly. This kind of renewable energy is popular to be used instead of conventional energy in developed countries as it contributes to about 40–50% energy supply (Ni et al. 2006). One drawback of energy from biomass is efficiency utilization. For instance, biogas, methanol, ethanol, biodiesel, etc., these sources of energy are mostly collected from crops, so they do affect the food market undeniably; a good example is ethanol produced from corn. Again, this energy cannot be used in intensive technology such as spacecraft fuels.

Science alongside technology grows simultaneously day by day; researchers are antsy; they keep exploring the other new alternative technologies as much as possible. In addition to being very clean, renewable energies are also very eco-friendly. We took biomass as an example, but the typical energies, including those in the family of biomass, have some hindrances in their production process or applications. Biodiesel is famous at the moment, which is extracted from palm oil, recycled grease, oils, soya, and animal fat by chemical processes, mostly used in car engines (Kapdan and Kargi 2006). America is among the top countries with the highest consumption of biodiesel; as of 2019, it was 118.25 barrels per day. The next countries are China, Indonesia, Brazil, and Thailand. Biodiesel does appear to be clean in some ways, but it isn't at the same level as bio-hydrogen. Some experiments and researches were conducted and indicated that burning biodiesel increases the NO_x effluents in the environment and, apart from that, affects the food market and increases deforestation.

Hydrogen fuel is a result of the combination of hydrogen with other elements to form compounds. Hydrogen is an element that is found abundant in nature and exists in its simplest form with just one electron and one proton. Over the years of research and invention, scientists have finally concluded that hydrogen can be used as a better alternative to conventional fuels. The concepts of ecological footprint and carbon footprint have grabbed the attention of today's world because of the increasing carbon emissions, especially from fossil fuels (Ivanova et al. 2009). On the other hand, hydrogen fuel is a zero carbon emission fuel that shall help in carbon neutrality for the future goals of reducing climate risks. The demand for hydrogen these days has been soaring as it is offering a good promise to substitute the use of fossil fuels. A global hydrogen review for 2021 states that the world's need for hydrogen is 90 megatons, an increase of 50% over the past decades. This demand was primarily fuelled by hydrogen being used in refining and industrial applications. For example, about 75% of it was used in methanol production. Hydrogen demand is not being shifted to other applications to replace fossil fuel demand. Currently, 79% of hydrogen comes from fossil fuel-based methods. About 21% comes from by-products of refining. This shows that even if hydrogen is a green fuel, indirectly it produces a high concentration of carbon dioxide. Hydrogen production by fossil fuels generates about 2.5% of global emissions in industries and energy. If this condition persists, the goal of net zero emissions by 2050 will be impossible. Researches are underway to produce low-carbon hydrogen, but government body and community supports are required.

There are many reasons why hydrogen is preferred rather than other biofuels like ethanol, biodiesel, biogas, and methanol. Hydrogen energy has a high efficiency that emits energy content about double (gravimetric energy of 143 MJ/kg) that of natural gas (gravimetric energy of 53.6 MJ/kg) (Martin et al. 2002; *Fossil and Alternative Fuels—Energy Content* 2008; Tashie-Lewis and Nnabuife 2021). As a result of its lower ignition temperature, it does not produce carbon dioxide when it burns, thus requiring less energy to ignite (Guo et al. 2010). Due to its high calorific value, it can be used to run intensive technology. In addition, the technology change on the engine is low because hydrogen has a wide range of flammability, and it has a low flash temperature of (−231 °C), so it is suitable for use in extremely cold conditions (*Fossil and Alternative Fuels—Energy Content* 2008). Besides fuels, hydrogen is used in many industries as a starting material, like the production of nitrobenzene from aniline, the production of synthetic gas, the production of ammonia and methanol, and hydrogenation processes. Hydrogen also has some challenging factors like storage and handling due to its high flammability. Production of hydrogen by electrolysis looks very clean, but one constraint is the huge amount of electricity consumed throughout the whole production process (36 KWh/kg); the electricity used might be renewable or nonrenewable. For cost minimization, researchers went far to find other alternatives that are cheap. Production of hydrogen from agro-wastes, livestock waste, crop wastes, and food wastes as it indicates that those are abundant, cheap, and highly biodegradable. By using enzymes, biological catalysts, microorganisms, and raw materials are fed into reactors at certain conditions to stay there at a specific time to produce hydrogen through the processes named photo-fermentation and dark fermentation. We can perform one process solely or combine them together because it is possible to work as a hybrid of two, found even to be more effective (Guo et al. 2010). Hydrogen is produced by thermochemical processes, electrolytic processes, photobiological processes, and biological processes. The thermochemical process is combined heat to remove hydrogen from raw materials that have hydrogen in their molecule or compound structures like hydrocarbons, natural gas, coal, and biomass that produce synthetic gas (a combination of hydrogen, carbon monoxide, and small amount of carbon dioxide) and a chemical reaction to remove hydrogen from the mixture by using water (Parthasarathy and Narayanan 2014). The thermochemical process is the majority of hydrogen produced by this process, whose by-product is carbon dioxide that increases the concentration of carbon dioxide that contradicts sustainable renewable energy production or low-carbon hydrogen (Martin et al. 2002; Guo et al. 2010). In another method, hydrogen is produced by splitting water into hydrogen and oxygen molecules by using electricity. Electrolytic processes include alkaline water electrolysis, water electrolysis, solid oxide electrolysis, and microbial electrolysis. Instead of using fossil fuels, this method can use renewable energy sources such as wind energy, solar energy, hydropower energy, and biomass to produce green hydrogen (low-carbon hydrogen) (Kumar and Himabindu 2019). However, it is not viable to produce hydrogen using electrolysis due to the high cost, low-performance efficiency, high energy consumption, and need for advanced technology (Parthasarathy and Narayanan 2014; Kumar and Himabindu 2019). Researchers and scholars have studied hydrogen production to produce hydrogen that does not contain carbon and is inexpensive to produce. This idea has been dubbed for bio-hydrogen (Harshita et al. 2022; Arimi et al. 2015; Kapdan and Kargi 2006) production.

The production of bio-hydrogen, a gas free of carbon, is strongly needed to achieve net zero carbon emissions in 2050. Thus, the production of hydrogen by biological methods is one of the solutions. A few of these methods include photo-electrochemical technology, microbial biomass conversion (dark fermentation), and photobiological methods (green microalgae or cyanobacteria). Photo-electrochemical technology uses sunlight energy with specialized semiconductors to split water into hydrogen and oxygen (*Hydrogen Production Processes* 2023). Photobiological methods split water by photosynthesis of green algae, and hydrogen is extracted indirectly or directly from ions formed (hydrogen ion and oxygen ion) as hydrogen gas (Holladay et al. 2009). Another photobiological method is called photo-fermentation, in which hydrogenase and nitrogenase enzymes break down organic matter in the presence of sunlight (Gaweł et al 2019). The challenge for hydrogen production by photo-electrochemical and photobiological processes is finding a novel reactor that is comfortable for microorganisms, has insufficient sunlight, and requires an optimum enzyme (*Hydrogen Production Processes* 2023).

An anaerobic microorganism decomposes biomass and uses it for metabolic purposes and produces hydrogen; this process is sometimes referred to as dark fermentation because no sunlight is required (Kapdan and Kargi 2006; Li et al. 2012a; Yadav et al. 2018). As a matter of fact, dark fermentation utilizes low-cost agricultural waste that is diverse and readily available all over the world. It is a sustainable resource of renewable energy and carbohydrate rich, and it is eco-friendly since it reduces environmental waste.

Agricultural residues contain lignocellulose composed of three components, cellulose, hemicellulose, and lignin (Łukajtis et al. 2018). The efficiency of bio-hydrogen production depends on the feedstock. Wheat straw, corn straw, rice bran, sugarcane bagasse, corn straw, maize leaves, and grass silage are the most common agricultural residues that produce various yields of hydrogen. Statistics show that over 220 billion tons of agricultural waste are produced every year (Ren et al. 2009). Utilizing these residues is therefore healthy for the environment, contributes to global energy resources, and promotes economic growth. However, lignocellulose cannot be used directly due to its complex structure. A chemical and biological process does not degrade cellulose and lignin. Due to these characteristics, pretreatment is required to increase the rate of production, for efficient conversion of hydrogen from biomass and for sustainable production of green hydrogen (Harshita et al. 2022).

The bio-hydrogen production yield does not only depend on the substrate but also on operating conditions, operational parameters (temperature, pressure, pH), plant design (reactor), and types of microbial use in fermentation. This book chapter will focus deeply on the production of bio-hydrogen harnessed from agricultural wastes, the recent technology which is being used, sustainability, and some advantages of this clean energy.

7.2 Different Types of Agriculture Waste Materials for Hydrogen Production

Agriculture is one of the most common economic driving factors in the world. Agricultural production has residues and wastes at the time of harvesting. Agricultural products may be classified as direct waste products from agricultural activity, animal manure, and food waste. Each year it generates waste which contains around 220 billion tons of wastes which are composed of carbohydrates and other simple sugars.

Most agricultural wastes contain starch and cellulose which are complex sugars and affect the biodegradability of waste materials. Starch can be separated into glucose and maltose by acidic or enzymatic processes. Glucose and maltose can be converted into carbohydrates and organic acids followed by hydrogen formation. Cellulose-rich agricultural wastes need mechanical or chemical pretreatment before fermentation. Cellulose and semi-cellulose content of materials will be hydrolyzed to carbohydrate and hydrogen gas after further process. If the material has a high amount of lignin content, it will be difficult for enzymatic hydrolysis (Kapdan and Kargi 2006).

Different agricultural products have different content of sugars, organic matter, and other chemicals. Bio-degradation and pretreatment are required for solid agricultural waste materials to separate some hazardous and nonbiodegradable materials. In addition to agricultural wastes, animal manure is also another agricultural product used in the generation of green energy including bio-hydrogen. Animal manure can be liquid or urinary manure, solid manure or farmyard manure, wastewater that is composed of liquid manure, disinfectants, and feedlot runoff. These animal products may cause environmental pollution and other health effects to humans and animals because they contain huge amounts of organic matter, nitrogen, and phosphorus (Rathore and Singh 2017). Therefore, proper management is required to reduce the adverse impacts on humans and the environment.

7.2.1 Corn Straw

The total production of corn in the world from worldwide statistics is 1.11 metric tons; the United States, China, and Brazil are the leading producers of corn in the globe (Gaweł et al. 2019). These large amounts of corn production result in the production of corn straws after maize/corn is plucked or harvested. Corn straws are rich in sugar; thus, utilizing the by-product of corn harvesting to produce hydrogen is a cheap and environmentally friendly process. Corn straw contains a large amount of cellulose which is the most useful input for the production of hydrogen. It contains 30–60% cellulose, 20–40% hemicellulose, and 15–25% lignin, which are complex sugars containing hydrogen. Fermentation of such products rich in glucose results in the production of hydrogen gas and other products, but the presence of different

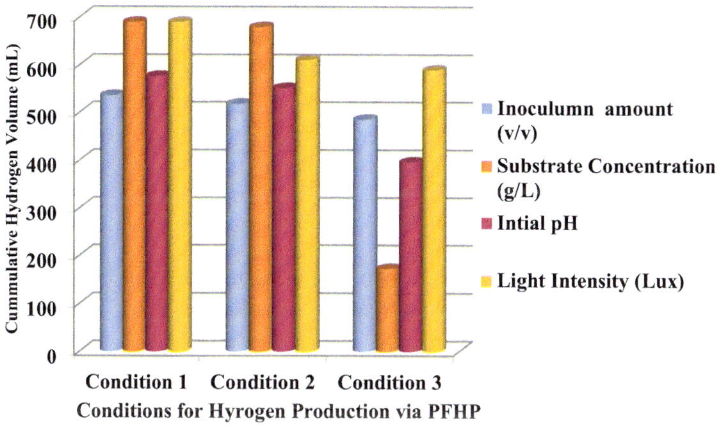

Fig. 7.2 Condition for hydrogen production via PFHP (Gaweł et al. 2019)

microbial lives and uncontrolled environments conducive to yield high amounts of hydrogen (basic fermentation) won't be effective and economical thus the use of Photo-Fermentative Hydrogen Production (PFHP) as the process that yields more hydrogen from the substrate produced. PFHP and a photocatalyst (TiO_2) improve the HY and HPR by 1.7 and 1.54, respectively, using photo-fermentation bacteria *Rhodobactersphaeroides* NMBL-02. Corn straw is most productive when PFHP and nano-TiO_2 are used under optimum parameters coupled with HAU-M1 (this bacterium contains *Rhodospirillum rubrum*, *Rhodobactercapsulatus*, and *Rhodopseudomonas palustris*) (Sivaramakrishnan et al. 2021) as a photo-bacterium yields 688.8 mL of CHV (Kapdan and Kargi 2006). Nano-TiO_2 is used to increase and maximize the production of photo-induced electrons in the system and thus increases the production of hydrogen. Nano-TiO_2 particles have been experimented with and found that it increases efficiency in the PFHP process (Gaweł et al. 2019). Production of hydrogen from corn straw depends on the dosage of the photocatalyst, concentration of the corn straw, pH value of the system, and the microbial organism present in the system. In different conditions, HY differs due to the factors set.

Figure 7.2 represents the conditions for hydrogen production via PFHP. These separate conditions detail how hydrogen can be yielded at maximum potential and conditions. Total HY is 688.8 mL of CHV, meaning an increase of CHV, HPR, and HC in percentage from the normal PFHP of 32.6, 27.9, and 8.3, respectively, all in percentage. More experiments are done using Box-Behnken design to evaluate the interaction between pH values, substrate concentration, and light intensity and its effect to the PFHP process while using corn straw and nano-TiO_2 present (Zhang et al. 2020a). For further testing, experiments have been done, and higher HY has been recorded up to 709.5 mL of CHV; the experiments were done using a mathematical (quadratic) relation:

$$Y = 686.88 + 118.03X_1 - 32.86X_2 + 32.04X_3 + 36.87X_1X_2 - 11.73X_1X_3 + 48.00X_2X_3 - (157.17X_1)^2 - (106.94X_2)^2 - (32.41X_3)^2$$

Y—cumulative hydrogen volume (CHV).
X_1—initial pH.
X_2—substrate concentration.
X_3—light intensity.

Through the above mathematical expression, more data can be obtained in the experimentation of the PFHP process (Kapdan and Kargi 2006).

7.2.2 Rice Bran

Many of the countries in the world produce rice as their agricultural products, especially India and Thailand that are the most famous producers where rice bran is easily accessible. Rice bran is majorly constituted of lignocellulose biomass which should be removed in pretreatment before fermentation. Phytic acid is another component of rice bran in addition to lignin. Phytic acid is anti-nutritional and causes environmental pollution through eutrophication.

Using rice bran for hydrogen production is an eco-friendly process which can reduce environmental impacts and has economic benefits. But the presence of lignin in rice bran may hinder fermentation to produce hydrogen. Therefore, appropriate pretreatment should be selected to degrade lignin. The degradation of lignin produces cellulose and hemicellulose which are favorable for fermentation and increases the accessibility of fermentation microorganisms.

There are different methods used for degradation of lignin and phytic acid from rice bran so as to enhance the fermentation process. Those are biological pretreatment, chemical, physical, and fungal pretreatment. Biological pretreatment is one of the best and most used methods due to its use of fewer chemicals for the process. In addition to this, biological pretreatment does not produce inhibitor which affects the fermentation process. Fungal treatment is also another recent method of lignin degradation where lignin will be more degraded relative to cellulose and hemicellulose (Sivaramakrishnan et al. 2021). Moreover, fungal treatment is also used for degradation of phytic acid from rice straw.

7.2.3 Sugarcane Bagasse

Sugarcane bagasse is waste material which is removed after the extraction of sugarcane in sugar industries and other operations. There is a huge amount of sugarcane bagasse left each year as bagasse contains 25% of sugarcane (Sangyoka et al. 2016). Combustion is one of the most used methods for the generation of energy from sugarcane bagasse.

Sugarcane bagasse is composed of cellulose, lignin, and hemicellulose where hemicellulose is the major content having 30–35% composition. Generation of energy from sugarcane bagasse using combustion creates many environmental hazards as it emits carbon dioxide and other toxic environmental pollutants. So to setback sugarcane bagasse should be pretreated. Mostly dilute acid pretreatment is recommended which produces xylose and glucose with very few amounts of arabinose. In this process, there is a formation of solid glucose and lignin due to strong bonds in cellulose. Even if production of energy using hydrolysate of sugarcane bagasse creates several inhibitory compounds, it is the most attractive way of producing hydrogen.

Hydrogen production from sugarcane bagasse may involve physical-chemical and biological processes. Physical-chemical processes require a high amount of energy, and they emit hazardous gases into the environment which contribute to global warming. This process produces greenhouse gases like carbon dioxide, methane, and carbon monoxide. However, the biological process is an eco-friendly and less energy-intensive process of bio-hydrogen production. There are many factors which affect the production of hydrogen from sugarcane bagasse such as temperature, pH, and substrate concentration, profile of organic acids, type of inoculum, and type of substrate. pH is one of the factors which mostly affect enzyme activity in the biological process of hydrogen production.

7.2.4 Wheat Straw

Wheat straw is another mostly produced biomass; agricultural production has a capacity of producing 3945 kg of hydrogen per day (Rezania et al. 2017). Wheat straw consists of 3.6% lignin, 19% hemicellulose, and 38.7% cellulose which makes it feasible for bio-hydrogen production. The composition of the hemicellulose and cellulose makes the substrate an important raw material to use in the bio-hydrogen production. Wheat straw is composed of biochemical components which not only are rich in glucose but also have a complex structure which when processed without pretreatment, then hydrogen yields would be low. Therefore, the utilization of wheat straw as raw material is dependent on the preprocessing technique due to the presence of lignified structure and other complex composition of wheat straw. Pretreatment is done so as to increase the permeability of the compounds broken down from sucrose (monosaccharide, disaccharides), so as to be used by the microbial life present in the reactors to convert the complex form of sugar to hydrogen molecules (Zhu et al. 2022). This practice of pretreatment enhances the delignification and the production of more carbohydrates which will produce hydrogen atoms after a reaction with enzymes. Wheat straw pretreatment can be done by ultrasound, lyophilization, and hydrothermal pretreatment (HPT). These techniques will have different products depending on their mode of operations. The important factor that constitutes the model at large is the amount of wheat straw being used and the type of pretreatment required. Treatment of concentrated substrate with alkali is to break the cell wall structure present in the raw material. This step is taken with further pretreatment techniques done to convert the concentrated substrate into a more likely reactive and high-yielding source. This alone can reduce lignin from the structure of wheat straw while leaving the carbohydrate content intact and add the amount of hydrogen collected after dark fermentation of the substrate. Relying on an alkali-treated substrate would yield a substantial amount of hydrogen. Zhang et al. (2020b) have experimented on wheat straw which has been alkali-cooked and had a hydrogen yield of 144 mL/g total solids (TS). Each pretreatment technique has a specific way of cell wall permeation in the substrate structure. Ultrasound uses its mechanical acoustic waves, heat, and its strong shear force ability to depolymerize wheat straw; He et al. (2014) experimented on the production of bio-hydrogen using a hydrothermal pretreatment technique and had a hydrogen yield of 28 mL/g volatile solids (VS) at 210 °C.

7.2.5 Maize Leaves

Maize leaves are produced in huge amounts every year from agricultural land all over the world as the maize is the main food item and source of ethanol production in many countries. Batch condition fermentation which stays for 48 days is necessary for hydrogen generation from maize leaves.

The rate of hydrogen generation in untreated maize leaves is low as compared to the treated ones. So pretreatment of maize leaves is done with *Bacillus amyloliquefaciens*, which is an anaerobic and fast-growing bacterium (Ivanova et al. 2009). Pretreated maize leaves are better for bio-hydrogen production, whereas untreated maize leaves are less applicable for hydrogen production. Heat-aided pretreatment did not produce enough sugar substrates for hydrogen generation. Maize leaves are considered for hydrogen production with the assistance of digesting microbial life which eases the breaking down of the sugar structure so as to be readily reactive and to produce bio-hydrogen, with pretreatment at 130 °C for 30 min before the digestion process is done at 70 °C in a batch reactor for a period of time sufficient for bio-hydrogen production (Zhang et al. 2020b). Maize leaves are categorized as crop residues, whereas the processing of these materials to prepare them for bio-hydrogen production is differentiated into mechanical, physical, chemical, and biological. This is done so as to increase the reactivity of the constituents of crop residues since they contain a microcrystalline structure of lignocellulosic component which is hard to be digested directly with the microbial organism (Mtui 2009). These processes enhance the surface structure of the material to be readily reactive and break bonds or structures which are complex to digest for microbial life.

7.2.6 *Grass Silage*

Grass silage is the one major cellulosic material which is used for biological hydrogen production. Silage is the degradation product which is produced by fermenting grass with lactic acid. Silage is composed of lignin, cellulose, and hemicellulose which can be hydrolysed to fermentable sugars. Earlier, silage was used for treating water which contains metals, as a source of electrons and carbon for sulfate-reducing bacteria (Li et al. 2012b). Grass silage has a complex structure (recalcitrant cellulosic structure) for which limits its usage in bio-hydrogen production using fermentation method. Pretreatment methods are used to prepare grass silage as a raw material for bio-hydrogen energy production. By performing the pretreatment, the microstructure of grass silage degradation increases.

The thermochemical treatment method helps in the reduction of the barrier present in grass silage (Deng et al. 2019). The thermochemical pre-treatment method increases the surface area of grass silage and favors the digestion of contents. Proper application of mild acids (H_2SO_4) and temperature at a certain time can be seen to increase the digestion of grass silage. With those parameters, the yield of biofuel increases along with two-stage processes of anaerobic digestion (production of CH_4 and CO_2) and fermentation (Deng et al. 2019). The ideal conditions for pretreatment of grass silage are 2% w/w of sulfuric acid, 135 °C, and 15 min, respectively. If any alteration occurs with these numbers, it effects the general yields of the production of hydrogen and methane. This can be explained by using the Severity Factor (SF) of which the yield of sugar reduces (Gonzales et al. 2016; Hsu et al. 2010). The optimum SF for reducing sugar yields is between 1.6 and 2.0, and this is because higher than 2.0 SF requires an extra process where the inhibitors are removed from the sugar to allow digestion to continue since they have been denatured by exposure to a high concentration of acids. With the data gathered from Ireland where grass is the predominant crop, grass silage can be used to produce 35 PJ approx. by 2035. This is to show that grass silage has potential in the production of biofuels from renewable energy resources and utilization of the readily available materials from different sectors to yield energy.

7.3 Biological Reactor Operation

7.3.1 *Dark Fermentation*

Fermentation is the act of transferring biological feedstock into energy by using bacteria and enzymes. Hydrogen production is heavily reliant on fossil fuels or natural gases. Using bio-based substrates will reduce the risk to the economy, health, and environment. Biomass feedstock with high levels of carbohydrates or starch, such as both agricultural and industrial residues, has been utilized for fermentation (Kumar et al. 2019). Therefore, dark fermentation turns organic compounds from industrial and agricultural wastes into hydrogen through anaerobic organisms without any need for light (Renaudie et al. 2021; Mona et al. 2020). Dark fermentation or heterotrophic fermentation under anaerobic conditions appears to be the most advantageous of several biological bio-hydrogen production methods due to its high rate of yield and low cost of hydrogen production (Mona et al. 2020).

A variety of factors influence the efficiency of substrate conversion and bio-hydrogen production by microbial biocatalysts, including fermentation conditions, process parameters, and reactor configurations. A decent bioreactor must really be capable of operating with a shorter hydraulic retention time, and the reactors must also be capable of reducing biomass washout with a short retention time. Bioreactor efficiency is evaluated not only by reactor configurations but also by reformation suitable for different application scenarios (Jung et al. 2011). CSTR (Continuously Stirred Tank Bioreactor) and UASB (Upflow Anaerobic Sludge Blanket Reactor) are widely used during dark fermentation, together with the other fermentative reactors (Mata et al. 2022). However, due to their efficiency and economics, they are limited to batch reactors. To operate the process continuously, it must be carried out in suitable bioreactors (Christopher et al. 2021).

Bio-hydrogen production can be boosted by using proper substrates, bacteria, and pretreatment. Microbes produce hydrogen by using enzymes such as metalloproteins as well as hydrogenases (Kumar et al. 2019). Anaerobes (*Clostridia* and thermophiles), facultative anaerobes (*Enterobacter*), and full aerobes are microbes used in dark fermentation (*Alcaligenes* and *Bacillus*) (Mona et al. 2020). Sugars, amino acids, glucose, wastewater effluent, dairy wastewater, distillery wastewater, and other materials are popularly used as hydrogen production substrates (Show et al. 2011). Furthermore, volatile fatty acids and alcohols can be used to boost hydrogen yield during the dark fermentation process. Researchers are intrigued in dark fermentation reactors due to their versatility, straightforwardness, and ability to tarnish waste in different manners. Light has no effect on dark fermentation reactors, allowing for greater volume utilization. Because of the anaerobic conditions, removing oxygen from these reactors is not a major issue. The researcher's main goal is to develop dark reactors that can fulfill the demand on a massive scale. The dark fermentation method facilitates bacterial growth and is less invasive to the environment. Dark fermentation processes produce hydrogen continuously without any light, and thus the process's essence, low energy input, and use of waste as raw materials are recognized as significant advantages over others (Banu et al. 2018; Mona et al. 2020).

7.3.2 Dark Fermentation Reactors

Multiple types of reactor designs have been tested in order to level up the amount of bio-hydrogen production. Nevertheless, every reactor type has merits and demerits. The following paragraphs entail a few of these bioreactors (Jung et al. 2011). UASB and CSTR are commonly used during dark fermentation, along with other fermentative reactors (Mona et al. 2020). Batch reactors are simple to operate but cumbersome, so they are mainly employed in laboratory tests. In laboratory or pilot-scale studies, UASB bioreactors have been widely used (Lin et al. 2012). Fixed-bed reactors have also been discovered to produce sufficient bio-hydrogen. Fixed-bed reactors, by contrast, do have downsides, as well as localized populations that range along the length of the reactor, channeling caused by inefficient mingling, and flawed substrate conversion due to poor dispersion efficiency (Jung et al. 2011).

7.3.2.1 Suspended Bioreactors

In the beginning of hydrogen fermentation gives the approval to use of suspended bioreactors, which have been originally designed to work similarly to anaerobic digestion for CH_4 fermentation. Although immobilized bioreactors have recently been developed, suspended bioreactors continue to be useful, particularly for feed-stock with high specific content such as municipal solid waste and food waste (Mtui 2009). Suspended bioreactors are used in industry to handle a mixture of wastewater and organic waste. Again in suspended reactor, microbial contact with the substrate is high, resulting in greater mass transfer. As a result, suspended bioreactors are now positioned to dominate the activities of bio-hydrogen production (Jung et al. 2011).

7.3.2.2 Continuously Stirred Tank Reactor (CSTR)

As one of the dark fermenter reactors, CSTRs are often used for bio-hydrogen production due to their simplicity, lower cost, and effortless operation. In CSTR, there is a continuous addition of substrate and a continuous withdrawal of product, accompanied by constant agitation, in order to ensure a complete mixing of the substrate and product. The CSTR operation has yielded the majority of significant control parameters, including hydrogen content, substrate concentration, pH, and solids retention time (Rajesh Banu et al. 2021; Jung et al. 2011; Christopher et al. 2021). Mixed cultures can be selected for fermentation by CSTR. A CSTR starts up quickly compared to an ASBR, and stirrer design and mixing speed are the main factors affecting bio-hydrogen yield (Banu et al. 2018). The inoculums, waste properties, and factors such as temperature, pH, and OLR all contribute to process instability in CSTR. Hydraulic retention time is a critical parameter that influences the microbial population and CSTR properties (Mtui 2009).

7.3.2.3 Anaerobic Fluidized Bed Reactor (AFBR)

The biocatalysts in these reactors form as biofilms and adhere to the substratum. The substrate flows horizontally containing the biomass suspension, and the biomass acts as a catalyst to increase the production of bio-hydrogen. In comparison to CSTR, anaerobic fluidized bed reactors have a high mass transfer efficiency due to better mixing and less shear stress. Despite having good mass transfer, these reactors are susceptible to biocatalyst washout, similar to CSTR. As a result, the bio-hydrogen yield is reduced. These reactors are capable of operating with both a quicker HRT and a higher substrate weight (Lin et al. 2012). For instance, after investigating the effect of biomass load on hydrogen generation by treating a high substrate volume, they observed that hydrogen yield was high and short HTR in AFBR (Banu et al. 2018). Effluent as a substrate user observed higher bio-hydrogen volume fraction at a fairly short HRT of 2 h and a higher OLR of 126 kg.-COD/m^3/d. The only limitation of AFBR is the high required energy for fluidization (Jung et al. 2011).

7.3.2.4 Upflow Anaerobic Sludge Blanket Reactor (UASBR)

In these categories of bioreactors, the layout of granular beds enhances bio-hydrogen output. The granules are composed of encapsulated microbes. HRT in bioreactors has higher substrate conversion efficiency while being less expensive (Rajesh Banu et al. 2021). All through bioreactor operation granular biosolids are suspended within the fermentor. The main advantage of UASB is its ability to reduce biomass blowout. An investigation proposed by Yang et al. (2006) has characterized a higher bio-hydrogen creation of 720 mL H_2/L.d in a UASB reactor biodegrading citric acid effluent, with butyrate being the massive fermentative substrate. Correspondingly cheese whey effluent was treated in the UASB reactor and produced 122 mL H_2/L.d. of bio-hydrogen. The disadvantages of UASB are the long setup time, the greater long retention time, and the development of multiple levels of biomass in the effluent (Jung et al. 2011). As a result, the suitable method is required for constructing the performance reliability and granule formation during the continuous process of hydrogen creation in UASB (Christopher et al. 2021). The UASB reactor is capable not only of treating organic material but also of utilizing this into bio-hydrogen (Lin et al. 2012).

7.3.2.5 Membrane Bioreactor (MBR)

For microbial retention within the bioreactor, these biofilters combine membrane technology with the activated sludge method. Process variables like retention time and SRT must be controlled in these reactors. Increased SRT resulted in increased biomass retention. As a result, the substrate utilization rate increased while bio-hydrogen production lowered (Rajesh Banu et al. 2021). The main drawbacks of membrane bioreactors are membrane fouling and high operating costs (Banu et al. 2018). With brewery wastewater, granulated biosolids membrane bioreactors can produce bio-hydrogen. Scientists found that increasing effluent substrate load from 30 g/L/d to 60 g/L/d led to a decrease in bio-hydrogen generation rate and yield (Rajesh Banu et al. 2021). The sections that follow will review and discuss major factors that affect bio-hydrogen production, such as operating conditions, feedstock, and reactor configuration, in order to gain a better understanding of how to improve it (Show et al. 2011).

7.3.2.6 Photo-Bioreactor

The photo-bioreactor is used in photo-fermentation to cultivate microbes by providing the necessary operating parameters using sunlight or artificial light as well as carbon dioxide to facilitate their metabolic activity. Carbon dioxide is either provided from an organic carbon source (agricultural residues like molasses, wheat leaves, etc.) or injected into the photo-bioreactor from the bottom of the reactor (Show et al. 2011). To produce hydrogen from agricultural residues efficiently and economically, the photo-bioreactors physical parameters must match the optimal conditions (temperature control, area/volume ratio, and transparency, durability of reactor, gas exchange system, and agitation system). Furthermore, the photo-bioreactor's physicochemical parameters (dissolved O_2 and CO_2, pH, temperature, and light intensity and agitation shear stress) have been maintained (Guler et al. 2019; Wang et al. 2021). Photo-bioreactors come in a variety of shapes and sizes, including flat plates, rectangular tubulars, vertical tubulars, and helical tubulars (Show et al. 2011; Zhang et al. 2020a). The advantages and disadvantages of different photo-bioreactor are discussed in Table 7.1.

7.3.2.7 Flat Plate Photo-Bioreactor

A flat plate photo-bioreactor is a bioreactor that has the shape of a rectangular box and a flat that directs light from one side of a reactor made of transparent glass. Because of its high surface area to volume ratio, large illumination, and low maintenance costs, this reactor is suitable for large-scale hydrogen production (Guler et al. 2019; Zhang et al. 2020a). These days, the flat plate photo-bioreactor has been carried out using modeling and simulation (computational fluid dynamics) and artificial intelligence to determine the optimum conditions such as light intensity, shear stress, fluid flow rate, gas flow rate, and aeration. For example, Wang et al. (2021) showed how flat plate aeration with inclined baffles was developed to improve aeration, light intensity, and mass transfer, resulting in a 25% increase in biomass. In another design, hexagonal shaped airlift flat plates are used with inclined baffles, yielding 61% more biomass than the old flat plate and enhancing dead zone, airflow, and shear stress (Yaqoubnejad et al. 2021).

7.3.2.8 Vertical Tubular Photo-Bioreactor

Vertical tubular photo-bioreactors, like flat plates, have thin glass walls that allow light to pass through and are used to grow photosynthetic microbes (Singh and Sharma (2012). A sparger at the bottom of this bioreactor creates tiny bubbles that

Table 7.1 Advantages and limitations of different types of photo-bioreactor

Photo-bioreactors	Advantageous	Disadvantageous	References
Vertical photo-bioreactor	• There is little chance of contamination • Productivity of biomass with efficiency • There is a large surface-to-volume ratio • Gas exchange and mixing are excellent • Reduction in material costs and relative affordability • An independent liquid circulation system	• Shear increases when bubbles burst • The problem of scaling up bioreactors • Hydrogenation of uptake • Inhibition of feedback • There is a gas hold-up • Hydrogen stream is diluted	Dasgupta et al. (2010), Mona et al. (2020) and Saeid and Chojnacka (2015)
Horizontal photo-bioreactor	• Highly efficient light conversion • Large illumination surface area • Appropriate for outdoor biohydrogen production	• Formation of photobleaching • Large illumination surface area • Exchanging gas mixture at a low rate • The input of energy is high • It is necessary to have a large ground area to scale up	Mona et al. (2020), Singh and Sharma (2012) and Dasgupta et al. (2010)
Flat plate photo-bioreactor	• The sterilization process is simple • Illumination with a wide range • A relatively low level of shear stress is developed • It is easy to design, maintain, and construct • Photosynthesis of high efficiency • There is a short light-dark cycle • It is easy to scale up • Photochemical efficiency and mixing rate are high	• The power consumption of aeration is high • Mixing always necessary • The length of tube is limited • The length of tube is limited • Biomass is attached to inside bioreactor wall • Low control in cultural conditions	Wang et al. (2021), Yaqoubnejad et al. (2021) and Akkerman et al. (2002)

allow gasses to diffuse into liquids more quickly. As a result, heat transfer, mass transfer, and fluid turbulence are accelerated (Singh and Sharma (2012; Christopher et al. 2021). A vertical tubular system reduces photobleaching and increases photosynthesis efficiency by removing oxygen from the inside bioreactor to the outside. A vertical tubular system has the advantage of producing a liquid that is utilized to control the heat generated in the tube, having low shear stress, and consuming less energy. It is, however, difficult to produce hydrogen in huge quantities because mixing by sparger is difficult and light is insufficient quantity (Mtui 2009; Akkerman et al. 2002). In terms of the way liquid flows inside a reactor, vertical photobioreactors are classified into bubble columns and airlift reactors (Akkerman et al. 2002; Singh and Sharma (2012).

A bubble column reactor is made up of cylindrical columns with a sparger that diffuses the gas mixture from the bottoms (Dasgupta et al. 2010; Singh and Sharma (2012). This causes the inside fluid to become turbulent, allowing for aeration and mixing to occur without causing shear stress. The product yield in bubble column photo-bioreactors is highly influenced by the flow rate of the mixture of gases. The optimal gas mixture flow rate results in more efficient aeration and mixing performance, as well as high productivity. According to Singh and Sharma (2012), turbulence does not develop inside the reactor in cases where the gas flow rate is below 60 m/s. Bubble column bioreactors have been widely used in the photofermentative production of bio-hydrogen due to their low capital costs, high illumination areas, high mass transfer, and efficient oxygen release. On a larger scale, however, it is necessary to have a novel design because sparger agitation is difficult and some areas are not fully illuminated (Saeid and Chojnacka 2015; Dasgupta et al. 2010).

An airlift photo-bioreactor is similar to others which is made of transparent tube glass and has a riser and a downcomer segment. A riser and a downcomer are connected to a vessel and the disengagement zone in an airlift photo-bioreactor. A riser connects to a sparger where the gas mixture is injected into the reactor, allowing it to rise to the reactor top. Downcomer liquid traps the gas mixture on the top of the reactor and returns it to the downcomer zone. Other gases produced by microbe metabolism, such as oxygen, are removed via the top reactor gas discharge region. It has advantages such as good aeration, no shear stress because there is no physical agitation, and high photosynthetic efficiency. However, the overall design is complex, and the cost of large-scale production is prohibitively expensive unless optimization is performed and a novel design is discovered (Mtui 2009; Singh and Sharma (2012; Dasgupta et al. 2010).

7.3.2.9 Horizontal Tubular Photo-Bioreactor

A horizontal tubular photo-bioreactor is made of transparent polyvinyl chloride or polypropylene acrylic, and spargers are used to circulate and dissolve the gas (Zhang et al. 2020b). The surface of the bioreactor is sprinkled with water, the feed temperature is kept constant before feeding, and cool jackets are provided to the reactor to maintain a fluid temperature. Depending on the illumination system and aeration, there are several types of horizontal photo-bioreactors, including alpha, helical, and loop patterns (Singh and Sharma (2012). This reactor is due to the use of a high surface area to volume ratio of artificial fiber optics in addition to natural light for illumination. Furthermore, because of the longer retention period of gas in the reactor, it has high aeration efficiency. It has high productivity when compared to other photo-bioreactors, but the amount of energy required to achieve an optimal turbulent condition is high. According to Singh and Sharma (2012), it requires approximately 2000 w/m^3 of energy, whereas flat and vertical photo-bioreactors require approximately 20 m/w^3. In addition, due to very low gas mixture flow rate and unequal lighting distribution throughout the reactor, horizontal photo-bioreactors face scaling-up issues.

7.3.3 *Reactor Operating Conditions*

Inoculums preparation and beginning, temperature, pH, hydraulic retention time (HRT), substrate concentration and liquid product inhibition, feedstock, hydrogen partial pressure, and essential minerals are thought to have a significant impact on DFHP behavior. Defining their optimal ranges, it would be useful in determining reactor and structure size, materials, equipment needed, chemical reagents, and so on (Jung et al. 2011). The operating conditions of dark fermentation of agricultural residues are shown in Table 7.2.

7.3.3.1 Types of Inoculums and Pretreatment

Heating, acid, and alkali pretreatment of naturally occurring mixed microflora enhances bio-hydrogen creators while inactivating non-sporulating hydrogen consumers such as methanogens. The production of bio-hydrogen in the thermal treatment process is primarily determined by the period of sludge thermal treatment (Li et al. 2012a). It appears from an engineering perspective that blended cultures are more practical than pure cultures since they are simpler and more controllable and can use a wide variety of feedstock. A variety of seeding sources such as anaerobic digester sludge, sewage sludge, compost, manure, and soil have been established for obtaining H_2-producing inoculums, *Clostridium* sp., which produces H_2 and produces spores that are commonly attacked by physicochemically (Jung et al. 2011).

7.3.3.2 Temperature

The production yield of hydrogen is influenced by fermentation temperature. The yield of hydrogen depends on the enzyme activity. The enzyme activity is affected by temperature because it contributes to the hydrolysis of the substrate. Depending

Table 7.2 An overview of the operating conditions of dark fermentation of agricultural residues

Substrate	Types of inoculum/pretreatment	Condition of the process		Hydrogen yield	Reactor mode	References
		pH	°C			
Rice straw	Heat-treated sludge	6.5	55 °C	24.8 mL $H_2 g_{TS}^{-1}$	Batch	Chen et al. (2012)
Wheat straw 1%	*Caldicellulosiruptor saccharolyticus*	–	70 °C	3.4 mol H (mol glucose)$^{-1}$	Batch	Ivanova et al. (2009)
Switchgrass (SWG)	Extreme thermophile *Caldicellulosiruptor saccharolyticus*	–	65 °C	11.2 mmol H_2/g SWG	Batch	Talluri et al. (2013)
Microcrystalline cellulose (MCC)	Extreme thermophile *Caldicellulosiruptor saccharolyticus*	–	65 °C	9.4 mmol H_2/g MCC	Batch	Talluri et al. (2013)
Cornstalk	*Thermoanaero bacterium thermosaccharolyticum* DD32	7.5	55 °C	2.17 mol H_2 /mol glucose	Batch	Shin et al. (2004)
Corn straw	*Thermoanaerobacterium thermosaccharolyticum* DD32	–	55 °C	6.38 mmol/g	Batch	Shin et al. (2004)
Wheat straw	*Caldicellulosiruptor saccharolyticus* DSM 8803	–	70 °C	1.58 mmol/g	Batch	Kamil et al. (2019)
Rice straw	*Thermotoga neapolitana*/chemically treated	7.5	75 °C	2.7 mmol/g straw	Batch	Nguyen et al. (2010)
Grass	Acid and alkaline pretreatment	7	35 °C	72.21 mL/g dry grass	Batch	Cui and Shen (2012)
Soybean straw	Anaerobic mixed (Cui and Shen 2012) culture	7	35 °C	60.2 mL/g dry soybean	Batch	Han et al. (2012)
Wheat straw	Enzymatic treatment	5.5	37 °C	19.63 mL H_2/g VS	Batch	Marianne et al. (2012)

on the types and growth rates of the bacteria and the substrate (monosaccharides, disaccharides, starches, lignocellulosic) used in the fermentation, the optimum temperature will be determined. In order to effectively hydrolyze substrates at different temperatures, different bacteria are needed, such as psychrophiles (below 25°), mesophiles (25–45°), thermophiles (45–65°), extreme thermophiles (55–80°), and hyperthermophiles (above 80°) (Łukajtis et al. 2018).

Mesophilic bacteria specifically require the optimal temperature of 35–37 °C and are frequently used in the production of bio-hydrogen by dark fermentation. As a result of this operating temperature, the operational cost of energy is lower compared to thermophilic and extra thermophilic conditions. In particular, mesophilic bacteria used substrates like lower sugar molecules (sucrose, fructose, glucose, and so on) and organic substances to reduce the denaturation of molecules (Łukajtis et al. 2018). However, the mesophilic condition (35 °C) is not the operating temperature when agricultural residues and waste foods are used as substrates. Instead, the thermophilic and extra thermophilic conditions are used as operating temperatures for a high yield of bio-hydrogen production (Guo et al. 2010; Pakarinen et al. 2008). The lignocellulosic materials of agricultural residues and starch in waste food need a high temperature for hydrolysis in fermentation and reported that the varieties of hydrogen production with temperature and found the highest yield at a high temperature of 70 °C (Pakarinen et al. 2008). In another study, Shin et al., using food waste as a substrate, found that rates at mesophilic temperatures (35 °C) were nine times lower than rates under thermophilic circumstances (55 °C) (Shin et al. 2004). The results showed that the temperature of 35 °C was too low to hydrolyze the lignocellulosic materials and starch of agricultural residues and waste food.

7.3.3.3 pH

Managing pH is essential to the dark fermentation of bio-hydrogen production, because it affects the activity of hydrogenase and metabolic pathways. It has been shown that a low pH in the fermentation medium inhibits the metabolic activity of the hydrogen-producing bacteria or triggers a metabolic pathway switch that results in bio-hydrogen generation stopping (Łukajtis et al. 2018; Pakarinen et al. 2008). A specific pH range is optimal for enzyme activity in bacterial metabolic processes (Łukajtis et al. 2018). During fermentation, it is also important to maintain pH levels at an optimal level. It is necessary to do so because hydrogen is produced by the synthesis of organic acids (acetic, lactic, butyric, and propionic), which decrease the pH value of the medium, thus inhibiting enzyme activity.

7.3.3.4 Hydraulic Retention Time (HRT)

A hydraulic retention time (HRT) is directly related to the amount of organic waste processed per unit time, which directly impacts economic operations. During the exponential growth phase of *Clostridium* sp., H_2-producing bacteria produce volatile fatty acids (VFAs), while during the stationary growth phase, they produce alcohol. Short HRT is preferred by the microbial community as well. After reducing the HRT from 8 to 6 h, propionic acid bacteria wash out, which consumes H_2. Short HRT is believed to prevent methanogen growth because methanogens grow so slowly (Show et al. 2011). The biodegradability of the substrate determines the HRT value for dark fermentation. Throughout continuous culture development, the HRT is typically gradually reduced from long to short distances to permit microorganisms to accumulate new surroundings and preclude bacteria involved from being washed away (Łukajtis et al. 2018). Comparing the H_2 fermentation performance of food waste over HRTs of 2, 3, and 5 d, the best H_2 (2.2 mol H_2/mol hexose) was obtained under a HRT of 5 d; a HRT of 2 d showed the worst of 1.95 mol H_2/mol hexose. A review of the literature revealed that the lowest HRT, the more stable and high effectiveness was obtained. Using a thermophilic (60 °C) packed bed reactor to treat waste food, and paper waste has taken 1.2 days to date. It is feasible to modify HRT in dark fermentation to confine or nullify the activity of bacteria utilizing hydrogen in their own metabolic pathways by attempting to make use of differences in the productivity growth of hydrogen consumers and producers (Łukajtis et al. 2018). Hydraulic retention time has an impact on the amount of hydrogen in the biogas produced.

7.3.3.5 Substrate Concentration and Liquid Product Inhibition

The concentration of the substrate must be taken into account while designing the fermentation. Most often, diluted substrates, such as 10 g/dm^3 or 1% TS, are used to provide the maximum hydrogen yields (Łukajtis et al. 2018; Pakarinen et al. 2008). The optimum substrate concentration in the batch experiments varied widely and was strongly impacted by other operational factors like pH. Because of the low pH condition when the pH was uncontrolled, hydrogen yield typically dropped as the substrate concentration raised (Show et al. 2011). The effects of pH and substrate concentration on the production of hydrogen and their correlation were reported by (Sung and Lay 2001). They used a fractional factorial design to determine that the optimal substrate concentration was 7.5 g COD/L at pH 5.5.

7.3.3.6 Feedstock

Agricultural wastes, organic wastes (industrial wastewater), and waste foods are all used as feedstocks in fermentation. The productivity of hydrogen production varies depending on the types of feedstock. Agricultural waste is more practical because it is abundant and more inexpensive as a raw material. They include cellulose and hemicellulose, which are made up of glucose and pentose or hexose, respectively. However, degradation of the material is challenging, and it contains lignin, though pretreatment is essential (Łukajtis et al. 2018; Bartacek et al. 2007). Different researchers investigate the productivity of hydrogen production without pretreatment or with various pretreatments. For instance, Zhang et al. (2020b) found that corn stover hydrogen production increased by 33% when it was treated with 2% NaOH. Using corn stover, they studied various pretreatments and found different hydrogen yields. In Sect. 7.2, (different types of agricultural waste materials for hydrogen production), it is generally discussed in details how the specific raw material is obtained and behaves.

7.4 Microbiology of Bio-Hydrogen Production

The improvement and stability of hydrogen production depend heavily on the development of bio-hydrogen dark fermentation, photo-fermentation, mixed fermentation technology, and bacterial breeding. There are many different ways that biogenic hydrogen bacteria are produced, including heat treatment of shrimp ponds to create bacteria that manufacture hydrogen, lots of light, and pig farms' high organic matter content that enriches photosynthetic bacteria in the sludge discharge. Hydrogen strains are isolated from water production wells and used in soil, commercial fertilizers, and sewage plant sludge to prepare cyanobacterium *Anabaena* algae and a community of hydrogen-producing bacteria (Singh and Wahid 2015).

7.4.1 Bio-Hydrogen Producers

The diversity in physiology and metabolism of microorganisms allows them to have different ways of producing hydrogen with different advantages and drawbacks. The main advantages of using microorganisms for production of hydrogen include lower cost catalyst and less energy needed for process than current industrial method for producing hydrogen. There are the main four approaches for the production of hydrogen using microorganisms (Hallenbeck et al. 2009).

1. Bio-photolysis of algae using cyanobacteria/algae.
2. Utilizing anaerobic bacteria, dark fermentation hydrogen production.
3. Utilizing photosynthetic microorganisms, photodegradation of organic molecules.
4. Bio-electrohydrogenesis.

In bio-photolysis, some types of cyanobacteria and green algae participate in the two photosystems of plant-type photosynthesis that work together to separate water with absorbed light and create reduced ferredoxin, which drives the reduction of protons to hydrogen. The most significant benefit of this process is the use of water as substrate which is easily obtainable everywhere. However, simultaneous production of oxygen and hydrogen causes severe problems: the production of combinations of these gases that could explode, as well as the hydrogenase inhibition, which is oversensitive to even slightly reduced oxygen levels. Production of hydrogen by using cyanobacteria is much less sensitive to oxygen than hydrogen production by nitrogenase in heterocysts. However, there is a metabolic cost to this, in aspects of heterocyst syntheses as well as maintenance, as well as nitrogenase's high ATP requirement. Low solar energy conversion efficiency results in additional issues by drastically raising the area needed for essential transparent, hydrogen-impervious materials.

Another process that requires light energy input is a generation of hydrogen powered by light by photosynthetic bacteria from a variety of materials, particularly synthetic acids, also known as photo-fermentation. Indeed, photosynthetic bacteria are known for their ability to generate huge amounts of hydrogen because of their superior substrate conversion efficiencies and capability of degrading a wide variety of raw materials. Mostly pure water is used as substrate in model studies, but for some trials, industrial water is demonstrated as the substrate. However, because of the effluent's high toxicity or its color/opacity, pretreatment may be required prior to the production of photosynthetic bio-hydrogen gas. For instance, low light passage into the bioreactor makes high biomass concentration undesirable. Regardless of the successes of hydrogen production through the organic compound photosynthetic disintegration, much more effort needs to be done to develop a large-scale, profitable technology. Although the rate of substrate conversion is often high, the hydrogen synthesis rate is slow, and the yields of hydrogen are still much below the maximum theoretical value. The important factors in increasing product yield are intensity of light and diffusion, much like in any other light-based production process. The rate of production and yield of hydrogen rise with an increase in light intensity (up to a specific threshold), but the efficiency of light conversion decreases. The need for huge reactor surface areas and expensive equipment are still significant limitations. It may be necessary to combine photosynthetic hydrogen generation with another method so as to make it commercially feasible.

Dark fermentation is a third strategy for biological hydrogen generation, which occurs without the presence of oxygen. These kinds of anaerobic systems have advantages because they are more affordable and easier and have a high rate of hydrogen production. The main disadvantage is that these bacteria cannot get past the barrier to intrinsic thermodynamic energy that prevents complete substrate breakdown. Therefore, fermentative systems generally have modest hydrogen yields. This is due to the fact that anaerobic metabolism has evolved to maximize biomass rather than hydrogen. Anaerobic organisms often produce gas at the time of exponential growth, and when the culture reaches the stationary growth phase, the metabolism switches from producing hydrogen or acid to solventogenesis. Reduced iron concentration, relatively high substrate concentration, high substrate partial pressure, and/or low pH have all been connected to low hydrogen yields. When compared to traditional reforming methods, the fermentative method is not economically viable at the current maximum hydrogen yields. This problem is now being under research in an effort to find a set of conditions that will maximize rate of production and yield.

It has been suggested that a two-step, hybrid biological process would be needed to produce hydrogen by dark fermentation for both commercially viable and sustainable. Higher total substrate conversion efficiency is feasible for the anaerobic and photosynthetic processes coming together because the photosynthetic bacteria can break down the fermented soluble metabolites using sunshine. The primary soluble breakdown products from the first phase are VFAs, which photo-heterotrophic bacteria prefer as substrates. The two-step procedure can be theoretically 1 mole of glucose yields and 12 moles of hydrogen. It should be noted that since there are species of photosynthetic bacteria that can utilize some sugars as substrate, it is theoretically possible for them to complete this process in a single step. Therefore, the only practical benefit of using a two-step procedure like this may be to shorten the period of time and volume needed for substrate conversion. In fact, complicated substrates, such as the majority of wastes containing carbohydrates, might be difficult for photosynthetic bacteria to break down and would likely call for the usage of mixed consortia. Several studies on two-stage methods have recently been published. Co-cultures of photosynthesis and fermentation organisms have been the subject of numerous studies. In comparison to pure culture production obtained separately, work using *C. butyricum* and *R. sphaeroides* co-cultures showed just a modest improvement in the hydrogen yield.

Finally, a new method of hybrid biological hydrogen production has been recently discovered. It is based on a microbial fuel cell (MFC) theory and application. The goal is a slight increase in electrical potential produced by a MFC in order to attain a force strong enough to convert protons to hydrogen, a process known as bio-electrohydrogenesis. Consequently, the cell might be referred to as a microbial electrohydrogenesis cell (MEC). The phrase "microbial electrolysis cell" is inappropriate because it suggests that water splitting produces the protons.

The benefit of MECs and MFCs is that waste stream energy will be recovered directly as electricity (in MFC) or hydrogen (in MEC). The underlying metabolic processes are unclear, and MEC research has only been conducted thus for using mixed cultures, frequently using those whose concentrated and active forms are currently used in microbial fuel cells (MFC). However, *Shewanella* bacteria and *Geobacter*, which are known to successfully link electrode surfaces to their metabolism, are typically found in MFCs. In essence, these reactions are anaerobic respirations in which an electrode is used as the external electron acceptor rather than the more typical oxidized molecule. Thus, it employs electrochemically active microorganisms that, with a low to moderate input voltage, use coupled anode-cathode processes to transform dissolved organic matter into hydrogen inside an electrochemical cell or MFC (Hallenbeck et al. 2009). Thus, both in theory and in fact, it is possible to supply enough energy to enable the conversion of substances like acetate, a by-product of dark fermentation, to hydrogen. Of course, typically, as previously noted, these bacteria cannot perform this conversion by themselves, unless in syntrophic association with an organism that consumes hydrogen and can endure very low partial pressures of hydrogen. Despite being an appealing notion, and undoubtedly one that might allow the full conversion of basic substrates, such as sugars or acetate, there are several ways to convert wastewaters or even hydrogen. Naturally, many of these are also encountered in the longer term. The power density of MFC at low volumetric hydrogen production is caused by low electrode surface areas (Hallenbeck et al. 2009; Reungsang et al. 2018).

7.4.2 Hydrogen Consumers and Metabolic Competitors

During the fermentation process of the organic compounds which results into yield of bio-hydrogen, a balance between the microorganisms, substrate, medium, and the environment has a crucial role in the overall hydrogen yield. Hydrogen production takes part in an anaerobic condition of which the microbes feed off using the organic material present as medium (some of these reactions take part under light and some in the dark). Fermentation that takes part in the light depends on light energy to convert the organic complex compounds into carbon dioxide, methane, and hydrogen gas. With that, process microbial life will generate these products as long as the presence of light energy is constant and at required sufficient amounts (Kapdan and Kargi 2006), on dark fermentation microbial organisms depend upon a medium to source their energy and convert complex organic compounds into hydrogen gas and other products. Therefore, the hindrance and competitors of hydrogen production depends upon the concentration of hydrogen hindering bacteria in the system acting as inhibitor. These three bacteria were experimented on and found to inhibit hydrogen production, sulfate-reducing bacteria (SRB), methane-producing bacteria (MPB), and homoacetogenic bacteria (HAB), and they directly or indirectly converge the bio-hydrogen potential of carbohydrates or any carbon source. Some bacteria operate within the range of 30–35 °C and optimal pH of 7.0 (Hung et al. 2011). So when bacteria live below these conditions, hydrogen production will drop due to shifts in the equilibrium of hydrogen production in the reactor. There are some main bacteria which commonly inhibit hydrogen production: homoacetogenic bacteria and methanogens (Hung et al. 2011).

7.4.2.1 Methanogens

Methanogens are considered to be among the most hydrogen-consuming organisms in anaerobic environment, and the production of methanogens is closely related to the system operating conditions. Methanogens tend to grow sustainably in pH between 6 and 8; thus, the initial acidic environment in the reactor is crucial for inhibiting the production of these bacteria and overall increase of hydrogen gas. To control these bacteria, chemical inhibitors (bromoethanesulfonate (BES), chloroform, and acetylene) are produced so as to decrease the growth of methanogen bacteria and the production of chemical inhibitors tend to be expensive toward the environment and not feasible for large-scale production of hydrogen. The most common way of inoculums treatment is by heating the medium at 100 °C for approximately 10 min so as to select spore-forming and hydrogen-producing bacteria helps in removing methanogens because they cannot withstand such high temperatures (Lay et al. 2005; Vijayaraghavan et al. 2006).

7.4.2.2 Homoacetogenic Bacteria

These are bacteria which catalyze the formation of methane from H_2 and CO_2. These bacteria are versatile anaerobic organisms with enzymes which can catalyze the formation of acetyl-CoA and are converted to either acetate in catabolism or to cell carbon in anabolism (Diekert et al. 1994). With such capability of the microorganism, changes in the system will occur (hydrogen concentration); thus, the calculated value will not match the experimental value because the homoacetogenic bacteria have produced methane using the hydrogen gas in the system. Experiments done by Kotsopoulos et al. (2009) showed the values anticipated in the CSTR production of hydrogen gas are different compared to the VFA accumulation; thus, hydrogen was assumed to be consumed by acetogenic organisms.

7.5 Challenges in Bio-Hydrogen Production Using Agricultural Waste

The production of hydrogen from agricultural waste for a clean energy source will reduce the amount of carbon footprint in the manufacturing and processing world. Core challenges of bio- hydrogen is the extraction of hydrogen from its agricultural source, since some agricultural wastes have a rigid structure and its bonds holding the hydrogen atoms are strong resulting in requirements of pretreatment that will prepare the hydrogen source to yield the maximum amount of hydrogen gas. The availability of technology to prepare the hydrogen source for hydrogen production and the system of hydrogen production itself requires some equipment, organisms, and special conditions. With such inputs to produce hydrogen gas as output, it can be analyzed and sought for an optimum way to acquire a sufficient amount of hydrogen. The agricultural wastes vary with locations and thus can affect the amount of hydrogen yield based on the agricultural waste.

7.6 Conclusion

A detailed study of bio-hydrogen production from agricultural waste materials has been discussed here. Various types of agriculture waste which can be used for hydrogen production are described at a large in this chapter. Different bioreactors and their operating conditions are explained over here. The circumstances of microbes which help in bio-hydrogen productions have been talked. Bio-hydrogen can be used as a source of energy and power up activities and processes based on the yield of hydrogen collected. More studies and research could be done to improve the yield of hydrogen from agricultural waste, which will definitely help to increase the chances of a carbon-free economy.

References

2021 at a glance—Statistical Review of World Energy 2022 (2022) BP. Accessed 16 Oct 2022

Akkerman I, Janssen M, Rocha J, Wijffels RH (2002) Photobiological hydrogen production: photochemical efficiency and bioreactor design. Int J Hydrogen Energy 27(11–12):1195–1208

Arimi MM, Knodel J, Kiprop A, Namango SS, Zhang Y, Geißen S-U (2015) Strategies for improvement of biohydrogen production from organic-rich wastewater: a review. Biomass Bioenergy 75:101–118. https://doi.org/10.1016/j.biombioe.2015.02.011

Banu JR, Kannah RY, Kumar MD, Gunasekaran M, Sivagurunathan P, Park J-H, Kumar G (2018) Recent advances on biogranules formation in dark hydrogen fermentation system: mechanism of formation and microbial characteristics. Bioresour Technol 268:787–796. https://doi.org/10.1016/j.biortech.2018.07.034

Bartacek J, Zabranska J, Lens L, P. N. (2007) Developments and constraints in fermentative hydrogen production. Biofuels Bioprod Biorefin 1(3):201–214. https://doi.org/10.1002/bbb.17

Chen C-C, Chuang Y-S, Lin C-Y, Lay C-H, Sen B (2012) Thermophilic dark fermentation of untreated rice straw using mixed cultures for hydrogen production. Int J Hydrogen Energy 37(20):15540–15546. https://doi.org/10.1016/j.ijhydene.2012.01.036

Christopher FC, Kumar PS, Vo D-VN, Joshiba GJ (2021) A review on critical assessment of advanced bioreactor options for sustainable hydrogen production. Int J Hydrogen Energy 46(10):7113–7136. https://doi.org/10.1016/j.ijhydene.2020.11.244

Cui M, Shen J (2012) Effects of acid and alkaline pretreatments on the biohydrogen production from grass by anaerobic dark fermentation. Int J Hydrogen Energy 37(1):1120–1124. https://doi.org/10.1016/j.ijhydene.2011.02.078

Dasgupta CN, Gilbert JJ, Lindblad P, Heidorn T, Borgvang SA, Skjanes K, Das D (2010) Recent trends on the development of photobiological processes and photobioreactors for the improvement of hydrogen production. Int J Hydrogen Energy 35(19):10218–10238. https://doi.org/10.1016/j.ijhydene.2010.06.029

Deng C, Lin R, Cheng J, Murphy D, J. (2019) Can acid pre-treatment enhance biohydrogen and biomethane production from grass silage in single-stage and two-stage fermentation processes? Energ Conver Manage 195:738–747. https://doi.org/10.1016/j.enconman.2019.05.044

Diekert G, Wohlfarth G, Leeuwenhoek A (1994) Metabolism of homoacetogens. Antonie Van Leeuwenhoek 66:209–221

Fossil and Alternative Fuels—Energy Content (2008) The Engineering Tool Box. https://www.engineeringtoolbox.com/fossil-fuels-energy-content-d_1298.html. Accessed 15 Oct 2022

Gaweł S, Izabela K, Adam C (2019) Perspectives of hydrogen production from corn wastes in Poland by means of dark fermentation. Ecol Chem Eng 26:255–263

Gonzales RR, Sivagurunathan P, Kim S-H (2016) Effect of severity on dilute acid pretreatment of lignocellulosic biomass and the following hydrogen fermentation. Int J Hydrogen Energy 41(46):21678–21684. https://doi.org/10.1016/j.ijhydene.2016.06.198

Guler BA, Deniz I, Demirel Z, Oncel SS, Imamoglu E (2019) Comparison of different photobioreactor configurations and empirical computational fluid dynamics simulation for fucoxanthin production. Algal Res 37:195–204. https://doi.org/10.1016/j.algal.2018.11.019

Guo XM, Trably E, Latrille E, Carrère H, Steyer J-P (2010) Hydrogen production from agricultural waste by dark fermentation: a review. Int J Hydrogen Energy 35(19):10660–10673. https://doi.org/10.1016/j.ijhydene.2010.03.008

Hallenbeck PC, Ghosh D, Skonieczny M, Yargeau V (2009) Microbiological and engineering aspects of biohydrogen production. Indian J Microbiol 49:48–59

Han H, Wei L, Liu B, Yang H, Shen J (2012) Optimization of biohydrogen production from soybean straw using anaerobic mixed bacteria. Int J Hydrogen Energy 37(17):13200–13208. https://doi.org/10.1016/j.ijhydene.2012.03.073

Harshita S, Sakshi T, Qureshi KA, Jaremko M, Rai PK (2022) Recent advances in biomass pretreatment technologies for biohydrogen production. Energies 15(3):999. https://doi.org/10.3390/en15030999

He L, Huang H, Lei Z, Liu C, Zhang Z (2014) Enhanced hydrogen production from anaerobic fermentation of rice straw pretreated by hydrothermal technology. Bioresour Technol 171:145–151. https://doi.org/10.1016/j.biortech.2014.08.049

Holladay JD, Hu J, King DL, Wang Y (2009) An overview of hydrogen production technologies. Catal Today 139(4):244–260. https://doi.org/10.1016/j.cattod.2008.08.039

Hsu T-C, Guo G-L, Chen W-H, Hwang W-S (2010) Effect of dilute acid pretreatment of rice straw on structural properties and enzymatic hydrolysis. Bioresour Technol 101(13):4907–4913. https://doi.org/10.1016/j.biortech.2009.10.009

Hung C-H, Chang Y-T, Chang Y-J (2011) Roles of microorganisms other than *clostridium* and *Enterobacter* in anaerobic fermentative biohydrogen production systems—a review. Bioresour Technol 108(18):8437–8444. https://doi.org/10.1016/j.biortech.2011.02.084

Hydrogen Production Processes (2023) https://www.energy.gov/eere/fuelcells/hydrogen-production-processes. Accessed 16 Oct 2022

Ivanova G, Rákhely G, Kovács KL (2009) Thermophilic biohydrogen production from energy plants by *Caldicellulosiruptor saccharolyticus* and comparison with related studies. Int J Hydrogen Energy 34(9):3659–3670. https://doi.org/10.1016/j.ijhydene.2009.02.082

Jung K-W, Kim D-H, Kim S-H, Shin H-S (2011) Bioreactor design for continuous dark fermentative hydrogen production. Bioresour Technol 102(18):8612–8620. https://doi.org/10.1016/j.biortech.2011.03.056

Kamil K, Andrzej L, Krystyna M, Qiao W (2019) Renewable energy sources biohydrogen dark fermentation wastes management biogas plants. J Ecol Eng 20(2):46–160

Kapdan IK, Kargi F (2006) Bio-hydrogen production from waste materials. Enzyme Microb Technol 38(5):569–582. https://doi.org/10.1016/j.enzmictec.2005.09.015

Kotsopoulos TA, Fotidis IA, Tsolakis N, Martzopoulos GG (2009) Biohydrogen production from pig slurry in a CSTR reactor system with mixed cultures under hyper-thermophilic temperature (70°C). Biomass Bioenergy 33(9):1168–1174. https://doi.org/10.1016/j.biombioe.2009.05.001

Kumar SS, Himabindu V (2019) Hydrogen production by PEM water electrolysis—a review. Mater Sci Energy Technol 2(3):442–454. https://doi.org/10.1016/j.mset.2019.03.002

Kumar G, Mathimani T, Rene ER, Pugazhendhi A (2019) Application of nanotechnology in dark fermentation for enhanced biohydrogen production using inorganic nanoparticles. Int J Hydrogen Energy 44(26):13106–13113. https://doi.org/10.1016/j.ijhydene.2019.03.131

Lay JJ, Fan KS, Hwang JI, Chang JI, Hsu PC (2005) Factors affecting hydrogen production from food wastes by clostridium-rich composts. J Environ Eng 131(4):595

Li Y-C, Liu Y-F, Chu H-Y, Chang P-L, Hsu C-W, Lin P-J, Wu S-Y (2012a) Techno-economic evaluation of biohydrogen production from wastewater and agricultural waste. Int J Hydrogen Energy 37(20):15704–15710. https://doi.org/10.1016/j.ijhydene.2012.05.043

Li Y-C, Kokko ME, Wu SY, Lin C-Y, Puhakka JA (2012b) Silage as source of bacteria and electrons for dark fermentative hydrogen production. Int J Hydrogen Energy 37(20): 15518–15524. https://doi.org/10.1016/j.ijhydene.2012.04.060

Lin C-Y, Lay C-H, Sen B, Chu C-Y, Kumar G, Chen C-C, Chang J-S (2012) Fermentative hydrogen production from wastewaters: a review and prognosis. Int J Hydrogen Energy 37(20):15632–15642. https://doi.org/10.1016/j.ijhydene.2012.02.072

Łukajtis R, Hołowacz I, Kucharska K, Glinka M, Rybarczyk P, Przyjazny A, Kamiński M (2018) Hydrogen production from biomass using dark fermentation. Renew Sustain Energy Rev 91: 665–694. https://doi.org/10.1016/j.rser.2018.04.043

Mafizur RM, Eswaran V (2020) Renewable and non-renewable energy consumption-economic growth nexus: new evidence from South Asia. Renew Energy 147:399–408

Marianne Q, Bittel M, Trably E, Dumas C, Fourage L, Ravot G, Steyer J-P, Carrère H (2012) Effect of enzyme addition on fermentative hydrogen production from wheat straw. Int J Hydrogen Energy 37(14):10639–10647. https://doi.org/10.1016/j.ijhydene.2012.04.083

Martin W, Anja H, Cecilia G, Anastasios M, Thomas H (2002) [Fe]-hydrogenases in green algae: photo-fermentation and hydrogen evolution under sulfur deprivation. Int J Hydrogen Energy 27(11-12):1431–1439. https://doi.org/10.1016/S0360-3199(02)00095-2

Mata TM, Martins AA, Calheiros CSC, Villanueva F, Alonso-Cuevilla NP, Gabriel MF, Silva GVI (2022) Air quality: a review of cleaning technologies. Environments 9:118

Mona S, Kumar SS, Kumar V, Parveen K, Saini N, Deepak B, Pugazhendhi A (2020) Green technology for sustainable biohydrogen production (waste to energy): a review. Sci Total Environ 728:138481. https://doi.org/10.1016/j.scitotenv.2020.138481

Mtui GYS (2009) Recent advances in pretreatment of lignocellulosic wastes and production of value added products. Afr J Biotechnol 8(8):1398–1415

Nguyen T-AD, Kim K-R, Kim MS, Sim SJ (2010) Thermophilic hydrogen fermentation from Korean rice straw by *Thermotoga neapolitana*. Int J Hydrogen Energy 35(24):13392–13398. https://doi.org/10.1016/j.ijhydene.2009.11.112

Ni M, Leung DYC, Leung MKH, Sumathy K (2006) An overview of hydrogen production from biomass. Fuel Process Technol 87(5):461–472. https://doi.org/10.1016/j.fuproc.2005.11.003

Nicoletti G, Arcuri N, Nicoletti G, Bruno R (2015) A technical and environmental comparison between hydrogen and some fossil fuels. Energ Conver Manage 89:205–213. https://doi.org/10.1016/j.enconman.2014.09.057

Pakarinen O, Lehtomäki A, Rintala J (2008) Batch dark fermentative hydrogen production from grass silage: the effect of inoculum, pH, temperature and VS ratio. Int J Hydrogen Energy 33(2): 594–601. https://doi.org/10.1016/j.ijhydene.2007.10.008

Parthasarathy P, Narayanan KS (2014) Hydrogen production from steam gasification of biomass: influence of process parameters on hydrogen yield—a review. Renew Energy 66:570–579. https://doi.org/10.1016/j.renene.2013.12.025

Rajesh Banu J, Mohamed Usman TM, Kavitha S, YukeshKannah R, Yogalakshmi KN, Sivashanmugam P, Bhatnagar A, Kumar G (2021) A critical review on limitations and enhancement strategies associated with biohydrogen production. Int J Hydrogen Energy 46(31): 16565–16590. https://doi.org/10.1016/j.ijhydene.2021.01.075

Rathore D, Singh A (eds) (2017) Biohydrogen production: sustainability of current technology and future perspective. Springer, Berlin. https://doi.org/10.1007/978-81-322-3577-4_4

Ren N, Wang A, Cao G, Xu J, Gao L (2009) Bioconversion of lignocellulosic biomass to hydrogen: potential and challenges. Biotechnol Adv 27(6):1051–1060. https://doi.org/10.1016/j.biotechadv.2009.05.007

Renaudie M, Dumas C, Vuilleumier S, Ernst B (2021) Biohydrogen production in a continuous liquid/gas hollow fiber membrane bioreactor: efficient retention of hydrogen producing bacteria via granule and biofilm formation. Bioresour Technol 319:124203. https://doi.org/10.1016/j.biortech.2020.124203

Reungsang A, Sittijunda S, Zhong N, Xia A (2018) Hydrogen from photo fermentation. In: Green energy and technology, pp 221–317. https://doi.org/10.1007/978-981-10-7677-0_7

Rezania S, Md Din MF, Taib SM, Sohaili J, Chelliapan S, Kamyab H, Saha BB (2017) Review on fermentative biohydrogen production from water hyacinth, wheat straw and rice straw with focus on recent perspectives. Int J Hydrogen Energy 42(33):20955–20969. https://doi.org/10.1016/j.ijhydene.2017.07.007

Saeid A, Chojnacka K (2015) Toward production of microalgae in photobioreactors under temperate climate. Chem Eng Res Des 93:377–391. https://doi.org/10.1016/j.cherd.2014.06.008

Sangyoka S, Reungsang A, Lin C-Y (2016) Optimization of biohydrogen production from sugarcane bagasse by mixed cultures using a statistical method. Sustain Environ Res 26(5):235–242. https://doi.org/10.1016/j.serj.2016.05.001

Shin H-S, Kim S-H, Paik B-C (2004) Characteristics of hydrogen production from food waste and waste activated sludge. J Water Environ Technol 1(2):177–187. https://doi.org/10.2965/jwet.2003.177

Show K-Y, Lee D-J, Chang J-S (2011) Bioreactor and process design for biohydrogen production. Bioresour Technol 102(18):8524–8533. https://doi.org/10.1016/j.biortech.2011.04.055

Singh RN, Sharma S (2012) Development of suitable photobioreactor for algae production—a review. Renew Sustain Energy Rev 16(4):2347–2353. https://doi.org/10.1016/j.rser.2012.01.026

Singh L, Wahid ZA (2015) Methods for enhancing bio-hydrogen production from biological process. J Ind Eng Chem 21:70–80. https://doi.org/10.1016/j.jiec.2014.05.035

Sivaramakrishnan R, Ramprakash B, Ramadoss G, Suresh S, Pugazhendhi A, Incharoensakdi A (2021) High potential of *Rhizopus* treated rice bran waste for the nutrient-free anaerobic fermentative biohydrogen production. Bioresour Technol 319:124193. https://doi.org/10.1016/j.biortech.2020.124193

Ginkel SV, Sung S, Lay JJ (2001) Biohydrogen production as a function of pH and substrate concentration. Environ Sci Technol 35(24):4726–4730

Talluri S, Raj SM, Christopher LP (2013) Consolidated bioprocessing of untreated switchgrass to hydrogen by the extreme thermophile *Caldicellulosiruptor saccharolyticus* DSM 8903. Bioresour Technol 139:272–279. https://doi.org/10.1016/j.biortech.2013.04.005

Tashie-Lewis BC, Nnabuife SG (2021) Hydrogen production, distribution, storage and power conversion in a hydrogen economy—a technology review. Chem Eng J Adv 8:100172. https://doi.org/10.1016/j.ceja.2021.100172

Vijaya Raghavan K, Ahmad D, Ibrahim MKB, Herman HNB (2006) Isolation of hydrogen generating microflora from cow dung for seeding anaerobic digester. Int J Hydrogen Energy 31(6):708–720. https://doi.org/10.1016/j.ijhydene.2005.07.002

Wang L, Wang Q, Zhao Z, Tao T, Ying K, Mao X (2021) Novel flat-plate photobioreactor with inclined baffles and internal structure optimization to improve light regime performance. ACS Sustain Chem Eng 9(4):1550–1558. https://doi.org/10.1021/acssuschemeng.0c06109

Yadav VS, Vinoth R, Yadav D (2018) Bio-hydrogen production from waste materials: a review. In: MATEC Web Conference, vol 192. https://doi.org/10.1051/matecconf/201819202020

Yang H, Shao P, Lu T, Shen J, Wang D, Xu Z, Yuan X (2006) Continuous bio-hydrogen production from citric acid wastewater via facultative anaerobic bacteria. Int J Hydrog Energy 31(10):1306–1313. https://doi.org/10.1016/j.ijhydene.2005.11.018

Yaqoubnejad P, Rad HA, Taghavijeloudar M (2021) Development a novel hexagonal airlift flat plate photobioreactor for the improvement of microalgae growth that simultaneously enhance CO2 bio-fixation and wastewater treatment. J Environ Manage 298:113482. https://doi.org/10.1016/j.jenvman.2021.113482

Zhang J, Kong C, Yang M, Zang L (2020a) Comparison of calcium oxide and calcium peroxide pretreatments of wheat straw for improving biohydrogen. ACS Omega 5(16):9151–9161. https://doi.org/10.1021/acsomega.9b04368

Zhang T, Jiang D, Zhang H, Jing Y, Tahir N, Zhang Y, Zhang Q (2020b) Comparative study on bio-hydrogen production from corn stover: photo-fermentation, dark-fermentation and dark-photo co-fermentation. Int J Hydrogen Energy 45(6):3807–3814. https://doi.org/10.1016/j.ijhydene.2019.04.170

Zhu J, Song W, Chen X, Sun S (2022) Integrated process to produce biohydrogen from wheat straw by enzymatic saccharification and dark fermentation. Int J Hydrog Energy 48:11153–11161. https://doi.org/10.1016/j.ijhydene.2022.05.056

Chapter 8
Conversion of Food Waste into Valuable Products

Ila Jogesh Ramala Sarkar, Steward Laishi, Michael C. Kabesha, Kakeeto Ismail, and Sanjay Kumar

Abstract The increasing amounts of food wastage generated per annum due to the growing human population worldwide are frequently associated with environmental pollution issues and shortage of natural resources and economic loss. Considering this, the researcher has worked toward finding sustainable approaches to replace the conventional practices for food waste management. Food waste serves as an excellent source of value-added products owing to high organic content. Effective conversion of food to valuable resources is often challenged by its heterogeneous nature. This book chapter will lay down the prospects and consequences associated with food waste management toward converting valuable products. The various social, economic, and environmental concerns associated with food waste management, especially in terms of greenhouse gas emission, will be discussed. The difficulties in proper collection, storage, and bioconversion of food waste to precious by-products will be pointed out in this chapter. Finally, the wide array of challenges to produce value-added products using food waste will be enlisted to emphasize the prospects of food waste management.

Keywords Food waste · Bioconversion · Value-added products · Utilization

8.1 Introduction

Food waste is a global challenge that has affected the vast majority for over 800 million years, while people don't have access to food in developing countries; in other words, this can mean that one fourth of the food wasted around the globe is able to provide for each person (TRVST LTD 2022). There are different reasons to as why food is wasted, among them include challenges in processing, unfavorable weather conditions, overproduction, unreliable marketplace, and many more. In other cases, overbuying, short on planning and unsureness, and high price of

I. J. R. Sarkar (✉) · S. Laishi · M. C. Kabesha · K. Ismail · S. Kumar
Department of Chemical Engineering, Faculty of Technology, Marwadi University, Rajkot, India

D. B. Pal, A. K. Tiwari (eds.), *Sustainable Valorization of Agriculture & Food Waste Biomass*, Clean Energy Production Technologies,
https://doi.org/10.1007/978-981-99-0526-3_8

commodities can also lead to food waste in homes and stores (foodprint organization 10/08/2018). It became a problem during the late 1800s and early 1900s. A greater number of food wastes end up as municipal solid waste (MSW). When disposed of in landfills, food waste produces harmful gasses such as methane and carbon dioxide generally greenhouse gasses, contributing to the global warming and climate change, and becomes a habitant of various harmful microorganisms (Ahmad et al. 2019). It also leads to depletion and contamination of natural resources such as freshwater, land, fossil fuels, and human natural resources. Therefore, to manage the wasted food, sustainable and innovative valorization technologies are required in waste recovery and recycling management.

Farmer factories, transports, storage, and retailing in supermarkets are some of the long-chain processes that are part of the system that induce food waste which has its aim to provide good service to the customers, cumulatively local shops, retail one stops, and others (bakery, butcher's, grocer's shops, fast foods), and local markets cost lots of muttered products for various reasons (Nitayavardhana et al. 2008).

Since the industrial revolution, many new manufacturing processes have been discovered and implemented. Waste has always been disregarded, but as time goes on, scientists and other profound intellects have realized that one man's trash is another's treasure. Waste was mostly managed by burning or dumping. After the knowledge of the ozone layer depletion, the idea of green chemistry was initially developed as the response to the pollution in 1990 and began its effectiveness practice in 1998 (Mazloumian et al. 2020). With green chemistry in mind knowing that food waste contributes to reduction in pollution, inventions and innovations of food waste to by-products are major steps to a greener environment with less industrial interest to it. Food waste management skills can lead to major production companies.

It is observed that the consumers are the ones who produce the majority of food waste. This is because of habits that are relatively part of human sensation such as eating too much, cooking too much, and buying too much. Food waste is now employed in industrial processes for the production of biofuels and biopolymers. Bio-fertilizers can also be produced from food waste just like biofuel, biodiesel and biochemicals, but this method is under looked upon when it comes to utilizing food waste for value-added products. Wasted food and food processing waste are rich in nutrients like fats with high moisture content, carbohydrates, and protein (Mazloumian et al. 2020; Sindhu et al. 2019); all the compounds have been studied and found to be convenient for treatment using processes like chemical hydrolysis, anaerobic digestion (AD), and aerobic composting. One of the biggest obstacles of the food waste processing industry is that food waste has calorific value, high moisture content, and its heterogeneous nature (Adam et al. 2012). Composition of food wastes varies depending upon the source. Therefore, a prevention technique cannot be adapted for every kind of food waste. Depending on source and composition, a certain treatment technique may be applied on the food waste for it to be accessible for growth of microorganisms and production of the craved finished good of interest in an eco-friendly and profitable manner. In this chapter, we are going to discuss things related to food waste management like classifying food waste according to its sources, physical nature and properties, the various creative ways in which we can convert food waste into value-added products, useful industrial products that can be obtained from food waste as well as the challenges that come along food waste valorization.

8.2 Classification of Food Waste

Variation of food waste is simply the failure to use potentially edible items for the satisfaction of human hunger, inefficient use of plants' energy content, and nutrients for human and livestock purposes. Allowancing waste, however, involves the recognition that waste occurs throughout a series of consecutive steps to produce well-defined goods and products and over to consumption and disposal. Knowing that food waste arises in different locations and for different reasons, the difference between food waste and food loss is that waste occurs in the consumption stage, while losses may be assumed from the perspective of postharvest but preconception waste.

Most commonly food waste is usually classified by type of foods like drinks, meat, fruits, cereal, fish, and vegetables. A classification such as this helps assess the amount of food waste based on mass which is most common, energy content, and economic cost. There are many examples on how to classify food waste for instance, according to its food confines, the Global Product Category (GPC) code or the United Nations Standard Products and Services Code (UNSPSC) as comprehensive codes (Somers et al. 2018). Also, food waste can also be classified according to its nutrient content (e.g., carbohydrate, protein, or fat content), chemical composition (e.g., C, H, N, O, S, and Cl content), or repository temperature (e.g., ambient, chilled, or frozen) (Alexander et al. 2013; Amicarelli et al. 2021). Either way, such counsel on the assumption that these examples outline is not satisfying to lay out some waste management alternatives against others; other researchers also recognize the left-overs and untouched food which go to waste.

Different authors have additionally classified food waste at the household level as cooked/uncooked, as unpackaged/packaged food waste (when waste is packaged, it is further sorted as opened/unopened packaging) and according to their reason of disposal. An inclusive and detailed attempt to classify food waste was uniquely approached by Cane et al., from a perspective where food waste is classified into the following categories: organic crop residue (including fruits and vegetables), packaging, catering waste, animal by-products, domestic waste, and mixed food waste (Cane and Parra 2020). The actual focus of food waste classification is to provide support for an extravagant alternative for food waste management; however, in this chapter, whether loss or waste, we classify it as food waste. It is intellectual to always identify the origin of a source product; food wastes can simply be classified based on their source of origin and based on physical nature and properties.

8.2.1 *Based on Source of Origin*

We must consider that if the product is produced by an animal, then it is animal based additionally as well as the parts of the animals (e.g., milk or any other dairy product, honey, eggs, meat, including fish); indifferently, the product is considered plant based or of natural occurrence (e.g., salt) (Akhtar et al. 2019). Some products may contain both animal- and plant-based materials like a ready meal can therefore be classified according to its principal ingredient, whereas if it be a plant ingredient, the product will be taken or considered as a mixed product. A product is single if it appears homogeneous of ingredient and it has not been made subject with other food materials simply that it has not mixed or been in contact with other materials. Furthermore, the product is heterogeneous; a product is unpackaged if it is not contained in any packaging material.

If there exists a technology that is able to unpack and separate the content of the package from its pack efficiently, then the material or content is considered unpackaged; apart from that, the product is packaged. Routinely, the ability of material to decompose or have biodegradable capability means that it can be worked on or processed into other products although the process may take longer that is possibly years or months by microorganisms. Therefore, in this concept, biodegradable packaging is simply the initiative of having material that can be subject to anaerobic respiration and is recyclable in a practical composting plant (e.g., the "OK compost" logo and "DIN CERTCO" logo) (Wang et al. 2011). Packaging that is made of bioplastic, paper, wood, or any plant-based material or product with exclusions of products like rubber and any other material made of metal, plastic, or glass is nonbiodegradable. Some wastes from food services and domestic disposal that are from schools, restaurants, hospitals, or prison facilities are known as catering waste while non-catering waste occurs in earlier stages of produce or supply chain which is the stages of primary production that is farming, secondary production which is manufacturing and tertiary production which wholesale and retailing. In context, the following are the major source categories of food waste.

8.2.1.1 Organic Crop Residues

They are the noneconomic plant parts that are brought about as a result of harvested vegetables, grains, and fruits as well as processing by-products which include husks, stones, peels, straw, pomace, stover, factory vegetable oil, and oleochemical residue. Residues are also from thrashing sheds as well as that which is eliminated during crop processing. Additionally, process wastes like groundnut shell, rice husks and cobs of maize, oil cake, cumber, and sorghum. The most considered likely to be great potential as a biomass resource appears to come from the field residues of maize, sorghum, cotton, soybean, sugarcane, and wheat. About 190 lakh tons of crop residue are estimated to be available for use in Tamil Nadu, which will add on to the production about 1.0 lakh tons of nitrogen, 2.0 lakh tons of potassium, and 0.5

lakh tons of phosphorus (TNAU AGRITECH PORTAL ORGANIC FARMING). These residues have an adequate value content of carbohydrates, lipids, sugars, and inorganic compounds (mainly silica); food processing wastes can result to biodegradable and biocompatible surfactants, such as sodium lauryl sulfate, which is used in home and personal care products, (Nα,Nε-dioctanoyl lysine), and ethyl-N-lauroyl-l-arginate, used by its high biodegradability. Furthermore, organic crop residues can include large proportions in phytochemicals, such as phenolic, carotenoids, and tocopherols, which show high potential in the fields of food, pharmaceutical industry, and cosmetics (Somers et al. 2018).

8.2.1.2 Catering Waste

Waste that results from kitchens, canteens, restaurants, pubs, coffee shops, or any other wastes related with food produced no longer intended for human consumption is known as catering waste. Examples include processed or cooked meats and vegetables as well as baked goods containing dairy, fish, or meat. There is a growing concern about the huge amount of waste that the food and hospitality sector is producing; about 90% of the catering waste can be reused or recycled, but it is hardly recycled due to a lack of awareness or logistical problems as well as the difficulties of separating the waste from its containers (Alexander et al. 2013). Catering waste has a tendency of containing mixed waste from food preparation which includes packaging and separated waste, glass, and organic, plastic and cardboard as well as used cooking oil.

8.2.1.3 Animal By-Products

Anything of animal origin or sourced from an animal by-product is further suggested not for human consumption only to mention a few such as animal carcasses and parts of animal carcasses, abdominal animal parts, fish, and manure from farm animals like chicken, cattle, and pigs. Depending on the type of animals, the by-products are the majority of which we can list on, for example, the semen, ova, and embryo which have not been used for breeding purposes. Feathers, hooves, blood, shells, wool, fur, hair, animal skins, and crustacean waste all don't qualify for human consumption and hence can be considered as food waste under animal by-product. Other food wastes of animal or fish origin regarded as longer appreciated for human consumption are eggs, milk, and cooking oil used to prepare animal products. Generally, anything that is of animal or fish origin becomes a by-product when it is deemed unworthy of human consumption and is regarded as edible (Northern Ireland Environmental Agency 2002).

8.2.1.4 Waste Packaging

Packaging waste is simply that part of food waste that consists of packaging and packaging material, also being the major contributing factor of waste around the globe. Almost all food packages are biodegradable which include plastic and cans; other waste packages include bottles, paper, and cardboard. Sourced from catering and domestic, this type of packaging waste needs to undergo pretreatment at the recycling or end up somewhere at a landfill site, and this packaging waste on average makes up about 73% of the household waste. On the other hand, domestic waste can also be used for biogas production by fermentation of the biodegradable waste content retrieved from it such as paper. Material such as plastic can be recycled, so is glass basically for reuse under aesthetics.

8.2.1.5 Domestic Waste

Domestic waste is waste that occurs as a result of the ordinary day-to-day use of a domestic premise. Domestic wastes include food waste, paper, glass, metals, plastics, textiles, and foam. A significant amount of domestic wastes consists of plant and animal wastes such as fruit peels, vegetables, leftover meat and bones, fish waste and chicken, and anything classified as wet wastes.

8.2.2 Based on Physical Nature and Properties

8.2.2.1 Fats and Used Cooking Oil

Used cooking oil often referred to as UCOs comes about when fats or oils have been used for frying or cooking and are still available, not changing in chemical structure but maybe in composition. UCOs are a major part of the food processing industry that is fast food chains, restaurants as well as homes and other food joints mainly sourced or found in these places. UCOs can be in the form animal fat or vegetable oil such as corn oil, canola oil, olive oil, palm oil, almond oil, and many more plant-based oils (Al Kamzari et al. 2021). Many are the creative and industrial positive ways that fats and used cooking oil can be processed into, and some of these are as follows:

Lamp oil: It is evident that oil is highly flammable; hence, with properties like this, the oil can be used for lamp lighting, vegetable oil being the most favorable.

Animal feed: This has just been abducted recently in the animal food production and is more complex than soap-making process as wherever consumption of something has an effect; hence, it requires basic knowledge in animal nutrition and veterinary administration.

Soap Making: A common type of soap is lye soap that has UCOs as the primary ingredient and is easy to produce using simple tools and equipment with a number of tutorial videos freely available on the internet on how to make them.

Biodiesel: The main advantage to this is that it has a high potential of replacing fossil fuels. It is sourced from used oils as UCOs are its main or key ingredient.

A number of technical applications have recently come into appreciation such as surfactants, coating, lubricants, and polymers (Boyde 2002). With dangers and caution about highly lubricant and being highly flammable, many attempts have been made to produce lubricants with high thermal/oxidative stability and are the potential UCOs and other oils have, as well as the synthesis of low operating temperature hydraulic fluids and cutting fluids, being highly biodegradable (Khanal et al. 2007; Saboya et al. 2017). It is also amazing to take note of the intellectual initiative to recycle animal by-products such as fats to be used as components of pet supplements, fuels, and cosmetic precursors. The most precise regulations of biofuel production from evinced fats have latterly increased in importance in the European Union; obtaining high biodiesel capitulates from animal by-products even with 30% water of triglyceride (Aravantinou et al. 2013).

8.2.2.2 Wasted Meat and Dairy Waste

Among the top groups of food wasted, meat, poultry, dairy, and fish account for the most considerable category representing 30% of wasted food. These occur due to transportation, overproduction, loss in market, and bad consumption habits.

8.2.2.3 Citrus Fruit Peels

The fruit might be sweet and sour for most of the citrus fruits, but the peel is often bitter; despite being bitter, its benefits are good having a favorable amount of flavonoid glycosides of a massive scope of biological effects such as being used as antioxidants, anti-carcinogenic, and anti-inflammatory, which are widely contributed in the pharmaceutical industry, cosmetics, and food production such as flavorings.

8.2.2.4 Roughage and Fiber-Filled Food Waste

Any part of plant food that cannot be digested or put into use by the human body can be referred to as roughage more technically as dietary fiber. It is often regarded as dry matter but contributes to land restoration and is said to have high cellulose content.

8.3 Methods of Conversion of Food Waste into Value-Added Products

8.3.1 Anaerobic Digestion

The management and handling of huge quantities of "food waste" have resulted into remarkable environmental issues and also financial issues across the globe. In comparison to the conventional methods of disposal and handling like composting, incineration (burning), and land filling, anaerobic digestion technique is a promising technology of food waste processing and management.

Anaerobic digestion is a sequential process by which microbes break down (decompose) biodegradable materials (biomass) in the absence of oxygen (Wang et al. 2011). The process is incorporated into energy production, fertilizer production, and other related processes. During anaerobic digestion, microorganisms degrade innumerable quantities of wastes and organic into biofuels such as biogas which comprises 60–70% methane and the rest being carbon dioxide and traces of other gases like hydrogen sulfide and hydrogen. The residual matter is rich in nutrients and can be incorporated into fertilizer production. In contrast with other technologies for biofuel production, anaerobic digestion uses a broader range of biomass (substrate with enormous impurities, high moisture content) and can be set up for both large- and small-scale productions.

Despite the fact that anaerobic digestion is already in use and widely applied for different processes of wastewater treatment, fertilizer production from the food waste management faces several economical, technical, social, and operational challenges such as process instability, low buffer capacity, foaming, volatile fatty acid accumulation, high operation, and transportation.

8.3.2 Fermentation

Fossil fuel sources being costly, limited nonrenewable resources, and detrimental to the environment research and different technologies are employed into transformation of biomass (waste) and different substrates into various valuable chemicals as alternative production means from fossil fuels. Through fermentation, chemicals such as acids like butyric acid, propanoic acid, and some traces of acetic acid are produced alcohols such as butyl alcohol and ethanol can also be obtained.

Fermentation is the process by which microbes break down substrates or biomass to produce energy under anaerobic conditions. The process allows the transformation and conversion of waste materials whose contents are carbohydrates, fats, and lipids plus proteins. These organic substrates act as nucleophiles and electrophiles during the process. The breakdown of the macro- and micronutrient results into the production of the central metabolites, which are biotransformed into a mixture of a number of value-added products including aromatic compounds.

Organic compounds in the food waste act as the electrophiles and nucleophiles and the process is done by microbes to hydrolyze organic compounds to produce energy anaerobically and other products such as alcohols and fatty acids. This biotic process is identified by reduced costs, reduced energy requirement, and reduced effluent generation and can be utilized to recycle food waste into value-added or more useful products.

Types of fermentation are classified on basis of the nature of substrate used:

- Liquid state fermentation (submerged and cultured fermentation).

This is a form of fermentation in which microbes use a substrate in liquid or gel form for their growth requirements which in turn they biotransform into useful products.

- Solid state fermentation (SSF).

In this form of fermentation, the microbes utilize substrate in solid state which they degrade in absence of water to form value-added products. This is applied in manufacturing processes such as pharmaceuticals, cosmetology production textile, and fuel production. Despite its application in different fields, it has limitations such as slow growth of microorganisms on solid substrate, control of the process, and difficulty in scale-up operations.

This has merits such as

- Cost effectiveness.
- Less or no water requirement.
- Low utilization of substrate.
- Easy aeration.

8.3.3 Enzyme Hydrolysis

Various by-products are generated which are rich in carbohydrates, lipids, and proteins by food processing industries and households; these are detrimental to the environment in different aspects, but when converted into value-added products like biofuels, important organic chemicals and some ingredients can be a great deal of revenue generation and also saving the environment in the due course. The huge carbohydrate and protein content of the food wastes makes them consummate substrate for enzymatic valorization. Although the micro- and macronutrient in the waste cause high biological oxygen demand, these can be converted into marketable products through processes which include acylation, phosphorylation of carbohydrates, hydrolysis, oxidation, deamination, glycosylation, hydrogenation, and esterification of lipids (Gupta and Suhas 2009). Through those processes, food and agricultural wastes are modified and transformed into several products of great value.

8.3.3.1 Esterification and Hydrogenation

This is a reaction in which an alcohol and organic acid combine to form an ester and water while hydrogenation is the process by which hydrogen is added to unsaturated oils, fats, and other products to improve their qualities such as making oils more spreadable.

8.3.3.1.1 Applications of Esterification

- Esterification of sugars and starch to form coatings, plastics.
- Esterification of oils using alcohols in the production of biodiesel.
- Esterification of sugars to form surfactants.

The above processes require catalysts to a reasonable amount of energy input, and the amount of by-products generated depends on the substrate used. The limitations of the conventional methods can be overcome by utilization catalysts.

For instance, the enzyme lipase limits the generation of by-products like soap along the biofuels produced as the yield is used and also other merits such as cost-effectiveness and reaction at mild environmental conditions. Thus, biocatalysts are better options than conventional catalysts in the recycling of food wastes. Their immobilization further improves their recovery, stability, and economic feasibility. The major limitation of the use of enzymes is maintaining their effectiveness under dynamic environments in the reactors like:

- Elevated temperatures.

  Since enzymes are proteins in nature, their shapes can easily be distorted at higher temperatures thus limiting their use in some of the processes. Different enzymes have temperatures for optimum activity below which their activity is low and beyond which they are denatured; for example, lipase enzyme activity increases with an increase in temperature up to about 45 °C beyond which it's denatured thus a reduction in activity.
- pH.

  Different enzymes require or operate best in a certain temperature range beyond or below which their action is affected any deviation from the optimum pH range results into the change of shape of the active site; thus, substrate cannot fit according to the lock and key theory; for example, during yogurt processing, the optimum pH value must be less than 5, while for glucose isomerizes, the optimum pH falls in the range of between 7 and 8.5.
- Substrates are denaturing the enzyme itself.

  For instance, during esterification, lipase enzyme is denatured by alcohols. This can be counteracted by enzyme immobilization. This does retain not only enzyme activity but also recoverability and reutilization of the biocatalyst over a number of cycles.

The following are the immobilization techniques:

- Adsorption.
  In this technique, there is movement of the biocatalyst occurring instantaneously, and this is enhanced by the varying pH values. Despite being inexpensive and simple, it is unsuitable for processes which involve waste valorization processes.
- Entrapment, encapsulation, and cross-linking.
  This offers greater shielding and conservation to the confined enzymes (microenvironment) of the enzyme. Size of the pore to control enzyme migration and allowing substrate accessibility are critical points to consider.

8.3.3.1.2 Limitations of Immobilization

1. Decreased enzyme activity.
2. High ph sensitivity.
3. Ionic strength.

8.3.4 Using High Pressure and Temperature

The thermochemical (heating at high temperatures) of biomass is also referred to as pyrolysis and proceeds in absence of limited oxygen. The main products of the process are biofuels, gasses, and semi-coke. The quantity and quality of the products are formed based on different conditions such as process temperature, heating rate, and nature of substrate used.

8.3.4.1 Heating Rate

On the basis of heating rate, pyrolysis is divided into the following:

1. Slow pyrolysis proceeds relatively for a long time that is several minutes or more to form secondary products like oil, gas, semi-coke, and biofuels.
2. Fast pyrolysis: the process occurs in a limited time span giving rise to bio-oil and gas. This has been successfully used in the production of syngas.
3. Microwave pyrolysis: this operates in a similar way to an oven (involves dielectric heating), so there is no need to splinter waste materials; the process is limited by gasification.

8.4 Various Creative Ways from Food Waste to Product

8.4.1 Make DIY Beauty Products

Undoubtedly, food waste has adverse effects across the globe. However, this same food waste can be utilized in almost all businesses to make value-added products.

The beauty industry is one of the many industries that are making use of food waste to make some of their beauty products. Using food waste in making beauty products will not only bring profit to these brands but also contribute to the fight against global warming because when this waste is dumped in landfills, they accumulate and eventually start emitting greenhouse gasses that are like methane that destroys the ozone layer in the atmosphere which is responsible for protecting us from ultraviolet radiation from the sun.

The beauty industry has identified the potential that food waste ingredients have for developing their products such as skin care and hair care products. Food waste like manna, lemon peels, used coffee, black currant palm, etc. can be used to make dyes, skin care, and hair care products. Olive groves are equally useful as the extract from their leaves is used in gel cleanser. There are some brands that have already started utilizing food waste for manufacturing as it is a sustainable way of producing products and, at the same time, preventing the emission of greenhouse gasses thereby fighting against global warming.

Apart from making skin care and hair care products, these food wastes can be used in the making of henna, deodorants, solid perfume, and natural sunscreen.

8.4.2 Utilize the Aromas of Fruits

In the latest decades, the way natural waste was dealt with in distinct parts or areas of the globe was different. Farmers and people living in rural areas used organic waste as livestock feed and mixed it with humid substances to make compose, while on the other hand, people in urban areas dumped the food wastes in landfills where it was due for burning, therefore posing some environmental problems like air pollution, groundwater pollution, and global warming.

The concept of waste valorization is entirely dependent on sustainable technology to help reuse and recycle or else it would also just lead to more environmental problems in the world. The main idea of waste valorization is to make value-added products by converting waste into other resources. Products to be obtained could include energy, chemicals, fragrances, fibers for making textile, and other products that could benefit local and global economies.

The process of fermentation has been in use for an exceedingly long time even by ancient civilization for manufacturing of products like wine, bread, milk, etc. Enzymatic conversions gave the products suitable properties, for instance, flavor, longer shelf life, and easier digestibility. Microorganisms can be directed to work on specific substrates. The biodiversity of microorganisms can be used to our advantage not just to make foods with special and cherished aromatic notes but also to convert precursors into smaller chemicals such as fragrances and aroma compounds by the process of biocatalysis.

As we all know, fermentation is a process that is carried out by enzymes to decompose carbon-containing compounds for energy in the absence of oxygen (anaerobic metabolism). This process is classified by low power consumption, low wastewater production, and inexpensive; it can also be exploited to recycle organic wastes into value-added products.

Flavor compounds or aromas can be synthesized chemically or extracted by biotechnological processes from compounds found in foods, oils, and fragrances. This makes food waste a qualified feedstock for extraction of aromas. This is an approach that will bring about the possibility of producing additives convenient for various significance in the industrial sector. The process is environmentally friendly compared to chemical synthesis, and aromas obtained through this process can find applications in various fields including pharmaceutical industry, cosmetic industry, chemical industry, and food ingredients. These aromas can either improve the original aroma in the product or completely change it; this allows the aromas to gain immense importance from the consumer market over the acceptance of products. Recently available biotechnological processes for producing aromas utilize enzymes, microbial cultures, or less frequently plant cell cultures (Bhattacharya et al. 2008).

8.4.3 Industrial Products

8.4.3.1 Fats, Oils, and Grease to Make Biodiesel Fuel

Biodiesel is a renewable type of diesel which is gaining more significance because of the exhausting resources of fossil fuel. Biodiesel is chemically a long chain of fatty acids obtained from feedstock such as animal or vegetable oils, etc. It is produced by transesterification which is a reaction for converting fats contained in oils into useful biodiesel in the presence of a catalyst. Waste from food can be utilized by transforming it into fatty acids and biodiesel. Many microorganisms can produce microbial oils naturally which can be used as a replacement for oils from plants because they have some similarities in fatty acid composition; this qualifies them for biodiesel production.

This process of transesterification is affected by temperature of reaction, concentration of catalyst, time taken by reaction, agitation speed, purity of reactants, etc.

A report on utilization of palm oil obtained from coconut residue for biodiesel production using catalysts that are not expensive for biodiesel production was made (Joshi and Gogate 2019; Chemat et al. 2017). Palm oil residue yielded 92.7% of biodiesel using the catalyst in an open reflux method. Catalyst showed properties of high stability and was active for another cycle or more. Fuel properties were found to be consistent. In addition, biodiesel was rapidly produced from waste pepper seeds (WPS) without isolating lipids (Gupta et al. 2003; Singh and Gupta 2006).

8.4.3.2 Biofuel

Biofuel is a renewable type of fuel synthesized in a brief time that is sourced from organic matter. The feedstock material used to produce biofuel is easily recharged and therefore making it a renewable energy source, contrary to petroleum, coal, natural gas, and other fossil fuels. Biofuel is said to be eco-friendly and cost-effective alternative to crude oil and other fossil fuels, especially that petroleum has led to an increase in carbon emission which in turn has resulted in global warming and not forgetting the soaring prices of petroleum. Biofuel is produced organic matter such as agricultural, plants, and industrial biowaste. There are two most common types of biofuels which include bioethanol and biodiesel. In the world, the United States of America is the largest producer of bioethanol and biodiesel.

8.4.3.2.1 Ethanol Production

Ethanol has a diverse application in the industry sector, and therefore, its demand keeps on increasing. Ethanol can be used as a fuel and this has been proven by researches, and it is one of the products that can be produced from food waste by several processes such as fermentation and distillation. One of the uses of ethanol is manufacturing of ethylene which has a high demand on the market; polyethylene and other plastics use ethylene as a key material for their production. Bioethanol production from cheap feedstock has also gained worldwide interest. Bioethanol comes from agricultural products that have starch and cellulose like sugarcane, potatoes, and rice; it is made by fermenting the glucose that is obtained from starch in the various kinds of agricultural harvests in the presence of some catalysts. Food waste tends to be hard to hydrolyze if the raw materials have massive amounts of cellulose. Utilization of food waste, for example, municipal and food wastes, was investigated as an alternative way of ethanol creation.

8.4.3.2.2 Hydrogen Production

Hydrogen has high energy yield and is used as a compressed gas. It is produced from food waste that has a good amount of carbohydrates. Hydrogen production depends upon lots of factors which may include the composition of food and waste pretreatments.

8.4.3.3 Paper and Cellulose

8.4.3.3.1 Cellulose

For food waste to be used as an alternative to renewable resource, it needs to be purified and separate the component containing cellulose and convert it into useful energy or standard feedstock. Cellulose is produced annually in enormous quantity through the process of photosynthesis, seed hairs of the cotton plant. In some green plants, cellulose is a constituent part of the cell, of various kinds of algae. Cellulose forms the cell wall of plants, hence making it the most common organic compound on the earth. A good amount of plant material is cellulose. Wood pulp and cotton are major sources of cellulose used for industrial purposes. Industrial products like paper, paperboard, card stock, and textiles made from cotton, linen, etc. are all made of cellulose as the main ingredient (Bond et al. 2013). Other applications of cellulose include production of biofuel like ethanol, in the process of chromatography, filter bed for liquid filtration, and pharmaceutical and commercial products. To separate cellulose from the plant or lignin which is another major constituent of plant matter, the kraft process is employed.

Fiber cellulose is a major ingredient of textiles made from cotton, linen, and some other plant fibers. Cellulose is regenerated to form cellophane, which is a thin transparent sheet and turns into a particularly important fiber rayon. Smokeless gun powders also make use of cellulose as feedstock, not in an ordinary way but rather nitrocellulose, and as the main substrate for celluloid a compound that was used in the past to make movies, cellulose is also used to synthesize adhesives and binders which are used in wallpaper paste (Cane and Parra 2020; Cecilia et al. 2019).

8.4.3.3.2 Paper

Most of the agricultural harvest ends up as food waste which is either burnt or deposited in landfills, while a small part of it is what is usually taken to be food. Some companies (paper companies) have produced a way of utilizing the waste as their raw materials rather than dumping or burning which indirectly contributes to global warming. This system of paper production is an effective way of reducing deforestation and is in line with one of the sustainable development goals.

Every paper either produced in industries or traditionally is also made from fibers that can be obtained from a different number of sources. Paper fibers are mostly made from cellulose which is obtained from trees on a large scale, but with the help of the latest technology and scientific research, paper manufacturers are now able to produce paper by utilizing the cellulose from food waste of coffee, mango, tobacco, bananas, and many more (Maitan-Alfenas et al. 2015). Oranges, palm hearts, and pineapples are under studies for new fibers as well. The fibers can come from a few sources including cellulose fibers from plants like hemp, flax, cotton, cloth rags, etc.

8.4.3.4 Textile

In the fashion industry, there is a strong push to make every phase of production more sustainable. Today, about 65% of fibers that are used in the world in textile production are obtained from natural deposits such as hemp, seed hair, stem fibers, leaf fibers, husk fibers, and many more (Bhatia et al. 2015). Despite this, an immense quantity of natural fiber is wasted every year in the form of food waste. Each year at least 270 million tons of waste is generated by the banana industry alone according to what the estimates show. Food peels and stems are dumped in landfills where they are left to rot or burnt, thereby giving rise to carbon dioxide, one of the greenhouse gasses which contribute to global warming. Now, concerned individuals and companies are working together to find sustainable ways of utilizing the natural fibers found in food waste into value-added products, one of which is using them for manufacturing textile.

Circular Systems use diverse kinds of technologies to make use of food waste, one of which is biorefinery, which utilizes biomass and changes into a wide range of products including textiles (Makanjuola et al. 2020). Cash crop farmers are the targets of this system, assisting in generating additional income streams and, at the same time, making food crops to become a considerable source of fiber. Circular Systems Social Purpose Corporation (CSSPC) has developed an eco-friendly technology platform that makes use of agricultural waste. Capitalizing on the natural fibers found in plants, Circular Systems Agraloop Technology (CSAT) converts crop waste into textiles (Khan et al. 2018). Their technology focuses on these five crops-flaxseed oils, pineapple leaves, hemp seed oil, banana trees, and sugar cane bark; these crops can produce roughly 250 million tons of fibers a year. These fibers can also be used to create materials like cardboard or eco-friendly forms for packaging. Turning food waste into usable fibers does not only make use of neglected resources, but it also minimizes the environmental impact of agriculture. When tropical crop by-products and remains are left to rot, their mass decomposition produces a significant amount of methane gas. The following are some of the crops that are used in the production of textile:

8.4.3.4.1 Banana

Since the thirteenth century, Japan has been making use of bananas as a natural source of textile fiber production, and in recent years, it has gained interest from all around the globe. This banana fiber is majorly produced in the Philippines; it is also named as "Musa," and it has been proven to be one of the strongest fibers across the globe (Ravindran et al. 2017). Banana trees are pseudo-stems that sprout out of the ground. The same pseudo-stems bear leaves, flowers, and fruits. This pseudo-stem becomes a waste by-product after harvesting of fruits and cutting it down. By making use of this by-product from the banana tree, banana fiber production is highly efficient using a widely available resource and dramatically reducing waste.

For producing fiber from bananas, the fibrous part of the banana pseudo-stem is separated and dried. Separation continues after drying of the pseudo based on quality of fiber. The soft fibers are left out, while the inner part of the pseudo-stem is said to contain good fibers which are considered superior in quality. The fiber is drawn out and twisted to convert them into yarn; it may also be mixed with other fibers according to the requirement of the product to be made from the said fibers. Treatment of the yarn with chemicals carries on until later it is dyed and read for use in textile production.

8.4.3.4.2 Corn

Comparatively corn fiber was just introduced to the textile industry; hence, it is new innovation. Cargill Dow Polymers LLC developed corn fiber (Wadhwa et al. 2015). Corn plants are known for their high starch content that is why it makes a reliable source to produce fiber. The starch in the corn is converted into dextrose sugar. To manufacture polymer in corn fiber, some processes are involved such as distillation, fermentation, and polymerization of maize dextrose. Sugar is fermented and converted to a high-performance polymer called polylactide by the process of additional polymerization (Brownstein 2014; Slater 2003). This high-performance polymer can now be spun or processed into corn fiber. This fiber can be mixed with other materials like cotton wool and silk. Corn fiber shows both natural and synthetic fiber qualities. Pure corn fiber has good stain resistance, odorless, is naturally retardant to flame, and can be washed or dry cleaned; it is also known for ultraviolet radiation resistance making it convenient to be used for clothing requirements.

8.4.3.4.3 Mushroom

Some research has shown that mushroom can be used as a source for leather production and this could save more than just animals. The tinder fungus or horse fungus (*Fomes fomentarius*) lives on dead or weak trees. The fibers of the fruit bodies were originally used as "tinder" for making fiber. The fibers of tinder fungus are currently being studied and experimented with to use it as a replacement for leather. It has a marbled, velvety surface and has visual similarities to animal leather, it is a sustainable alternative, and it possesses antiseptic properties and feels amazingly comfortable because of its high air content. It will be used to make leather products such as wallets, shoes, watch straps, and many more (Khanal et al. 2007; Malik et al. 2014).

8.5 Challenges to Produce Value-Added Products Using Food Waste

8.5.1 Difference in Characteristics of Waste Food Material from Variable Sources

The sophisticated nature of different food waste materials from variable sources renders their different characteristics. The change in attics is due to the source of the food waste, process by which the food waste is produced and environmental factors.

8.5.1.1 Environmental Factors

Environmental parameters of light and temperature, maturity stages, and agricultural parameters such as nutrients, water, and minerals greatly change the nature of the food wastes making their conversion into value-added products even harder. Light and temperature greatly result in differential nutritional content even for waste of materials from the same source.

8.5.1.2 Process by Which the Food Waste Is Produced

During different production stages in food processing, food wastes are given of as by-products; these differ in quality nutrient and moisture content and value, thus requiring differential treatment methods before being incorporated into different production processes. The differential treatment methods mean an extra cost to be incurred making the process expensive.

8.5.2 Need for Densification of By-Products

Fruit and vegetable by-products are bulky, and their density is low and so the low density makes the practical conversion of these by-products (food wastes) into value-added products difficult, especially when there is a need for transportation and storage of the food wastes for a long time. Densification also facilitates controlling particle size and distribution, the scaling up, and extraction of the value-added products at the end of the process.

8.5.3 Need for Size Reduction of the Food Waste

Food wastes materials may be too large, and so there is a need to increase the surface area for efficient conversion and transformation of the food waste into value-added products. Using equipment such as crushers, mills, cutters, and shredders, the food waste material size is reduced.

8.5.4 Requirement for Physical Pretreatment

There is a need to prepare the food waste materials before being fed into convertors, techniques such as sieving and drying to reduce the moisture content; sorting may be involved.

8.6 Conclusions

A detailed study of conversion of food waste into valuable products has been discussed here.

Classification of food waste based on their sources of origin and physical nature and property is described at a large in this chapter. Various creative ways and methods of converting food waste into value-added products are explained here. The various social, economic, and environmental concerns associated with food waste management, especially in terms of greenhouse gas emission are discussed. Numerous challenges to produce value-added products using food waste have been talked over here.

References

Adam F, Appaturi JN, Iqbal A (2012) The utilization of rice husk silica as a catalyst: review and recent progress. Catal Today 190:2–14

Ahmad T, Aadil RM, Ahmed H (2019) Treatment and utilization of dairy industrial waste: a review. Trends Food Sci Technol 88:361–372

Akhtar MN, Mushtaq Z, Ahmad N, Khan MK, Ahmad MH, Hussain AI, Imran M (2019) Optimal ultrasound-assisted process extraction, characterization, and functional product development from flaxseed meal derived polysaccharide gum. PRO 7:189

Al Kamzari SM, Nageswara Rao L, Lakavat M, Gandi S, Reddy PS, Kavitha Sri G (2021) Extraction and characterization of cellulose from agricultural waste materials. Mater Today Proc (article in press)

Alexander C, Gregson N, Gille Z (2013) Food waste. In: Murcott A, Belasco W, Jackson P (eds) The handbook of food research. Bloomsbury, London, pp 471–484

Amicarelli V, Lagioia G, Bux C (2021) Global warming potential of food waste through the life cycle assessment: an analytical review. Environ Impact Assess Rev 91:106677

Aravantinou AF, Theodorakopoulos MA, Manariotis ID (2013) Selection of microalgae for wastewater treatment and potential lipids production. Bioresour Technol 147:130–134

Bhatia S, Sharma K, Dahiya R, Bera T (2015) Modern applications of plant biotechnology in pharmaceutical sciences. Academic Press, Cambridge, MA, pp 121–126

Bhattacharya AK, Naiya TK, Mandal SN, Das SK (2008) Adsorption, kinetics and equilibrium studies on removal of Cr (VI) from aqueous solutions using different low-cost adsorbents. Chem Eng J 137:529–541

Bond M, Meacham T, Bhunnoo R, Benton TG (2013) Food waste within global food systems. A global food security report. www.foodsecurity.ac.uk

Boyde S (2002) Green lubricants. Environmental benefits and impacts of lubrication, uniqema lubricants. Green Chem 4:293–307

Brownstein AM (2014) Renewable motor fuels: the past, the present and the uncertain future. Butterworth-Heinemann, Oxford

Cane M, Parra C (2020) Digital platforms: mapping the territory of new technologies to fight food waste. Br Food J 122:1647–1669

Cecilia JA, García SC, Maireles-Torres P (2019) Industrial food waste valorization: a general overview. In: Biorefinery, pp 253–277

Chemat F, Rombaut N, Sicaire A-G, Meullemiestre A, Fabiano-Tixier A-S, Abert-Vian M (2017) Ultrasound assisted extraction of food and natural products. mechanisms, techniques, combinations, protocols and applications. A review. Ultrason Sonochem 34:540–560

Gupta VK, Suhas (2009) Application of low-cost adsorbents for dye removal—a review. J Environ Manage 90:2313–2342

Gupta VK, Jain CK, Ali I, Sharma M, Saini K (2003) Removal of cadmium and nickel from wastewater using bagasse fly ash—a sugar industry waste. Water Res 37:4038–4044

Joshi SM, Gogate PR (2019) Intensifying the biogas production from food waste using ultrasound: understanding into effect of operating parameters. Ultrason Sonochem 59:104755

Khan MI, Shin JH, Kim JD (2018) The promising future of microalgae: current status, challenges, and optimization of a sustainable and renewable industry for biofuels, feed, and other products. Microb Cell Fact 17:36

Khanal SK, Montalbo M, van Leeuwen J, Srinivasan G, Grewell D (2007) Ultrasound enhanced glucose release from corn in ethanol plants. Biotechnol Bioeng 98:978–985

Maitan-Alfenas GP, Visser EM, Guimarães VM (2015) Enzymatic hydrolysis of lignocellulosic biomass: converting food waste in valuable products. Curr Opin Food Sci 1:44–49

Makanjuola O, Arowosola T, Du C (2020) The utilization of food waste: challenges and opportunities. J Food Chem Nanotechnol 6(4):182–188

Malik A, Erginkaya Z, Ahmad S, Erten H (2014) Food processing: strategies for quality assessment. Springer, Berlin

Mazloumian A, Rosenthal M, Gelke H (2020) Deep learning for classifying food waste. Cornell University, Ithaca, NY, pp 1–5

Northern Ireland Environmental Agency (2002) Environment (Northern Ireland) Order 2002 (S.I. No. 3153 (N.I. 7) of 2002)

Nitayavardhana S, Rakshit SK, Grewell D, van Leeuwen J, Khanal SK (2008) Ultrasound pretreatment of cassava chip slurry to enhance sugar release for subsequent ethanol production. Biotechnol Bioeng 101:487–496

Ravindran R, Jaiswal S, Abu-Ghannam N, Jaiswal AK (2017) Evaluation of ultrasound assisted potassium permanganate pretreatment of spent coffee waste. Bioresour Technol 224:680–687

Saboya PP, Bodanese LC, Zimmermann PR, Gustavo AD, Macagnan FE, Feoli AP, Oliveira MD (2017) Lifestyle intervention on metabolic syndrome and its impact on quality of life: a randomized controlled trial. Arq Bras Cardiol 108:60–69

Sindhu R, Gnansounou E, Rebello S, Binod P, Varjani S, Thakur IS, Nair RB, Pandey A (2019) Conversion of food and kitchen waste to value-added products. J Environ Manage 241:619–630

Singh AK, Gupta AK (2006) Metalworking fluids from vegetable oils. J Synth Lubr 23:167–176
Slater K (2003) Environmental impact of textiles, 1st edn. Woodhead Publishing, Cambridge
Somers MH, Azman S, Sigurnjak I, Ghyselbrecht K, Meers E, Meesschaert B, Appels L (2018) Effect of digestate disintegration on anaerobic digestion of organic waste. Bioresour Technol 268:568–576
Wadhwa M, Bakshi MP, Makkar HP (2015) Wastes to worth: value added products from fruit and vegetable wastes. CABI Rev 2015:1–25
Wang S-L, Liang T-W, Yen Y-H (2011) Bioconversion of chitin-containing wastes for the production of enzymes and bioactive materials. Carbohydr Polym 84:732–742

Chapter 9
Food Waste Materials for Bioenergy Production

Shraddha Awasthi, Ambneesh Mishra, Rajeev Singh, and Dan Bahadur Pal

Abstract Globally, one-third of food production each year is either lost or deteriorated. Management of this wasted food is essential as it will discontinue the symmetry of the environment and majorly affect public health. On the other side, fossil fuel exploitation is at an all-time high, which results in increased carbon emissions, global warming, and climate change. As a result, converting food waste into energy sources not only makes money but also addresses the issues of trash management and the overuse of fossil fuels, boosting environmental viability. However food waste is a potential source; no such things are required to convert it into valuable products. The chapter discussed several conversion technologies like fermentation, incineration, pyrolysis, etc. An in-depth analysis of the most significant bioenergy sources including biofuel, biohydrogen, biogas, etc. is also mentioned in this chapter. The advantages and potential uses are determined.

Keywords Food waste · Conversion technologies · Anaerobic digestion · Incineration · Pyrolysis · Gasification · Biogas · Biofuel · Biohydrogen

9.1 Introduction

The usage of fossil fuels as energy sources is dwindling. Due to the excessive use of nonrenewable energy resources and GHGs release, there is an increasing need for more energy, and environmental contamination is getting worse (Zhang et al. 2011; Gurung et al. 2013). The subject of global warming is no longer novel to humanity.

S. Awasthi (✉) · A. Mishra
Department of Environmental Science, Ramanujan College, Delhi University, New Delhi, Delhi, India

R. Singh
Department of Chemistry, Atma Ram Sanatan Dharma College, Delhi University, New Delhi, Delhi, India

D. B. Pal
Department of Chemical Engineering, Harcourt Butler Technical University, Kanpur, India

D. B. Pal, A. K. Tiwari (eds.), *Sustainable Valorization of Agriculture & Food Waste Biomass*, Clean Energy Production Technologies,
https://doi.org/10.1007/978-981-99-0526-3_9

These are now some of the issues people facing today. Due to the ongoing usage of fossil fuels, contamination of the soil, air, and water has increased. It has compelled humanity to seek alternative fuel sources that would not harm the environment when utilized or combusted (Osuagwu 2014). The accumulation of food waste, especially trash from kitchens, canteens, restaurants, and food-processing facilities, has become a global issue (Ren et al. 2017). Approximately 1.3 billion tons of food, comprising fruits, bakery goods, bread, vegetables, dairy products, and meat, are wasted annually along the global food supply chain, according to FAO (2011, 2012). Every year, 1.4 billion hectares of arable land—28% of the total area used for agriculture—are used to produce food that is lost or squandered. Aside from wasting food and land resources, it is the prediction that food waste adds around 3.3 billion tons of CO_2 to the atmosphere annually, which plays a considerable role in contributing to the emissions of greenhouse gases (GHGs). Earlier, this food waste that is a part of municipal solid waste was burned or discarded in an open environment, which might have substantial cynical effects on human health and the environment (Talyan et al. 2008; Agarwal et al. 2005).

Conventionally, the trash food and other combustible municipal garbage to produce heat or electricity. The food waste contains a high degree of moisture that could produce dioxins when burned with other wastes with low humidity and high calorific values (Katami et al. 2004). Moreover, smoldering food waste can contaminate the air and disrupt its bioactive components. These indicate that appropriate food waste management is required to advance (Ma et al. 2019).

9.2 Defining Food Waste and Its Global Trend

9.2.1 Defining Food Waste

The term "food waste" refers to the loss of food at the end of the food supply chain, along with the loss of resources like labor, water, energy, and land, as well as suppliers and consumers (FAO 2011). Although "food waste" does not possess a definite meaning, it is essential to consider how different factions perceive the term. For example, "food waste" and "food loss" are referred to as the same thing by Ostergren et al. (2014) and Bellemare et al. (2017), neither of which establishes a distinction between the terms. The terms "food waste" and "food loss" will be used interchangeably because there is no explicit distinction made between the two in the literature (Coleman-Jensen et al. 2012). Contrarily, food waste was defined by Dou et al. (2016) as the quantity of food produced and then abandoned at any point along the food supply chain. According to Parfitt et al. (2010), "food waste" is defined as food spoilage at the end of the food distribution chain (i.e., retail, consumer) (Gustavsson et al. 2011). Nevertheless, Gustavsson et al. (2011) proceed by defining food waste as any product designed for human consumption that is not eventually consumed, including items that get utilized for other activities (e.g., animal feed, industrial by-products).

The Food and Agriculture Organization of the United Nations (FAO) defines "food loss" and "food waste" in the context at present (2019). According to the FAO, "food waste" originates at the consumer, and retail level "food loss" happens along the food supply chain up to but excluding the retail level.

9.2.2 *Global Trends*

Excessive food waste is absolutely an outcome of regional disparities in agricultural production, distribution, and consumption. Perpetual waste production and the migration of people from rural to urban areas have cumulative effects. Academics predict that by 2050, 68% of the world's population will live in cities, leaving only 30% of the community capable of generating the enormous quantities of fruits, vegetables, animal products, and so on that are necessary for themselves and the urban population (UN 2018).

9.2.2.1 United States of America

A contemporary analysis of the National Resources Defense Council, 40% of food manufactured in the United States of America gets misplaced during processing and transportation to stores, restaurants, and consumers (NRDC 2012; Gunders 2012). About one-fifth of the municipal solid waste at landfills from across the country is primarily composed of discarded food (United States Environmental Protection Agency 2011).

9.2.2.2 Japan

Even Japan, one of the most advanced economies, had one of the worst rates of food waste in the world in 2016, with 27.59 million tons (business waste: 19.70 million tons; home waste: 7.89 million tons) (MAFF 2020).

9.2.2.3 The United Kingdom

Similar trends are noticed in the United Kingdom, which annually discards around 30 and 40 per cent of its food production (Kosseva 2009).

9.2.2.4 South Africa

As per Oelofse and Nahman (2013), food waste of nine million tons weight is thrown away in South Africa annually.

9.2.2.5 Singapore

An analysis by Singapore's National Environmental Agency says the thickly urbanized, commercial city of Singapore generated 542,720 tons of food waste in 2006 and roughly 703,200 tons in 2012 (NEA 2012).

9.2.2.6 European Union

A report by the European Commission (2010), Food waste in the European Union is forecasted to rise from 89 million tons to 126 million tons in 14 years, i.e., between 2006 and 2020. The sector involved in processing of food in Europe generates enormous amounts of aqueous waste each year. According to a study by Kosseva (2011), these wastes comprised surpluses and discarded fruit and vegetable commodities, bagasse and molasses from sugar refining, skeletons, blood and flesh from preparing fish and meat, stillage and by-products which are generated from distilleries, wineries and dairy wastes breweries such as cheese whey, and wastewaters from washing, blanching, and cooling operations.

9.3 Methods/Technologies for Conversion of Food Waste to Energy

9.3.1 Anaerobic Digestion

The remediation of waste with a high moisture content could be accomplished via anaerobic digestion, one of the most efficient and sustainable waste management techniques. Microorganisms break down biodegradable materials through a series of biological processes called anaerobic digestion without the presence of oxygen. In landfills, organic waste is anaerobically digested and results in biogas production, which primarily consists of methane and carbon dioxide and other gases like nitrogen, oxygen, and hydrogen sulfide are also generated in small amounts, which evade the atmosphere as a result of which the ecosystem is impaired (Zhu et al. 2009). Research by Chanakya et al. (2007) as well as by Guermoud et al. (2009) says that a similar process can transform organic wastes into usable products like biofuels which is known as biogas and nutrient-rich leftover that is applied as fertilizers under regulated circumstances without oxygen.

The technology is pricier, ending with less residual waste, and eventually the food waste ends up to a rational source of energy the statement was practically proven by Nasir et al. (2012) and Morita and Sasaki (2012). The biological mechanisms include hydrolysis, acidogenesis, acetogenesis, and methanogenesis (Supaphol et al. 2011; Christy et al. 2014).

9.3.1.1 Hydrolysis

In the initial stage, substrates are hydrolyzed to create soluble molecules. To carry out hydrolysis reactions, extracellular enzymes referred to as hydrolase are used. Esterase, glycosidase, and peptidases are some examples of hydrolases (Elefsiniotis and Oldham 1994). The polymers are converted into oligomer or monomeric molecules through hydrolysis. The given equation depicts the synthesis of glucose molecules by starch hydrolysis. Oligosaccharides and monosaccharides are the breakdown products of polysaccharides. Peptides and amino acids are synthesized from proteins, while glycerol and fatty acids are produced from lipids (Dahiya et al. 2015).

$$nC_6H_{10}O_5 + nH_2O \rightarrow C_6H_{12}O_6$$

According to Mittal (1997), depending on the substrate type, bacterial concentration, pH range, and bioreactor temperature, the hydrolysis rate is often slower under anaerobic conditions than the rate of acid formation. The rate of hydrolysis is further influenced by considerable factors like pH, enzyme synthesis, and enzyme adsorption on substrate particles. According to Bryant (1979), the anaerobe genera *Streptococcus* and *Enterobacter* are in charge of hydrolysis.

9.3.1.2 Acidogenesis

The fermentative bacteria result in conversion to hydrolyzed soluble molecules into volatile fatty acids, lactate, alcohol, and CO_2 in the second phase (Silva et al. 2013). Acetic acid, propionic acid, and ethanol are the primary by-products in this phase (Zhou et al. 2018). Facultative anaerobic bacteria use carbon and oxygen during acidification to generate an anaerobic environment necessary for methanogenesis. In phase two, whenever a group of bacteria transforms the substrates into organic acids, the monomers produced in phase one serve as substrates for the microbes. Direct utilization of acetate, hydrogen, and carbon dioxide can result in the production of methane. However, syntrophic acetogenic bacteria subsequently deteriorate propionate, butyrate, valerate, and isobutyrate to produce acetate and hydrogen (Mittal 1997; Schink 1997).

9.3.1.3 Acetogenesis

VFAs and alcohols are metabolized into acetic acid and H_2 via acetogenic bacteria during the acetogenesis process (Zinder 1990). The products of the acid phase are altered into acetates and hydrogen using acetogenic bacteria of the genera *Syntrophomonas* and *Syntrophobacter*. While leveraging hydrogen as an electron source to eliminate carbon dioxide, a small proportion of oxalate molecules obtain.

Pathogenic organisms will continue to use acetates in subsequent processes. However, microbes get restricted by the hydrogen produced during the process. Because of this, hydrogenotrophic methanogens remove the hydrogen by using it to generate methane live in syntrophic interaction with acetogenic bacteria in anaerobic digesters. Additionally, because 70% of methane production takes place when acetate declines, the acetogenesis phase represents the effectiveness of the biogas production process. During the procedure, 11% hydrogen generates synchronously (Schink 1997).

$$nC_6H_{12}O_6 \rightarrow 3nCH_3COOH$$

9.3.1.4 Methanogenesis

Methanogenic bacteria produce CH_4 and CO_2 from the results of acetogenesis at the end of the process, which itself is known as methanogenesis (Eryasar and Kocar 2004). Methanogens, a group among *Archaea*, are accountable for this process. Methane can be generated by reducing carbon dioxide or by fermented acetic acid. Acetic acid, hydrogen, and carbon dioxide, the by-products of the previous step, thereby serve as a precursor for the synthesis of methane. Only 30% of the methane generated in this process is the result of methanogens' carbon dioxide reduction (Griffin et al. 1998; Karakashev et al. 2005).

Two distinct types of methanogens can produce methane in two different ways:

(a) Acetoclastic methanogens that convert acetic acid to methane.
   $CH_4 + CO_2$ from CH3COOH.
(b) Hydrogenotrophic methanogens that disintegrate carbon dioxide while using hydrogen.
   $CO_2 + 4H_2O \rightarrow CH_4 + 3H_2O$.

9.3.2 Ethanol Fermentation

A distinct method for transforming waste into energy is essential for ethanol generation from wasted food. Various food wastes as banana peel by Oberoi et al. (2011b) sugar beet pulp used pineapple waste, grape pomace (Socas-Rodríguez et al. 2021), potato peel waste, citrus waste (Oberoi et al. 2011a; Pourbafrani et al. 2010) as well as cafeteria food waste and household food waste by Matsakas et al. (2014) are used for the production of bioethanol. Due to the complicated structure of the lignocellulosic part of food wastes, numerous pretreatment procedures, comprising alkali, acid, enzymatic treatments, and thermal, came to be used to enhance digestibility of cellulose (Arapoglou et al. 2010; Vavouraki et al. 2013). The extensively used technique for pretreatment step during synthesis of ethanol from food waste is probably enzymatic hydrolysis Moon et al. (2009) can use carbohydrates and amyloglucosidases to produce 29.1 g/L of ethanol from food waste.

Even though the pretreatment can speed up the synthesis of ethanol by making cellulose more digestible, the soluble sugars can break down and produce inhibitors like furfural, especially when the treatment is carried out with alkali under extreme conditions (Matsakas et al. 2014). Even though Ban-Koffi and Han (1990) also employed *Zymomonas mobilis* another scientist used *Pichia rhodanensis*, *Saccharomyces cerevisiae* was typically used for fermentation. According to Balat (2011), the limitation of *S. cerevisiae*'s ability to utilize just hexose sugars is a drawback, in spite of the fact other organisms which perform fermentation can be drawn on ethanol synthesis using pentose carbohydrates. In a study by Kim an Kim (2011), production of 0.43 g ethanol/g TS and 0.31 g ethanol/g TS respectively was procured for the process of separated hydrolysis and fermentation in addition simultaneous saccharification and fermentation. Consequently, ethanol produced from food waste might just be estimated to possess an amount of energy in average 8.3–11.6 kJ/g TS based on the 26.9 MJ/kg energy of ethanol content.

9.3.3 Incineration

Incineration is the controlled direct burning of waste at temperatures of 800 °C and more than it in the environment having oxygen, releasing energy in the form of heat, gases, and inert ash. The density and makeup of the waste determine the net energy yield. Factors include the proportion of inert materials and moisture that contribute to heat loss, the ignition temperature, the size and the shape of the elements, the layout of the combustion system, etc. In reality, 65–80% of the energy contained in the organic matter can be recovered as heat energy, which can be used for direct thermal applications, to run steam turbines for energy production, or to heat up process streams in industrial settings using heat exchangers (Autret et al. 2007; Stillman 1983; Patil et al. 2014). Combustion, or confined and regulated burning, can lower the amount of solid waste treated in landfills while simultaneously recovering energy from the burning of garbage. Eliminating the requirement for energy from fossil sources and landfill methanogenesis generates a renewable energy source and lessens energy consumption (US-EPA 2022).

The solid waste that needs to disposed its volume can be drastically reduced using incinerators, which can lessen the volume by 80–85%. Although the incineration of solid wastes is an ancient technique, certain European Member States still don't fully acknowledge it as a legitimate waste management option. Due to the dioxins and heavy metals released into the air by antiquated machinery and technology, several nations are reluctant to rely on garbage incineration (Katami et al. 2004). As a result, garbage combustion received a bad reputation and even became outlawed in some nations. A new generation of facilities can also be manufactured that significantly adhere to a stricter environmental regulatory regime because of advancements in air pollution control technology, drastically reducing potential negative health consequences for humans (WHO 2007). The advancements in combustion technology made it possible for numerous new plants to build in various nations (Grosso et al. 2010). The installation of the energy recovery portion increased the importance of these plants. Solid waste burning was regarded as an element of energy plans that aimed to lessen reliance on fossil fuels by recovering heat or providing electricity.

However, a very less literature is there which says that concentrate on energy recovery by direct methods from food waste by incineration process. Due of high moisture content and fire-resistant contents, organic waste alone seems unsuitable for incineration (Mardikar and Niranjan 1995). It is not always feasible to obtain energy through the incineration of solely food wastes since it requires heat in evaporation of water available in this biowaste. Caton et al. (2010) investigated the possibility for energy recovery by the direct burning of food waste after use of consumers. They used the heat loss from combustion systems, which was released by the exhaust or the buildings of the devices, to dehydrate the food waste. The findings demonstrated that recovering energy from food waste could provide cost reductions by offsetting traditional fuel use and lowering disposal expenses.

9.3.4 Pyrolysis

Pyrolysis is an endothermic process that employs thermal degradation to produce biogas, charcoal, and bio-oil from pretreated (dry) food waste or bio-resistant digestate. At industrial pyrolysis temperature (500 °C), information on energy distribution yield and results from characterization out of these products was previously presented (Opatokun et al. 2015a, b). During pyrolysis, the heating rate can be defined to affect the product selection. Fast heating rates result in higher yields of bio-oil or biogas, whereas slow heating rates guarantee a higher yield of biochar.

As a result of the waste's degradation during the pyrolysis process, syngas, pyrolysis oil or tar (condensable hydrocarbon vapor that emerges from the solid matrix as gas and liquid in the form of mist), and char (the final leftover solid in nature and devolatilized) are all procured (Knoef 2005). Pyrolysis oil, also known as bio-oil, can be transformed into transportation fuels and highly specialized chemicals or used as a renewable industrial fuel to produce heat and electricity. In addition to being a solid fuel source, biochar has recently utilized to enhance agricultural land. In most cases, the non-condensable gases are recycled into the process to provide process heat and extra fuel for dual-fuel generator systems (W2E 2022). The generated gas is anticipated to be cogenerated (CHP) into electricity, with the resulting heat earmarked for the pretreatment of food waste (Opatokun et al. 2017).

9.3.5 *Gasification*

In gasification process, organic wastes or biomasses are heated, typically in the range of 500 and 1000 °C, in the partial presence of oxidative atmosphere in the presence of gasifying agents such as steam, air, or carbon dioxide (Jeguirim and Khiari 2022). Through partial oxidation at a high temperature (more than 700 °C), gasification converts food waste into simulated gas which is a combination of CO, CO_2, H_2, N_2, and CH_4. All the chemical properties of syngas must be determined before food waste is subjected to gasification, which forbids its direct use in internal combustion engines (Vimala Ebenezer et al. 2020).

Hydrogen-rich syngas is produced in gasification process which is very advantageous, which may be used as a fundamental requirement for developing valuable goods, like chemicals and fuels, under a controlled proportion of oxygen (Young et al. 2017). The ability to combine the operational conditions along with the characteristics of the particular reactor to gain a syngas which is worthy to execute in various executions are two main benefits of gasification over traditional incineration technology (Arena 2012). Additionally, there are evident benefits to the gasification process when employing steam to prevent the production of dioxins in current incineration facilities. Furthermore, by requiring little to no oxygen, these processes help reduce emissions of air pollutants.

9.3.6 *Hydrothermal Carbonization*

One thermal conversion method receiving more interest from researchers is hydrothermal carbonization (HTC), especially for waste streams having moisture content in very high amount (80–90%). To produce coal from cellulose, HTC was first experimented in 1913 (Libra et al. 2011). HTC is a wet process that works in autogenous pressures and relatively low temperatures (180–350 °C) to transform food wastes into a profitable, energy-rich resource (Berge et al. 2011).

HTC provides several advantages over alternative waste-to-energy conversion techniques that use biological processes, including less treatment spoor, better reductions of waste volume, and no odor related with process (Li et al. 2013). This process leads to the development of highly activated carbon and an energy-densified substance known as hydrochar, whose composition has been compared to that of lignite coal (Berge et al. 2011). According to Liu et al. (2010) and Parshetti et al. (2014), the patterns of surface functionalization of produced char make it corrigible to advantageous eventual uses such as an adsorbent for detrimental pollutants, stuff for carbon fuel cells as well as soil improvement (similar to char from pyrolysis/gasification) (Spokas and Reicosky 2009).

9.4 Various Energy Sources Produced from Food Waster

9.4.1 *Biogas*

An energy source which is renewable, lower manufacture costs, with very less generation of leftover trash make biogas production among the most favorable methods for managing waste which is organic (Kiran et al. 2014). Additionally, the digestate manufactured during anaerobic digestion is a nutritious substance which is applied as a biofertilizer and soil conditioning agent. Anaerobic digestion, the process by which microorganisms break down organic materials such as plant and animal products without the presence of oxygen, results in the production of biogas. These organic wastes are recycled by anaerobic digestion in biogas systems, creating biogas, which comprises energy (gas), and efficient soil by-products (liquids and solids) (EESI 2017).

Anaerobic fermentation, a complex metabolic conversion process, produces biogas from food waste. Fermentation of the anaerobic process is classified into three steps: first, the substrate—which includes fat, protein, and carbohydrate—is hydrolyzed by bacteria into organic units which are simple in structure like fatty acids, amino acids, and sugars and; second, by the action of acidogenic bacteria, the substrate is acidified into lesser fatty acids like propionic acid and butyric acid; and third, hydrogen gas and acetic acid obtained. Eventually, methane is produced from acetic acid and hydrogen in the presence of methanogenic bacteria. Food waste is categorized into organic waste that adapts itself well to anaerobic fermentation. One-phase and two-phase fermentation systems can be distinguished based on the fermentation process (Zhang 2017). Since the process is driven by a mixed culture of bacteria in the absence of oxygen, it doesn't require significant energy expenditures. The biogas fuel can be collected and burned to provide power, heat, or both. Biogas is regarded as carbon neutral because, unlike fossil fuels, which have been storing carbon for millions of years, all of the carbon released during combustion has just recently been removed from the atmosphere through photosynthesis. As a result, biogas is a greener fuel than natural gas. The remaining nutrients (nitrogen, phosphorus, and micronutrients) are kept in the affluent because anaerobic digestion only releases carbon to the gas phase. The effluent from anaerobic digestion serves as organic fertilizer of very high quality and soil modification since it only releases carbon to the gas phase and leaves the other nutrients (nitrogen, phosphorus, and micronutrients) behind (Wilkie 2008).

9.4.2 *Biomethane*

Biomethane referred to as "renewable natural gas" is a nearly pure form of methane that can be created by either "upgrading" biogas (a procedure that eliminates any CO_2 and other pollutants present in the biogas) or by gasifying solid biomass and then methanation it (IEA 2020). The solid by-products of enzymatic hydrolysis and fermentation of food waste were used to fuel the anaerobic digestion process, which has been chosen to produce biomethane. This is because the procedure has been demonstrated to be one of the most effective ways to extract and transform any leftover essential components into energy (Bampenrat et al. 2021). Due to cheaper in cost, low creation of leftover trash, and use as a inception of renewable energy, the biogas formation, in particular, methane produced by anaerobic processes, is a viable option for waste control (Morita and Sasaki 2012; Nasir et al. 2012). The digestate having nutrient richness that is obtained is used as fertilizer or soil conditioner in addition to producing biogas. Fruit and vegetable wastes anaerobic digestion which is a two-stage process was studied (Mata-Alvarez) (Viturtia et al. 1989), and 95.1% of the volatile solids were converted, yielding 530 mL/g of volatile solids of methane. In a study by Lee et al. (1999), food waste was turned into methane using a continuous digester fed with volatile solids, producing a methane yield in the process. On the basis of source of the wastes, Gunaseelan (2004) observed that 54 distinct fruit and vegetable wastes had methane generation capabilities ranging from 180 to 732 mL/g volatile solids.

Although the words "biogas" and "methane" are sometimes used interchangeably, biogas is the direct by-product of a methanization unit. It is created when organic waste in the gut ferments. It is made up of 40% CO_2 and 60% methane. It is capable of producing both heat and electricity simultaneously. Biomethane is biogas that has had water, carbon dioxide, and hydrogen sulfide removed. Biomethane has the same properties as natural gas as a result of that purification procedure and can be put into the grid. By increasing the percentage of green gases in our gas consumption, biomethane production supports the energy transition and lowers greenhouse gas emissions (Terega 2022).

9.4.3 Bioethanol

Another prominent biofuel is bioethanol. Bioenergy, a type of liquid biofuel created by digesting and converting biomass raw materials, includes bioethanol. Bioethanol has no sulfur and is safe for the environment. A research by Mannberg et al. (2014) says that pure ethanol automobiles emit far less carbon dioxide (CO_2) than comparable gasoline vehicles. Ethanol from biofuels is also a kind of renewable energy. Global pressure to reduce carbon emissions is growing, fossil energy resources are becoming rare, and the condition of oil security is getting worse. When nations attempt to make a switch from dirty, high-carbon fuel to clean, low-carbon fuel, bioethanol emerges as the top option. Global production of ethanol-based biofuels has expanded in the twenty-first century. Currently, the United States, Brazil, the European Union, China, and Canada are the leading nations as well as areas that support use of biofuel ethanol globally. The nations that have the biggest biofuel ethanol industries among them are the United States and Brazil (Adewuyi 2020, pp 77–88).

As food waste also contains protein in addition to carbs, these ingredients are frequently used in the industrial fermentation of yeast to produce bioethanol. Fuel ethanol, which may be made from refined ethanol and added to fuel oil, is crucial for addressing the energy crisis (Wang 2015). Food waste is mainly classified to liquid and solid components. The liquid gets divided into wastewater and lipid. After being crushed, a solid component is transformed into ethanol in two different methods. In the first approach, the solid contents are hydrolyzed in an environment which is low-acid in nature; the pH is adjusted up to approximately 5.5, after that yeast in powder form is added for fermentation. To control PH, the further approach involves sterilizing solids in a microwave reactor before adding yeast in powder form for fermentation. Subsequent to fermentation, liquid separation is used to recover the oil layer. For crude ethanol, the fermentation broth is distillate. The ethylene direct water technique, which can be separated into a liquid-phase method and a gas-phase method, makes up the majority of the commercial synthesis of ethanol. The gas phase process is currently used to create the majority of ethanol. The principle is that with the action of phosphoric acid and catalyst which is solid, the ethylene gas reacts with water (Guo 2015).

9.4.4 Biobutanol

Air pollution issues like haze and acid rain are getting more significant, and people are paying greater attention to them as a result of the global energy sparsity along with combustion of various nonrenewable fuels. Biobutanol is classified among raw constituents which are chemical and green fuels that come under the second generation of biofuels. During the actual combustion, no hazardous gases are produced. Another renewable fuel source is biobutanol. Similar to bioethanol, biobutanol comes under biofuel. Alike bioethanol in terms of raw ingredients and manufacturing methods, it is corrosive and has a low steam pressure when combined with gasoline. It also has a high tolerance for impure water. It gains a higher mixing ratio with gasoline than the presently offered biofuels without demanding vehicle modification. As a result, it also underscores the significance of biological butanol research and development (Stoeberl et al. 2011).

Similar to how ethanol is manufactured, butanol can also be obtained by fermentation. Butanol production is more expensive than ethanol because it utilizes facilities for evaporation, heating, cooling, and other processes, as well as higher initial investment. Therefore, enhancing the raw components' conversion to butanol and hastening the process of conversion which is crucial for the commercialization of biobutanol. It is controlled by the synthesis of high-performance biocatalysts and the design development of the manufacturing process. A study in 2017 was done by Shi et al. Butanol yield vary depending on the substrate's composition and content when using different raw materials. The generation of butanol is significantly impacted by the use of various strains (Shi et al. 2017). *Clostridium acetobutylicum*, *Clostridium beijerincki*, *Clostridium saccharobutylicum*, and *Clostridium saccharoper-*

butacetonicum are the four types of clostridiums acetobutanol that can create butanol in industry. Among them, *Clostridium acetobutyricum*, also known as clostridium starch degrading, is primarily made of cereal, corn, and different starch components. The other three species, sometimes termed saccharified clostridium, are typically fermented with cellulose hydrolysate, molasses and different sugar sources (Wang et al. 2019).

9.4.5 Biodiesel

The energy crisis and environmental degradation have grown to be two of the biggest issues facing humanity as we enter the twenty-first century. The imbalance between the supply and demand for oil is getting worse. The study of new alternative energy is now the most crucial endeavor. The demand for oil is rising quickly along with the economy's rapid development, and the supply-demand mismatch is widening. The widespread use of fossil fuels has increased environmental pollution, which is now one of the biggest problems facing the entire planet. In this setting, the globe is extending the progress of petrochemical fuel alternatives, with biodiesel garnering the attentiveness of all countries due to its superior eco-friendly nature. Biodiesel is manufactured by esterification or transesterification of replenishable oil sources (such as animal and vegetable oil, microbial oil, and used cooking oil), a type of liquid fuel (Dong 2020).

The raw ingredients for biodiesel can be exploited by any source of fatty acids, such as flora or fauna fats. Biodiesel's source materials are therefore highly diverse. The preparation processes for biodiesel can be further broken down into physical methods and chemical methods depending on the primary physical change or chemical reaction that occurs during the process. High-quality biodiesel is however still produced via chemical processes (Ishola et al. 2020).

(a) The foremost technique is Physical process, which entails modifying the oil's physical characteristics, such as its viscosity and freezing point, by emulsification and dilution in such a manner that it can fulfill the fundamental needs of fuel oil. Vegetable oil and diesel oil have equal calorific values and phase formation, but the prominent consistency of vegetable oil causes nozzle blockage when being used. Mixed dilution and microemulsion were utilized to reduce viscosity. These techniques can lessen the heaviness of oils, but they have major consequences for engine carbon build-up and are susceptible to nozzle choking. So, high-quality biodiesel cannot be produced physically (Ishola et al. 2020).
(b) The chemical procedure is the second technique. Transesterification is the process by which glycerides combine with tiny alcohols in the presence of a catalyst to create new esters and alcohols. The two primary components of oil used in biodiesel production are triglycerides and short-chain alcohol, which, when combined with a catalyst, produce biodiesel and glycerin. Transesterification is currently the best method for making biodiesel from

vegetable and animal oil because of its quick reaction time, moderate reaction environment, high conversion rate, and cost-effectiveness. Transesterification is a process utilized in the biodiesel business today. Utilizing a catalyst can speed up the chemical process and enhance biodiesel production (Ishola et al. 2020).

Hydrodeoxygenation cracking is an additional technique. Under the influence of a catalyst, animal and vegetable lipids get transformed to fuels which are liquid having 15–18 carbon chain length. Glycerides can create hydrocarbons which have improved performance at low temperature, high temperatures along with pressure through hydrodeoxygenation process. But the economic situation is dire, the reaction circumstances are severe, and the energy consumption is maximum (Ishola et al. 2020).

9.4.6 Biohydrogen

The development of sustainable and ecologically friendly energy has received more attention during the last few decades (Balat 2011; Demirbas 2008). Since it has a higher mass-energy density than other fuels and doesn't release any greenhouse gases when it burns, hydrogen has been termed the "perfect" fuel (Hosseini et al. 2015; Hallenbeck and Liu 2016). With a high energy production of 142.35 kJ/g, biohydrogen (H_2) is used as compressed gas. For the synthesis of H_2, carbohydrate-rich FW is appropriate. Several biotechnological procedures can turn food waste into biohydrogen (H_2). These procedures include photo-fementation and dark fermentation combined with AD, two-stage H_2CH_4 fermentation, and one-stage H_2 fermentation (Alibardi and Cossu 2016). According to a report by Tawfik et al. (2011) and Alibardi et al. (2014). Due to its low energy requirements, dark fermentation is used in the biorefinery idea as the primary method for producing H_2 from food waste (). Dark fermentation produces H_2 more quickly than photo-fermentation. Likewise, using readily accessible feedstock at a low cost, such as food waste, could increase the process's financial advantages. Utilizing food waste to create H_2, therefore, addresses both the waste problem and the need for sustainable energy.

However, several parameters, including the need for pretreatment procedures, raw material type, inoculum source and variety, reactor architecture, heat, and the accessibility of micronutrients, limit the conversion of food waste to hydrogen. The yields of H_2 are also influenced by several other elements, including moisture content, organic matter, nutrient concentration, particle size, chemical oxygen demand (COD), and compo stability (Zhang et al. 2007a, b, c).

9.4.7 Syngas

The food waste is gasified to create syngas, a useful synthesis gas. The distinction between gasification and incineration is the syngas produced during gasification.

The waste is used as a feedstock for a high-temperature chemical conversion process during the gasification process, not as a fuel. The syngas generated by gasification gets converted into higher value commercial products such as transportation fuels, chemicals, fertilizers, and even substitute natural gas, as opposed to only producing heat and electricity, as is done in a waste-to-energy plant utilizing incineration. This cannot be accomplished by incinerating (GSTC n.d.).

9.5 Advantages of Conversion of Food Waste to Energy

9.5.1 Reduction of Waste at Landfills

Many individuals comprehend the benefits of not disposing of plastic garbage in landfills. After all, it takes hundreds of years for it to degrade or break down during its whole existence. However, a lot of individuals may not understand how food waste poses issues when it gets dumped in landfills. There is more to food spoilage than just that; it eventually vanishes. Food that disposes of in landfills will decay and eventually disappear completely. Gases that are dangerous to our environment will be produced throughout the entire procedure. The biggest issue here is methane. Compared to CO_2, this gas is more powerful. Therefore, methane continues to enter the atmosphere when decomposing food is piled on top of it in a landfill (Adhikari et al. 2006).

Therefore, if we stop throwing our food waste in landfills, the harm will be lessened. Additionally, landfills are an eyesore in our countryside. They harm our environment and emit terrible odors. In general, expanding landfills and their utilization is not a sustainable course of action. Instead, we should strive to lessen the amount of waste that would otherwise end up in landfills (TRVST 2022).

9.5.2 Reduction in Carbon Emission

We can approach this in a variety of ways. Traditionally, garbage trucks pick up our rubbish from our houses when we dispose of food waste. After that, this will be delivered to landfills. To make this all work, more machinery will pump more CO_2 emissions into the atmosphere. This cycle of collecting waste, moving it, and disposing of it contributes to the greenhouse impact. Therefore, we can lower emissions if we take our food waste to processing facilities that generate their energy. Of course, garbage transportation is still necessary, but many facilities for processing food waste are self-sufficient. Here is where we can reduce CO_2 emissions (TRVST 2022).

9.5.3 *Electricity and Heat Production*

Traditionally, we have produced power using hydrocarbon deposits and unreplenishable resources. We will sooner or later come to an end of coal and other fossil fuels if we continue on our current use trajectory. We can move away from using nonrenewable sources of energy when we generate energy from food waste. As a by-product of processing food waste, the renewable natural gas that is stored can assist in producing heat and energy. As a result, we can use renewable resources to power homes with heat and electricity (TRVST 2022).

9.5.4 *Cost-Effective*

While there is a cost associated with turning food waste into energy, it may ultimately result in lower fuel expenditures. Despite attempts to reduce, reuse, and recycle, there will always be some amount of food waste because it is so prevalent. As a result, we can always generate energy using it. There are no labor-intensive procedures or coal mining. The only thing left to do is transmit the energy source from residences, eateries, and stores to the plants. Because of the huge cost reduction, consumers may save money (TRVST 2022).

9.5.5 *Job Opportunities*

The sector of converting food waste into electricity is relatively new. It will continue to grow as long as the technology is available, and it is established. There's a good chance we'll require extra processing facilities. In addition, we will require more personnel to oversee the procedures. Consequently, as this industry expands, more jobs will be created (TRVST 2022).

9.6 Application of Major Energy Products from Food Waste

9.6.1 *Biogas*

Biogas has an extensive range of applications. For electricity and heat, biogas is appended to combined heat and power (CHP) gas engines. The exhaust heat from these engines can be employed for a wide range of applications including process heating. Once it has been treated, biogas can potentially be utilized to power vehicles. Vehicles like biogas trains can be powered by purified biogas by using it as fuel. Sweden is home to the world's first biogas train. Since October 2005, the train has been in operation (Yağlı et al. 2016; Laval 2006).

9.6.2 *Bioethanol*

Gasoline engines can be replaced by bioethanol. It has characteristics to be mixed with any percentage of gasoline. A blend of approximately 15% bioethanol as well as oil is used by majority of current petroleum engines. A quality of having higher octane number of bioethanol than regular gasoline makes it responsible for an increase in engine's ratio of compression and also its efficiency. Additionally, it serves as the fuel for fireplaces that burn bioethanol. It is ideal for home use because it is free of smoke and does not require a chimney. Bioethanol has numerous other uses, such as a fuel for cogeneration systems, a raw material for the chemical industry, and fuel for thermochemical reactions in cells (Shi et al. 2017).

9.6.3 *Biodiesel*

Biodiesel is a renewable fuel that shares many characteristics with diesel. In contrast to diesel, biodiesel has a number of benefits. The environmental benefits of biodiesel are considerable. Biodiesel emits very little sulfur when used in diesel engines, compared to petrochemical diesel, which almost entirely lacks sulfur. Particles present in tail gas make up nearabout 20% of diesel and emissions by CO make up approximately 10% in petrochemical diesel that is why the index of emission can provide required level of emissions. The lubricity of biodiesel is good. Biodiesel has a higher viscosity in comparison to petrochemical diesel that can lower the wear rate on injection pump of fuel, engine block, and connecting rod and increase their useful lives. The adaptability of biodiesel is good. The addition of biodiesel to a diesel engine does not require any engine modifications. Biodiesel is highly climatically adaptable. Biodiesel has a superior ignition performance than regular diesel (Ishola et al. 2020).

9.6.4 *Biobutanol*

The transportation of biobutanol is safer and less corrosive. Strong compatibility between biobutanol and gasoline means that no engine modifications are necessary. Since biobutanol and gasoline share a long carbon chain, they are highly compatible. When it gets combined with gasoline, biobutanol has a high tolerance for water and is less volatile. It also stores better and is more tolerant of humidity and low water vapor pressure. Biobutanol has a lot of energy. About 30% of ethanol has a pretty high energy of burning. The waste produced is ethanol-like, and there is good combustion efficiency (Zhang 2018).

9.6.5 *Biomethane*

With a purity threshold of over 95%, biomethane produced and processed in plants using the latest technologies meets high-quality criteria. Natural gas (CH_4) can be completely replaced chemically by biomethane, which enables its introduction into gas distribution and transportation systems. Because it is a renewable resource, biomethane is used to generate energy, heat homes and businesses, and power automobiles (SNAM 2020).

9.6.6 *Biohydrogen*

To produce electricity, a fuel cell system uses the H_2 gas produced by the conventional process as fuel. However, there are currently no commercial biohydrogen generation systems available, and there have only been a few researches into their use (Rahman et al. 2016).

9.7 Conclusion

In compliance with rapidly increasing costs accompanied to energy production and waste disposal and rising public involvements with environmental quality, the transformation of food wastes to energy is now an environment-friendly and cost-effective practice. Compositions of food waste vary significantly based on their sources. In this chapter, we have discussed the current methodologies for management of food waste by incineration, composting, anaerobic digestion, pyrolysis, landfill, and biochemical methods on the basis of their composition, although sustainable technologies are required to be employed to save the environment which is largely neglected. Acknowledging to this, the chapter culminates the viable approaches like hydrothermal carbonization as this holds a verge pertaining to price, quality of product and all-embracing effectiveness. Various energy sources were generated from food waste like biomethane, biogas, biohydrogen, bioethanol, biobutanol, and biodiesel as green energy sources and utilization.

References

Adewuyi A (2020) Challenges and prospects of renewable energy in Nigeria: a case of bioethanol and biodiesel production. Energy Rep 6:77–88

Adhikari BK, Barrington S, Martinez J (2006) Predicted growth of world urban food waste and methane production. Waste Manag Res 24(5):421–433. https://doi.org/10.1177/0734242X06067767

Agarwal A, Singhmar A, Kulshrestha M, Mittal AK (2005) Municipal solid waste recycling and associated markets in Delhi, India. Resour Conserv Recycl 44(1):73–90

Alibardi L, Cossu R (2016) Composition variability of the organic fraction of municipal solid waste and effects on hydrogen and methane production potentials. Waste Manag 36:147–155

Alibardi L, Muntoni A, Polettini A (2014) Hydrogen and waste: illusions, challenges and perspectives. Waste Manag 34:2425–2426

Arapoglou D, Varzakas T, Vlyssides A, Israilides C (2010) Ethanol production from potato peel waste (PPW). Waste Manag 30(10):1898–1902

Arena U (2012) Process and technological aspects of municipal solid waste gasification. A review. Waste Manag 32(4):625–639

Autret E, Berthier F, Luszezanec A, Nicolas F (2007) Incineration of municipal and assimilated wastes in France: assessment of latest energy and material recovery performances. J Hazard Mater 139(3):569–574

Balat M (2011) Production of bioethanol from lignocellulosic materials via the biochemical pathway: a review. Energ Conver Manage 52(2):858–875

Bampenrat A, Sukkathanyawat H, Prommuak C (2021) Enhanced energy recovery from food waste by co-production of bioethanol and biomethane process. Fermentation 7:265. https://doi.org/10.3390/fermentation7040265

Ban-Koffi L, Han YW (1990) Alcohol production from pineapple waste. World J Microbiol Biotechnol 6:281–284

Bellemare MF, Çakir M, Peterson HH, Novak L, Rudi J (2017) On the measurement of food waste. Am J Agric Econ 99:1148–1158. https://doi.org/10.1093/cce/aax034

Berge ND, Ro KS, Mao J, Flora JR, Chappell MA, Bae S (2011) Hydrothermal carbonization of municipal waste streams. Environ Sci Technol 45(13):5696–5703

Bryant MP (1979) Microbial methane production: theoretical aspects. J Anim Sci 48(1):193–201

Caton PA, Carr MA, Kim SS, Beautyman MJ (2010) Energy recovery from waste food by combustion or gasification with the potential for regenerative dehydration: a case study. Energ Conver Manage 51(6):1157–1169

Chanakya HN, Ramachandra TV, Vijayachamundeeswari M (2007) Resource recovery potential from secondary components of segregated municipal solid wastes. Environ Monit Assess 135(1–3):119–127

Christy PM, Gopinath L, Divya D (2014) A review on anaerobic decomposition and enhancement of biogas production through enzymes and microorganisms. Renew Sustain Energy Rev 34:167–173

Coleman-Jensen A, Nord M, Andrews M, Carlson S (2012) House-hold food security in the United States in 2011. Economic research report number 141. USDA-Economic Research Service, Washington, DC

Dahiya S, Sarkar O, Swamy Y, Mohan SV (2015) Acidogenic fermentation of food waste for volatile fatty acid production with co-generation of biohydrogen. Bioresour Technol 182:103–113. https://doi.org/10.1016/j.biortech.2015.01.007

Demirbas A (2008) Biofuels sources, biofuel policy, biofuel economy and global biofuel projections. Energ Conver Manage 49:2106–2116

Dong S (2020, June) Food waste as a raw material for biofuel production. Thesis. Centria University of Applied Sciences, Environmental Chemistry and Technology

Dou Z, Ferguson J, Galligan D, Kelly A, Finn S, Giegengack R (2016) Assessing US food wastage and opportunities for reduction. Glob Food Secur 8:19–26. https://doi.org/10.1016/j.gfs.2016.02.001

EESI (2017). https://www.eesi.org/papers/view/fact-sheet-biogasconverting-waste-to-energy

Elefsiniotis P, Oldham W (1994) Substrate degradation patterns in acid- phase anaerobic digestion of municipal primary sludge. Environ Technol 15(8):741–751

Eryasar A, Kocar G (2004) The practicability of solar heated biogas reactors in Turkey conditions. In: Bioenergy Symposium, Ege University, Izmir

European Commission (2010) Final report–preparatory study on food waste across EU27. Bio Intelligence Service, Paris

FAO (2011) Global food losses and food waste e extent, causes and prevention by Gustavsson J, Cederberg C, Sonesson U, Otterdijk R, Meybeck A. Food and Agriculture Organization, Rome. http://www.fao.org/docrep/014/mb060e/mb060e00.pdf

FAO (2012) Towards the future we want: end hunger and make the transition to sustainable agricultural and food systems. Food Agriculture Organization of the United Nations, Rome

FAO (2019) The state of food and agriculture 2019. Moving forward on food loss and waste reduction. http://www.fao.org/3/ca6030en/ca6030en.pdf

Griffin ME, McMahon KD, Mackie RI, Raskin L (1998) Methanogenic population dynamics during start-up of anaerobic digesters treating municipal solid waste and biosolids. Biotechnol Bioeng 57(3):342–355

Grosso M, Motta A, Rigamonti L (2010) Efficiency of energy recovery from waste incineration, in the light of the new waste framework directive. Waste Manag 30(7):1238–1243

GSTC (n.d.). https://globalsyngas.org/syngas-technology/syngas-production/waste-to-energy-gasification/#:~:text=In%20a%20waste%2Dto%2Denergy,gasification%20so%20different%20from%20incineration

Guermoud N, Ouadjnia F, Abdelmalek F, Taleb F, Addou A (2009) Municipal solid waste in Mostaganem city (Western Algeria). Waste Manag 29(2):896–902

Gunaseelan VN (2004) Biochemical methane potential of fruits and vegetable solid waste feedstocks. Biomass Bioenergy 26(4):389–399

Gunders D (2012) Natural Resources Defense Council Issue Paper. Wasted: How America is losing up to 40 percent of its food from farm to fork to landfill (IP:12–06-B). www.nrdc.org/food/files/wasted-food-IP.pdf

Guo W (2015) Study on the efficient synthesis of syngas to fuel ethanol. Zhejiang University, Zhejiang

Gurung A, Karki R, Cho JS, Park KW, Oh SE (2013) Roles of renewable energy technologies in improving the rural energy situation in Nepal: gaps and opportunities. Energy Policy 62:1104–1109

Gustavsson J, Cederberg C, Sonesson U, van Otterdijk R, Meybeck A (2011) Global food losses and food waste: extent, causes and prevention. http://www.fao.org/3/a-i2697e.pdf

Hallenbeck PC, Liu Y (2016) Recent advances in hydrogen production by photosynthetic bacteria. Int J Hydrogen Energy 41:4446–4454

Hosseini SE, Wahid MA, Jamil MM, Azli AAM, Misbah MF (2015) A review on biomass-based hydrogen production for renewable energy supply. Int J Energy Res 39:1597–1615

IEA (2020). https://www.iea.org/reports/outlook-for-biogas-and-biomethane-prospects-for-organic-growth/an-introduction-to-biogas-and-biomethane

Ishola F, Adelekan D, Mamudu A, Abodunrin T, Aworinde A, Olatunji O, Akinlabi S (2020) Biodiesel production from palm olein: a sustainable bioresource for Nigeria. Heliyon 6(4): 2405–8440

Jeguirim M, Khiari B (2022) Chapter Nine—Thermochemical conversion of tomato wastes. In: Jeguirim M, Zorpas A (eds) Tomato processing by-products. Academic Press, New York, pp 285–332., ISBN 9780128228661. https://doi.org/10.1016/B978-0-12-822866-1.00008-9

Karakashev D, Batstone DJ, Angelidaki I (2005) Influence of environmental conditions on methanogenic compositions in anaerobic biogas reactors. Appl Environ Microbiol 71(1): 331–338

Katami T, Yasuhara A, Shibamoto T (2004) Formation of dioxins from incineration of foods found in domestic garbage. Environ Sci Technol 38(4):1062–1065

Kim M-H, Kim J-W (2011) Comparison through a LCA evaluation analysis of food waste disposal options from the perspective of global warming and resource recovery. Sci Total Environ 408(19):3998–4006. https://doi.org/10.1016/j.scitotenv.2010.04.049

Kiran EU, Trzcinski AP, Ng WJ, Liu Y (2014) Bioconversion of food waste to energy: a review. Fuel 134:389–399

Knoef H (2005) Practical aspects of biomass gasification. In: Knoef H (ed) Handbook biomass gasification. BTG-Biomass Technology Group. Enschede, The Netherlands. ISBN: 90-810068-1-9

Kosseva MR (2009) Processing of food wastes. In: Taylor SL (ed) Advances in food and nutrition research, vol 58, pp 57–136

Kosseva MR (2011) Management and processing of food wastes. In: Reference module in earth systems and environmental sciences comprehensive biotechnology, 2nd edn, pp 557–593

Laval P (2006) Sweden: the World's first biogas-powered train. Rail and public. Transport 12

Lee JP, Lee JS, Park SC (1999) Two-phase methanization of food wastes in pilot scale. Appl Biochem Biotechnol Pt A Enzyme Eng Biotechnol 77-79:585–593

Li L, Diederick R, Flora JR, Berge ND (2013) Hydrothermal carbonization of food waste and associated packaging materials for energy source generation. Waste Manag 33(11):2478–2492

Libra JA, Ro KS, Kammann C, Funke A, Berge ND, Neubauer Y, Titirici MM, Fuhner C, Bens O, Kern J, Emmerich KH (2011) Hydrothermal carbonization of biomass residuals: a comparative review on the chemistry, processes and applications of wet and dry pyrolysis. Biofuels 2(1): 71–106

Liu Z, Zhang F-S, Wu J (2010) Characterization and application of chars produced from pinewood pyrolysis and hydrothermal treatment. Fuel 89(2):510–514

Ma Y, Yu L (2019) Turning food waste to energy and resources towards a great environmental and economic sustainability: an innovative integrated biological approach. Biotechnol Adv 37(7): 107414. ISSN 0734-9750

Mannberg A, Jansson J, Pettersson T (2014) Do tax incentives affect households' adoption of 'green' cars? A panel study of the Stockholm congestion tax. Energy Policy 74:286–299

Mardikar SH, Niranjan K (1995) Food processing and the environment. Environ Manage Health 6(3):23–26

Matsakas L, Kekos D, Loizidou M, Christakopoulos P (2014) Utilization of household food waste for the production of ethanol at high dry material content. Biotechnol Biofuels 7:4–13

Ministry of Agriculture, Forestry and Fisheries of Japan (2020) Outline of the food loss reduction promotion Law and efforts to reduce food loss: new basic policy of the food recycling Law and amendment of relevant government ordinances. (in Japanese). https://www.maf.go.jp/j/chikusan/sinko/lin/l_siryo/attach/pdf/ecofeed-87.pdf

Mittal KM (1997) Biogas systems, policies, progress and prospects. New Age International, New Delhi

Moon HC, Song IS, Kim JC, Shirai Y, Lee DH, Kim JK, Chung SO, Kim DH, Oh KK, Cho YS (2009) Enzymatic hydrolysis of food waste and ethanol fermentation. Int J Energy Res 33:164–172

Morita M, Sasaki K (2012) Factors influencing the degradation of garbage in methanogenic bioreactors and impacts on biogas formation. Appl Microbiol Biotechnol 94(3):575–582

Nasir IM, Ghazi TIM, Omar R (2012) Production of biogas from solid organic wastes through anaerobic digestion: a review. Appl Microbiol Biotechnol 95(2):321–329

NEA (2012) Waste statistics and overall recycling. https://www.nea.gov.sg/energy-waste/waste-management/waste-statistics-and-overall-recycling

NRDC (National Resources Defense Council) (2012) In: NRDC (ed) Wasted: how America is losing up to 40 percent of its food from farm to fork to landfill, New York

Oberoi HS, Vadlani PV, Nanjundaswamy A, Bansal S, Singh S, Kaur S, Babbar N (2011a) Enhanced ethanol production from Kinnow mandarin (Citrus reticulata) waste via a statistically optimized simultaneous saccharification and fermentation process. Bioresour Technol 102(2): 1593–1601

Oberoi HS, Vadlani PV, Saida L, Bansal S, Hughes JD (2011b) Ethanol production from banana peels using statistically optimized simultaneous saccharification and fermentation process. Waste Manag 31(7):1576–1584

Oelofse SHH, Nahman A (2013) Estimating the magnitude of food waste generated in South Africa. Waste Manag Res 31(1):80–86

Opatokun SA, Kan T, Al Shoaibi A, Srinivasakannan C, Strezov V (2015a) Characterization of food waste and its digestate as feedstock for thermochemical processing. Energy Fuel 30:1589–1597

Opatokun SA, Strezov V, Kan T (2015b) Product based evaluation of pyrolysis of food waste and its digestate. Energy 92:349–354

Opatokun S, López-Sabirón A, Ferreira Ferreira G, Strezov V (2017) Life cycle analysis of energy production from food waste through anaerobic digestion, pyrolysis and integrated energy system. Sustainability 9:1804. https://doi.org/10.3390/su9101804

Östergren K, Gustavsson J, Bos-Brouwers H, Timmermans T, Hansen OJ, Møller H et al (2014) FUSIONS definitional framework for food waste. https://www.eufusions.org/phocadownload/

Osuagwu U (2014) Bio-hydrogen production from food waste through anaerobic fermentation. Sains Malaysiana 43:1927–1936

Parfitt J, Barthel M, McNaughton S (2010) Food waste within food supply chains: quantification and potential for change to 2050. Philos Trans R Soc Lond B Biol Sci 365:3065–3081. https://doi.org/10.1098/rstb.2010.0126

Parshetti GK, Chowdhury S, Balasubramanian R (2014) Hydrothermal conversion of urban food waste to chars for removal of textile dyes from contaminated waters. Bioresour Technol 161: 310–319

Patil A, Kulkarni A, Patil B (2014) Waste to energy by incineration. J Comput Technol 3:12–15

Pourbafrani M, Forgacs G, Horvath IS, Niklasson C, Taherzadeh MJ (2010) Production of biofuels, limonene and pectin from citrus wastes. Bioresour Technol 101(11):4246–4250

Rahman SNA, Masdar MS, Rosli MI, Majlan EH, Husaini T, Kamarudin SK, Daud WRW (2016) Overview biohydrogen technologies and application in fuel cell technology. Renew Sustain Energy Rev 66:137–162., ISSN 1364-0321, https://www.sciencedirect.com/science/article/pii/S1364032116303756. https://doi.org/10.1016/j.rser.2016.07.047

Ren Y, Yu M, Wu C, Wang Q, Gao M, Huang Q, Liu Y (2017) A comprehensive review on food waste anaerobic digestion: research updates and tendencies. Bioresour Technol 247:1069–1076

Schink B (1997) Energetics of syntrophic cooperation in methanogenic degradation. Microbiol Mol Biol Rev 61(2):262–280

Shi S, Gao M, Wang Q, Zheng J, Sonomoto K, Tashiro Y (2017) Study on bio-butanol production from food waste by ABE fermentation. Environ Eng 35(2):118–121

Silva F, Serafim L, Nadais H, Arroja L, Capela I (2013) Acidogenic fermentation towards valorisation of organic waste streams into volatile fatty acids. Chem Biochem Eng Q 27(4): 467–476. https://doi.org/10.1504/ijewm.2020.10026612

SNAM (2020). https://www.snam.it/en/energy_transition/biomethane/benefits_and_usage/#:~:text=Biomethane%20is%20chemically%20a%20complete,a%20fuel%20for%20motor%20vehicles

Socas-Rodríguez B, Álvarez-Rivera G, Valdés A, Ibáñez E, Cifuentes A (2021) Food by-products and food wastes: are they safe enough for their valorization. Trends Food Sci Technol 114:133–147. ISSN 0924-2244

Spokas KA, Reicosky DC (2009) Impacts of sixteen different bio-chars on soil greenhouse gas production. Ann Environ Sci 3:179–193

Stillman GI (1983) Municipal solid waste (garbage): problems and benefits. Ann N Y Acad Sci 403(1):1–26

Stoeberl M, Werkmeister R, Faulstich M, Russ W (2011) Biobutanol from food wastes—fermentative production, use as biofuel an the influence on the emissions. Proc Food Sci 1:1867–1874

Supaphol S, Jenkins SN, Intomo P, Waite IS, O'Donnell AG (2011) Microbial community dynamics in mesophilic anaerobic co-digestion of mixed waste. Bioresour Technol 102(5): 4021–4027. https://doi.org/10.1016/j.biortech.2010.11.124

Talyan V, Dahiya RP, Sreekrishnan TR (2008) State of municipal solid waste management in Delhi, the capital of India. Waste Manag 28(7):1276–1287

Tawfik A, Salem A, El-Qelish M (2011) Two stage anaerobic baffled reactor for biohydrogen production from municipal food waste. Bioresour Technol 102:8723–8726

Terega (2022). https://www.terega.fr/en/lab/what-is-the-difference-between-biogas-and-biomethane#:~:text=Biomethane%20is%20a%20biogas%20from,be%20injected%20into%20the%20grid

TRVST (2022). https://www.trvst.world/waste-recycling/food-waste/converting-food-waste-to-energy/

UN (2018) World urbanization prospects: the 2018 revision, methodology. In: UN-Department of economic and social affairs, population division. Working Paper No. ESA/P/WP.252. United Nations, New York

United States Environmental Protection Agency (2011) Food recovery challenge data. www.epa.gov/smm/food recovery/

US-EPA (2022). https://www.epa.gov/smm/energy-recovery-combustion-municipal-solid-waste-msw

Vavouraki AI, Angelis EM, Kornaros M (2013) Optimization of thermo-chemical hydrolysis of kitchen wastes. Waste Manag 33(3):740–745

Vimala Ebenezer A, Dinesh Kumar MD, Kavitha S, Uan DK, Rajesh Banu J (2020) Chapter 14. State of the art of food waste management in various countries. In: Banu JR, Kumar G, Gunasekaran M, Kavitha S (eds) Food waste to valuable resources. Academic Press, New York, pp 299–323., ISBN 9780128183533. https://doi.org/10.1016/B978-0-12-818353-3.00014-6

Viturtia A, Mata-Alvarez J, Cecchi F, Gazzini G (1989) Two-phase anaerobic digestion of a mixture of fruit and vegetable wastes. Biol Wastes 29(3):189–199., ISSN 0269-7483. https://doi.org/10.1016/0269-7483(89)90130-4

W2E (2022). https://www.waste2energy.co.nz/pyrolysis-waste-to-energy/

Wang H (2015) Study on utilization of waste starch in food waste. Wuhan Institute of Technology, Wuhan

Wang Y, Gao M, Bodjui O, Wang F, Wang Q (2019) BIO-BUTANOL production from direct fermentation of kitchen wastes. Environ Eng 37(4):138–141

WHO (2007) Population, Health and Waste Management: Scientific Data and Policy Options. Report of a WHO Workshop, 29–30 March 2007, Rome, Italy

Wilkie AC (2008) Biomethane from biomass, bio- waste and biofuels. In: Wall JD, Harwood CS, Demain A (eds) Bioenergy. American Society for Microbiology Press, Washington, DC, pp 195–205

Yağlı H, Koç Y, Koç A, Görgülü A, Tandiroğlu A (2016) Parametric optimization and exergetic analysis comparison of subcritical and supercritical organic Rankine cycle (ORC) for biogas fuelled combined heat and power (CHP) engine exhaust gas waste heat. Energy 111:923–932

Young W, Russell SV, Robinson CA, Barkemeyer R (2017) Can social media be a tool for reducing consumers' food waste? A behaviour change experiment by a UK retailer. Resour Conserv Recycl 117:195–203. https://doi.org/10.1016/j.resconrec.2016.10.016

Zhang G (2017) Study on anaerobic digestion of kitchen waste and pilot scale-up. China University of Petroleum, Beijing

Zhang C (2018) Research in the green process of fuel butanol production from straw. Beijing University of Chemical Technology, Beijing

Zhang ZY, Jin B, Kelly JM (2007a) Production of lactic acid from renewable materials by *Rhizopus fungi*. Biochem Eng J 35:251–263

Zhang R, El-Mashad HM, Hartman K, Wang F, Liu G, Choate C, Gamble P (2007b) Characterization of food waste as feedstock for anaerobic digestion. Bioresour Technol 98(4):929–935

Zhang R, El-Mashad HM, Hartman K, Wang F, Liu G, Choate C, Gamble P (2007c) Characterization of food waste as feedstock for anaerobic digestion. Bioresour Technol 98:929–935

Zhang N, Lior N, Jin HG (2011) The energy situation and its sustainable development strategy in China. Energy 36:3639–3649

Zhou M, Yan B, Wong JW, Zhang Y (2018) Enhanced volatile fatty acids production from anaerobic fermentation of food waste: a mini-review focusing on acidogenic metabolic pathways. Bioresour Technol 248:68–78. https://doi.org/10.1016/j.biortech.2017.06.121

Zhu B, Gikas P, Zhang R, Lord J, Jenkins B, Li X (2009) Characteristics and biogas production potential of municipal solid wastes pre-treated with a rotary drum reactor. Bioresour Technol 100(3):1122–1129

Zinder SH (1990) Conversion of acetic acid to methane by thermophiles. FEMS Microbiol Rev 6(2–3):125–137. https://doi.org/10.2172/10158582

Chapter 10
Biochar for Sustainable Crop Production

Neerja Sharma, Shalini Dhiman, Jaspreet Kour, Tamanna Bhardwaj, Kamini Devi, Nitika Kapoor, Amandeep Bhatti, Dhriti Kapoor, Amrit Pal Singh, and Renu Bhardwaj

Abstract Increased demand of growing population on crops puts down immense pressure on agricultural productivity by reduction of cultivable area and the depletion of natural resources. Inadequate soil management, i.e., excessive addition of chemical fertilizers, pesticides, and herbicides degrades the agricultural soil quality and hampers the quality of the produce. A sustainable approach in agricultural crop production is therefore necessary for the development of eco-friendly, ethically generous, and economically beneficial environment. Due to high porosity, alkalinity, and adsorption capacity, biochar application in agriculture decreases the soil acidity and results in mitigating the soil pollutants and hence enhances the crop growth and yield. This chapter summarizes the impact of biochar on soil stability and soil properties, its effect on plant growth and productivity, remediation of soil pollutants and resistance against plant diseases, and its role in lowering of greenhouse gas emission from the soil.

Keywords Biochar production · Pyrolysis · Global population · Soil amendment · Greenhouse gases

---

N. Sharma (✉) · S. Dhiman · J. Kour · T. Bhardwaj · K. Devi · R. Bhardwaj
Department of Botanical and Environmental Sciences, Guru Nanak Dev University, Amritsar, Punjab, India

N. Kapoor
PG Department of Botany, Hans Raj Mahila Maha Vidyalaya, Jalandhar, Punjab, India

A. Bhatti
S.R Government College for Women, Amritsar, Punjab, India

D. Kapoor
School of Bioengineering and Biosciences, Lovely Professional University, Jalandhar, Punjab, India

A. P. Singh
Department of Pharmaceutical Sciences, Guru Nanak Dev University, Amritsar, Punjab, India

D. B. Pal, A. K. Tiwari (eds.), *Sustainable Valorization of Agriculture & Food Waste Biomass*, Clean Energy Production Technologies,
https://doi.org/10.1007/978-981-99-0526-3_10

10.1 Introduction

The continuous use of various fertilizers and pesticides in order to improve the crop yields worsens the soil fertility because of leaching necessary nutrients and increases the soil acidification. The application of these agricultural chemicals threatens the quality and also reduces the crop produce. Due to bioaccumulation and long-range transport, pesticides have great impact on the human health and other organisms (Kuranchie-Mensah et al. 2012). So to improve the soil superiority and reduce greenhouse gas emission, the biochar an ecofriendly approach is used for sustainable crop production. Biochar is a carbonaceous material obtained from an extensive variety of organic materials such as wood remains, agricultural wastes, animal litters, etc. by pyrolysis or thermochemical conversion in oxygen-limited controlled conditions (Lian and Xing 2017; Zhao et al. 2013; Luo et al. 2015). Generally the advantage of biochar application to soil, to production schemes, and hence to the environment depends upon three factors: use of sustainable organic feedstock, sustainable production procedures, and sustainable final use (Elad et al. 2011). By decreasing soil bulk density and increasing porosity, biochar endorses the creation of aggregates and mends the structure of soil (Fungo et al. 2017; Pituello et al. 2018). Since most of the biochars are alkaline in nature, they enhance the pH of acidic soils and also increase the water holding and cation exchange capacity of soil (Chintala et al. 2014; Van Zwieten et al. 2010). As a soil amendment, the large content of recalcitrant carbon in biochar helps to sequester the carbon and therefore decreases the release greenhouse gas emission (Borchard et al. 2019; Singh et al. 2012; Han et al. 2018; Sui et al. 2016; Zimmerman 2010). High sorption affinity of biochar against heavy metals lessens their bioavailability in crops and also support to retrieve crop quality and yield (Bian et al. 2014a, b; Puga et al. 2015; Li et al. 2016; Xu et al. 2016; Yin et al. 2017). Biochar-treated soils with modified microbial communities decompose the organic matter effectively and decrease the nutrient leaching by increasing nutrient supply to plants (Kuzyakov et al. 2009; Liang et al. 2010).

10.2 Effect of Biochar on Soil Properties

Carbon-rich property of biochar makes it good soil conditioner and also recovers the fertility of degraded soils. Soil-water relationships are improved by biochar through increasing soil water penetration, aggregation power, and water holding capacity (Qambrani et al. 2017; Purakayastha et al. 2019). It has been reported that soil porosity is increased by stable carbon content and low bulk density of biochar which then reduces the infiltration resistance and also bulk density of soil (Gwenzi et al. 2015).

The application of biochar increases the cation exchange capacity (CEC) of soil and thereby enhances the fertility of acidic soils by the discharge of ions like K, Ca, Mg, and Na which contribute to enhanced soil pH. At the time of pyrolysis, the mineral elements of organic biomass are converted into oxides and carbonates that combine with hydrogen ion and Al in acidic soils by exchange capacity and increased pH (Ahmad et al. 2016; Dai et al. 2017a, b; Purakayastha et al. 2019).

10.2.1 Effect of Biochar on Physicochemical Properties and Nutrient Status of Soil

Physicochemical properties and nutrient status of the soil can be improved with the application of biochar in soil either alone or with mineral or organic fertilizers (Sänger et al. 2017). It has been reported that when biochar along with composting mixture is added in the soil, it enhances the retention of nutrients in the soil by coating the biochar (Joseph et al. 2018). Biochar also helps in improving the nutrient status of the soil due to presence of various micro- and macronutrients as Ca, P, K, and N in it (Manolikaki and Diamadopoulos 2017) (Fig. 10.1).

It was observed by Cao et al. (2018) that supplementation of biochar with compost enhances the nutrient level in the soil. Kammann et al. (2015) found that the biochar which is based on compost is rich in nitrates and phosphates and thereby biochar acts as a source of phosphorus in soil. Water content and organic carbon of soil are also enhanced by the biochar treatment. Agegnehu et al. (2015) had found that when fertilizer was applied in the soil alone, the soil organic carbon (SOC) increased by 0.93%, but when fertilizer was applied with biochar, the SOC increased

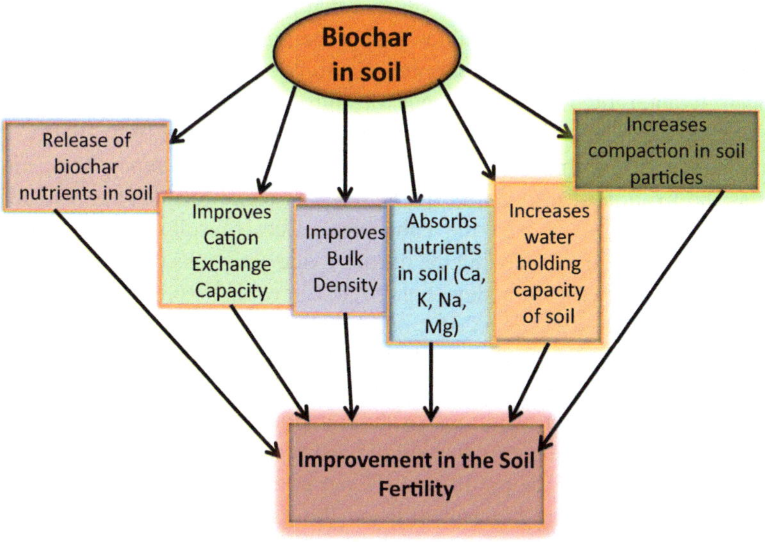

Fig. 10.1 Improvement in soil fertility by biochar

from 0.93% to 1.25%. Similarly, the soil water content was also increased from 18% to 23%, in both cases when fertilizer was applied alone and when biochar was used with fertilizer.

Biochar loaded with ammonia, nitrates, and phosphates is also known to upsurge the soil fertility by acting as fertilizer (Kammann et al. 2015). It has been reported by Liu et al. (2014) that due to application of rice husk biochar, the aggregation of soil particles increased from 8 to 36%. Due to this aggregation, there is enhancement in the water holding capacity of soil, and nutrient absorption also increases there by increasing the soil fertility. Different biochars produced from different waste materials show different properties in variety of soil. They help in improving the fertility of soil by enhancing the nutrient content of the soil and also improve the physico-chemical properties of the soil. Besides these applications the biochar also enhances the crop yield and productivity (Table 10.1).

10.2.2 Effect of Biochar on Biological Nitrogen Fixation

Nitrogen is the key element for the growth and productivity of the crop and is immensely depleted due to leaching, denitrification, and volatilization. The application of biochar reduces the loss of nitrogen from soil and enhances the confinement of NO_3 in the soil (Yao et al. 2012; Liu et al. 2018; Sui et al. 2021). The adsorption of nitrate (NO_3) on the surface of biochar (contain functional groups) produced from sugarcane bagasse at 800 °C reduces its leaching in dark red soil (Kameyama et al. 2012). Other reports (Clough et al. 2013; Wang et al. 2017; Liu et al. 2017, 2018) revealed the decreased the leaching of nitrogen via adsorption, curbing, and exchange of ions of nitrate upon the biochar that results in increasing the efficiency of crop productivity. Plant residues which derived biochar application in soil improves the growth of soil microorganisms, nitrogen cycling, and hence biological nitrogen fixation by regulating genes of nitrogen transformation, reduction in ammonia production, and mitigation of nitrous oxide emission (Fig. 10.2).

It was also stated that the elevation in the deposition of atmospheric nitrogen reduces the fixation of nitrogen in the legumes of some forest types (Zheng et al. 2016). Many studies showed the enhancing effect of soil microflora after the application of biochar by reducing the nitrous oxide emissions resulting in increased yield and growth of crops (Pandey et al. 2016; Liu et al. 2016; Ameloot et al. 2016; Abid et al. 2017; Feng and Zhu 2017; Ahmad et al. 2021). The influence of biochar on emission of nitrous oxide may vary with type of soil and crop and also the type of feedstock used for biochar production (Yoo et al. 2018). Some reports showed the mechanism of mitigation of nitrous oxide emission from the soil through biochar supplementation which elevates the soil pH and hence regulates the toxicity on soil microflora for nitrous oxide production (Harter et al. 2016; Cayuela et al. 2015). It has been studied that the biochar enriched the displaying of the nosZ gene via stimulating the alteration of nitrous oxide and nitrogen (Van Zwieten et al. 2014; Harter et al. 2017; Shaaban et al. 2018; Krause et al. 2018). It was studied by

Table 10.1 Effect of biochar on physicochemical properties and nutrient status of soil

S. no.	Biochar type	Effects	References
1.	Sludge biochar, straw biochar, and quicklime	Yield of the wheat was improved and the improvement in the physicochemical properties of the soil was caused due to application of biochar	Malik et al. (2018)
2.	Rice husk biochar	Water retention power, moisture content of soil, and total nutrient content as total C, N, and P was enhanced due to application of biochar as compared to control	Singh et al. (2018)
3.	Rice husk biochar	Water holding capacity and total nutrients as N, K, and P were enhanced in soil after application of biochar along with inorganic nitrogen fertilizer	Oladele et al. (2019)
4.	Waste mushroom substrate biochar	Ash content, nutrient content as P and K, and various minerals as Ca, Zn, Fe were increased by the biochar that was produced at 600 °C and 700 °C	Sarfraz et al. (2019)
5.	Peanut shell biochar	Biochar produced at 220 °C helped in maintaining the fertility of soil by improving the nutrient status as well as physical and chemical properties of soil	Wang et al. (2020)
6.	Coconut shell biochar	Increased nitrogen and potassium content and pH and EC were also improved due to application of biochar on sandy loam soil	Arunkumar and Thippeshappa (2020)
7.	Wheat straw biochar	Increase in water holding capacity, total potassium, and total nitrogen and nitrate in the soil after the application of biochar produced at 600 °C as compared to control	Ren et al. (2020)
8.	Wheat straw biochar	Improved pH, total nitrogen and phosphorus, CEC, water status, and total organic carbon	Futa et al. (2020)
9.	Coffee husk biochar	pH of soil, organic carbon, total nitrogen, CEC, iron, copper, manganese, and zinc content were increased in the soil as compared to the control plants	Getahun et al. (2020)
10.	Corn straw biochar	Increased porosity, pH, water holding capacity, total carbon, total nitrogen, and CEC were enhanced when applied in the alkaline soil	Liu et al. (2020)
11.	Tobacco straw biochar	Physicochemical properties of soil were improved due to the application of biochar	Zheng et al. (2020)
12.	Rice husk biochar	Nutrient contents of soil as N, P, and K were increased, and other physic chemical properties of the soil were also improved due to biochar application	Choudhary et al. (2021)
13.	Peanut shell biochar	Biochar when applied to saline sodic paddy soil improves nutrient status and physicochemical property. Rice yield and biomass were also enhanced due to biochar application	Yao et al. (2021)

(continued)

Table 10.1 (continued)

S. no.	Biochar type	Effects	References
14.	Cow manure	Plant productivity was increased. Total carbon and water holding capacity of soil were increased when biochar was applied along with the compost	Zahra et al. (2021)
15.	Residual biochar	Organic content, organic phosphorus, and the plant yield were significantly increased due to the biochar application	Mian et al. (2021)

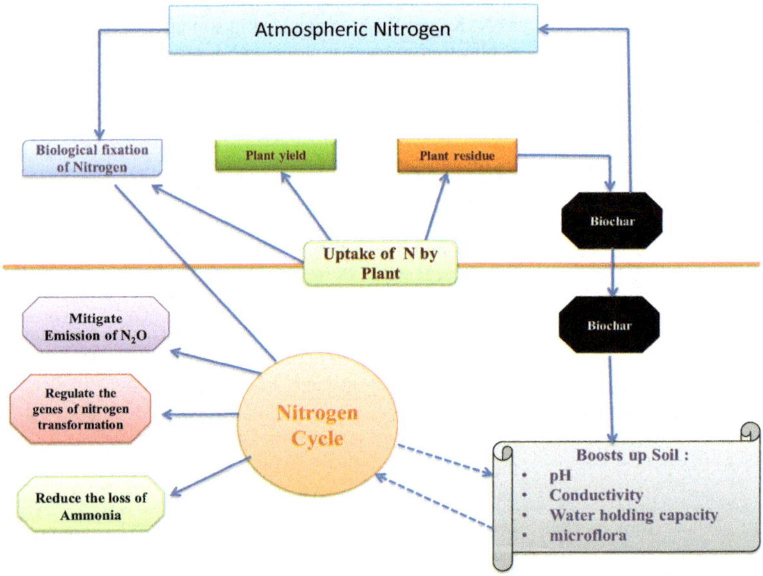

Fig. 10.2 Effect of biochar on nitrogen fixation

Cayuela et al. (2015) that hydrogen/carbon ratio of biochar reveals the polymerization of carbon within the biochar that plays an important role in lowering the nitrous oxide emission from the soil. Some reports also documented the aging effect of biochar on nitrous oxide emissions (Hawthorne et al. 2017; Hagemann et al. 2017; Oo et al. 2018). Supplementation of biochar to the soil enhances the biological nitrogen fixation (Azeem et al. 2019). Microorganisms' niches were directly provided by biochar that ultimately leads to change in soil pH, moisture, and adsorption of signalling molecules (Gul et al. 2015; Gul and Whalen 2016). The complex structure and porosity of the biochar help in regulating the transformation of nitrogen that lead to decrease in the loss of ammonia from the soil (Sánchez-García et al. 2015; Hale et al. 2015; Jain et al. 2020; Malinowski et al. 2019). Table 10.2 summarizes the effect of biochar on biological nitrogen fixation.

Table 10.2 Effect of biochar on biological nitrogen fixation

S. no.	Soil type	Biochar concentration	Effect of biochar	Reference
1	Calcareous soil	2%	Enhanced the activity of *Azotobacter* which further boosted up the fixation of biological nitrogen	Zhao et al. (2021)
2	Albic soil	10–40 g/kg soil	Enhanced the capability of rhizobacteria to improve the nutrient and thereby increased the fixation of the nitrogen in the soil resulting in the	Xiu et al. (2021)
3	Soil	10%	Transformed nitrogen via altering of some diazotrophs after the application of the biochar	Wu et al. (2020)
4	Tropical Arenosol	5%	Reduced the deprivation of nitrogen	Beusch et al. (2019)
5	Agricultural	2.25 t/ha	Increased the availability of phosphorous	Si et al. (2018)
6	Legume crop soil	11.4 g/kg soil	Increased the pH of the soil	Mia et al. (2018)
7	Agricultural	20 t/ha	Hydrochar has more potential than pyrochar to enhance the biological nitrogen fixation	Scheifele et al. (2017)
8	Agricultural	4%	Significantly maintained the nitrous oxide emission	Lin et al. (2017)
9	Soil	–	Maintained the fixation of biological nitrogen	Yuan et al. (2016)
10	Agricultural oxisols and cambisols	–	Boosted up the transformation of ammonia to nitrate	He et al. (2016)
11	Agricultural soil	–	Increased the soil aeration and further enhanced the total nitrogen in the soil	Abujabhah et al. (2016)
12	Soil/water	–	Ameliorated contaminated soil, improved the aeration of the soil, and also increased the efficiency of biological nitrogen fixation	Laghari et al. (2016)
13	Agricultural	5 t/ha	Enhanced the formation of nodules	Partey et al. (2016)
14	Agricultural	–	Increased the fixation of biological nitrogen	Van Zwieten et al. (2015b)
15	Sandy soil	–	Fixed the nitrogen loss by maintaining the emission of nitrous oxide	Yao et al. (2012)
16	Agricultural	0–10%	Enhanced the retention time of nitrate in the soil and further increased its availability for crops	Kameyama et al. (2012)

10.2.3 *Effect of Biochar on Soil Microflora and Soil Enzymatic Activity*

For the stability of soil ecosystem and ecology, its microflora plays a key role in maintaining the decomposition of carbon, uptake of nutrients, and transformation of nitrogen by the plants (Karhu et al. 2014; Castrillo et al. 2017). Biochar is one of the sustainable sources of carbon that sheltered the growth of microbes and result in boosting the biomass of microbial community (Zhang et al. 2016; Zhou et al. 2017). Liu et al. (2018) reported that soil amendment of biochar enhances the biomass of microbes by 12%. Biochar prepared form tea leaves proved to be an effective carrier of nutrients, sustaining shelter for growth of microbes (Azeem et al. 2021). In recent report, the activity of soil engineers including urease and catalase increases after the addition of biochar by enhancing the pH of the soil (Bashir et al. 2019; Jin et al. 2021). High adsorption capacity of biochar makes it superior by decreasing the leaching of nutrients and remediating the stressed soil (Blanco-Canqui 2021). The biochar application and its responses were broadly studied, but the interaction process between soil microflora and biochar still remains unclear (Liao et al. 2016; Luo et al. 2017; Jing et al. 2018; Hardy et al. 2019). Some reports showed the boosting effect of biochar that increases the activity if soil microflora (Xu et al. 2015; Liao et al. 2016). Depending upon the biochar feedstock, production and its properties decided the interaction between biochar and microflora (Biederman and Harpole 2013; Jing et al. 2018). Low temperature-pyrolyzed biochar contains mineralized carbon which is easily oxidized by microbes (Keiluweit et al. 2010; Luo et al. 2011; Dai et al. 2017a, b). 5–20 t/ha application of biochar enhanced the soil carbon, whereas the high rates of biochar application showed negative effect on mineralization of carbon. Biochar with high porosity considered the best habitat and source of nutrients for the growth of soil microflora and other enzymes (Ali et al. 2019). In some reports it was studied that after the biochar supplementation, there is an increase in the growth of mycorrhizae fungi which helped in the aggregation of soil, and nutrient cycle of the soil ultimately strengthened its physiochemical properties of the soil (Lone et al. 2015; Yu et al. 2019). It was reported by Cayuela et al. (2014) that any noxious compound present in biochar badly affects the biological activities of the soil. Biochar recovers the ecology depending upon its properties and soil as well as environmental conditions. The application of biochar can affect the flow of soil microbial assistance which results to elevate or restrict the degeneration of organic carbon of the soil (Zimmerman et al. 2011; Guo and Chen 2014; Bamminger et al. 2014; Xiao et al. 2014; Cely et al. 2014). Moreover, the application of biochar stabilized the sequestration and decaying of carbon contributing to the fractious carbon source. Figure 10.3 shows that the biochar application enhances soil fertility by providing nutrients, carbon source, habitat, and stability to the microflora which lead to blockade of carbon that is ultimately increasing the growth of plants. Biochar supplementation boosts up the properties of the soil which includes reduced loss of nutrients by slashing the filtration, escape, and emissions of nitrous oxide, inhabiting the microflora of the soil and adsorption of toxic

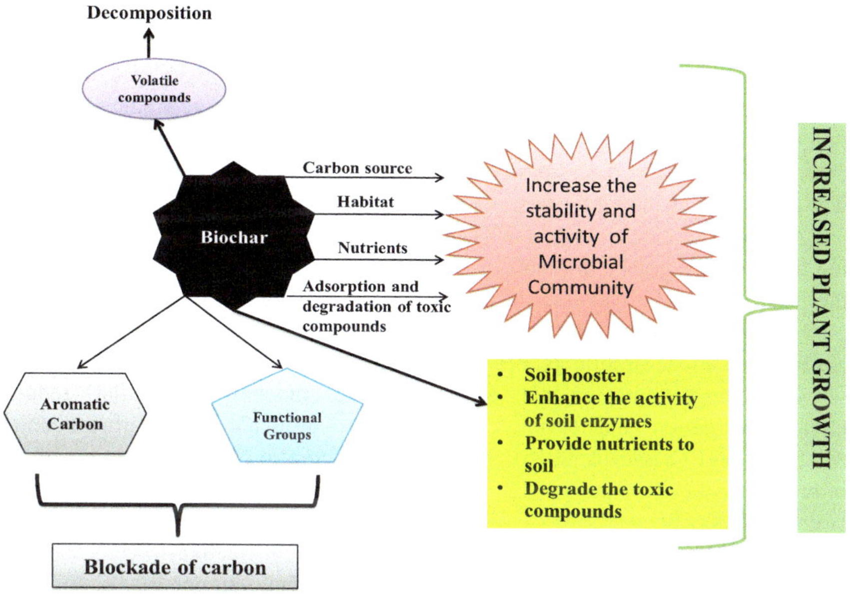

Fig. 10.3 Effect of biochar on soil microflora and soil enzymatic activity

compounds to provide micro-rich environment resulting in the reduction of acidity of the soil, enhances the confinement and availability of water in the soil, and provides suitable carbon and other nutrients to the microbial community that nourished the soil flora.

10.3 Biochar for the Remediation of Soil Pollutants

Remarkable characteristics of biochar especially its porosity property which facilitates sorption, desorption, degradation, bioavailability, and immobilization of various soil pollutants like pesticides, fungicides, herbicide, polychlorinated dibenzo-*p*-dioxins/dibenzofurans (PCDD/Fs), and polycyclic aromatic hydrocarbons (PAHs) hinder its leaching into the ground water. It not only has a potential to remediate the heavy metals from soil but also reduces heavy metal uptake inside the plants grown in the contaminated soil.

10.3.1 Biochar for Pesticide Remediation

Pesticides are often used to enhance productivity of crops. However, parts of pesticides applied in the field get accumulated in various parts of the environment and, thus, cause direct toxic hazards and health risk to the human population and other animals. Biochar is an ecofriendly method used for immobilizing and degrading pesticides and, thus, minimizing its risk in soil (Khalid et al. 2020). Biochar application affects pesticides leaching, bioavailability, degradation, desorption, sorption, hydrolysis, and oxidation processes inside the contaminated soil. Such characteristics of biochar minimized the risks of pesticide exposure to humans. It was capable of changing the pesticide fate in the soil. Various reports show reduced and degraded pesticide inside soil (Table 10.3). Biochar was also capable of reducing the pesticide uptake in plants grown inside the contaminated soil. Reduction in the pesticide uptake was due to the presence of biochar in the soil that has high porosity (i.e., having macro- and micropores) which supply more sites for pesticide adsorption and desorption processes. In one of the reports, it was revealed that sewage sludge-grown plants amended with biochar application reduced phytoaccumulation and toxicity of polycyclic aromatic hydrocarbon content (Zielińska and Oleszczuk 2015). Thus, higher adsorption and desorption capacity of biochar reduced the pesticide inside the soil and finally its uptake inside the plant system. Physiochemical process like sorption causes the pesticide molecule to be retained on the solid surface of biochar through the interactive forces caused by various bondings like hydrophobic, hydrogen, covalent, van der Waals, and ionic between pesticides and biochar surface. Biochar also reduced the pesticide leaching. Delwiche et al. (2014) reported reduction in atrazine leaching from soil having pine chip-derived biochar. Desorption of pesticide is an essential process as pesticide desorption directly impacts the other physiological process like degradation, bioavailability, and bioaccessibility processes of pesticides. The biochar thus becomes an essential sink for improving pesticide contamination.

10.3.2 Biochar-Induced Immobilization of Heavy Metals

Various investigatory reports revealed the magnificent immobilization effect of biochar on soil contaminated with heavy metals (Inyang et al. 2016). Physiochemical properties of biochar and environmental condition of soil are the two main factors on which immobilization potential of biochar on heavy metal depends. Various environmental factors constantly altered the properties of biochar (Rechberger et al. 2017). High pH causes increase in biochar immobilization efficiency, whereas steady surface oxidation and alkaline mineral leaching cause inhibition of biochar immobilization potential toward heavy metal toxicity (Cui et al. 2016). Heavy metal immobilization inside the contaminated soil occurs either through direct adsorption or through indirect absorption mainly through modification of adsorption capacity of the soil particle through altering the physiochemical characteristic of soil and then followed by biochar adsorption (He et al. 2019). Direct adsorption of heavy metals by biochar occurs through ion exchange, electrostatic forces, and complex formation. For instance, amendment of phosphorus-rich biochar in lead-contaminated soil facilitates lead immobilization through complexation of

Table 10.3 Effect of biochar on pesticide remediation of soil

S. no.	Biochar used	Pesticide remediated	Biochar response in polluted soil	References
1	Wheat straw biochar	2-methyl-4-chlorophenoxyacetic acid (MCPA)	Causes pesticide degradation	Tatarková et al. (2013)
2	Wood chip biochar	2-methyl-4-chlorophenoxyacetic acid (MCPA)	Causes pesticide degradation	Muter et al. (2014)
3	Wheat straw biochar	2-methyl-4-chlorophenoxyacetic acid (MCPA)	Causes pesticide degradation	Muter et al. (2014)
4	Poultry biochar	Atrazine	Causes pesticide sorption in biochar	Martin et al. (2012)
5	Paper mill biochar	Atrazine	Causes pesticide sorption in biochar	Martin et al. (2012)
6	Eucalyptus wood chip biochar	Isoproturon	Causes pesticide sorption in biochar	Sopeña et al. (2012)
7	Composed olive waste	Tricyclazole	Causes pesticide sorption in biochar	García-Jaramillo et al. (2014)
8	Wood chip biochar	Aminocyclopyrachlor	Causes pesticide sorption in biochar	Cabrera et al. (2014)
9	Wheat straw and rice straw biochar	Atrazine and imidacloprid	Causes pesticide sorption in biochar	Jin et al. (2016)
10	Rice hull biochar	Fomesafen	Causes pesticide sorption in biochar	Khorram et al. (2017)
11	Rice husk	Methamidophos, phorate, terbufos, parathion, isocarbophos	Causes pesticide desorption	Zheng et al. (2018)
12	Maize and pig manure-derived biochar	Thiacloprid	Causes pesticide sorption	Zhang et al. (2019)
13	Bamboo chip-, rice straw and husk-, eucalyptus bark-, and corn cob-derived biochar	Atrazine	Causes pesticide adsorption	Mandal et al. (2017)
14	*Azadirachta indica*-derived biochar	Bentazone	Causes sorption of pesticide	Ponnam et al. (2020)
15	Coconut shell	Diazinon	Pesticide adsorption	Baharum et al. (2020)

(continued)

Table 10.3 (continued)

S. no.	Biochar used	Pesticide remediated	Biochar response in polluted soil	References
16	Rice straw biochar	Carbamate and organophosphate	Decreases uptake of pesticide in *Allium cepa* var. aggregatum	Hemowng et al. (2021)

lead (Netherway et al. 2019). One more investigatory report shows that various types of biochar application such as sewage sludge-derived biochar (SSB), wood charcoal powder (i.e., wood biochar WB), raw sewage sludge (SS), and their blending (WB/SS) improves soil pH and immobilize of Cd, Pb, and Zn (Penido et al. 2019). Similarly, dairy manure-derived biochars (BCs) positively affect bioavailability and mobility reduction for Pb, Cd, Zn, and Cu (Chen et al. 2018). Holm oak-derived biochar were reported to immobilize Cd and Zn in the soil (Egene et al. 2018). Biochar not only improves contaminated soil quality but also causes decreased uptake of metal ions by plants grown in that heavy metal-contaminated soil (as shown in Table 10.4).

10.4 Potential of Biochar Against Plant Diseases

Global warming, climate change, greenhouse gases, and food security are some consequences of overexploitation of natural resources. Overindulgence of pesticides and fertilizers has been witnessed to meet the food demand. Lack of substantial efforts toward the soil health has led to proliferation of plant diseases. Biochar, a porous and alkaline medium with high adsorption capacity, is a promising alternative and a sustainable approach to such emerging issues. It also enhances nutrient availability to plants and water retention capacity of the soil (Frenkel et al. 2017) and thus the best substitute for soil amendment. Endurance of biochar in the soils is considered to be responsible for such positive effects. Studies were conducted to find the mechanism that helped biochar to induce SAR (systemic acquired resistance) and ISR (induced systemic resistance) pathways. A study was conducted to check if root zone application of biochar was capable of systemic induced resistance (SAR) in a foliar fungal infection and foliar mite infestation caused by biotrophic (*Leveillula taurica*) and necrotrophic (*Botrytis cinerea*) foliar pathogens in tomato and pepper plants. Induced systemic resistance was found at 1 to 5% biochar concentration in both the plants. But at most instances no significant results were recorded. A large-scale research on several other pathosystems especially foliar and soil pathogens has been done (Elad et al. 2011; Graber et al. 2014a, b).

Molecular evidences for systemic induction of plant diseases by biochar via SAR and ISR pathways have been reported in strawberry plants grown with biochar (Meller et al. 2012).

Table 10.4 Heavy metal uptake in plants under the effect of biochar

S. no.	Plant used	Biochar used	Effect on heavy metal	Effect on plant	References
1	Ryegrass (*Lolium perenne* L. var. Cadix)	Biochar derived from British oak, ash, sycamore, and birch	Causes the immobilization of both the heavy metals	Reduction in Cu and Pb in the ryegrass	Karami et al. (2011)
2	*Brassica juncea* L.	Quercus ilex wood-derived biochar	Causes the reduction in Cu, Pb, Ni, and Zn in soil	Reduction in the phytoavailable metal concentration	Forján et al. (2018)
3	Blue wildrye (*Elymus glaucus* cv. "Elkton")	Miscanthus (*Miscanthus giganteus*) biochar	Reduction in Cu, Zn, and Al content	After lime addition there was significant increase in the growth of blue wildrye	Novak et al. (2018)
4	Wheat (*Triticum aestivum* L.)	Wheat straw biochar	Reduced heavy metal uptake (Cd and Pb)	Increase in grain yield	Sui et al. (2018)
5	*Miscanthus* × *giganteus* rhizomes	Poplar biochar (P-BC)	Reduced $NH_4NO_3^-$ extractable Cd, Pb, and Zn concentrations	*Miscanthus* is an potential metal excluder	Karer et al. (2018)
6	Maize (*Zea maize*)	Wheat straw biochar	Reduction in available Pb and Cd in soil	Reduction in the Pb and Cd content in maize plant	Zhan et al. (2019)
7	*Oryza sativa*	Rice straw (RB) and coconut by-product-derived biochar (CB)	Remediation of Cd from soil	Increased rice yield and reduced Cd accumulation in rice	Chen et al. (2018)
8	Palmarosa (*Cymbopogon martini* (Roxb.)	*Cymbopogon flexuosus* (lemongrass) waste produced after distillation process	Reduced soil acidity and bioavailability of Al, Cr, Cu, and Pb	Improved soil health, plant growth, and yield	Jain et al. (2019)
9	*Brassica juncea* L.	Manure waste (M) collected from rabbit farm biochar	Causes reduction of As, Cu, Pb, and Se in soil	Increase biomass production and soil quality	Gascó et al. (2019)
10	Soybean	Sugarcane bagasse-derived biochar (SBDB)	Immobilization of Cd	Increases soybean biomass	Mohamed et al. (2019)

(continued)

Table 10.4 (continued)

S. no.	Plant used	Biochar used	Effect on heavy metal	Effect on plant	References
11	*Oryza sativa*	Water-washed rice straw biochar (W-BC)	Immobilization of Cd	Reduced Cd content in root and shoots	Li et al. (2019)
12	*Oryza sativa*	Iron-modified biochar	Reduction in mobility of Cd and As	Decreased Cd and As accumulation in rice	Pan et al. (2019)
13	Cabbage and wheat	Rice straw biochar	Reduced Pb and Cu mobility	Reduced Pb and Cu uptake in plant	Salam et al. (2019)
14	*Oryza sativa* (rice)	Wheat straw biochar	Supplemented with Si-enhanced immobilization of Cd	Decreased Cd uptake in rice plants	Sui et al. (2020)

ISR develops in response to the colonization of plant roots by plant growth-promoting rhizobacteria (PGPR) and plant growth-promoting fungi (PGPF), whereas SAR seems to be induced both by microorganisms and various chemical agents like residual tars present in biochar (Graber et al. 2010). It was assumed that plants respond to the stress of low levels of phytotoxic compounds in the root zone via induced resistance mechanism as evident in *Arabidopsis thaliana*. The plant exhibited an inverted u-shaped growth response to catechins. At low levels of catechins, leaves developed restricted lesions only at the site of inoculation, whereas at high concentrations these are phytotoxic 15 pathogens and 30 different pathosystems were tested for the highest concentration of biochar on disease severity. It was found that 70% of the pathosystems and 60% of the pathogen's response to the said dose (0.5–5%) were negative or neutral effect on the plant disease in comparison to the control (Bonanomi et al. 2015).

Out of the six pathosystems, only 33% suggested a u-shaped relationship in case of foliar pathogens, but it was 83% in soilborne pathogens (Elad et al. 2010, 2011; Harel et al. 2012).

Jaiswal et al. (2015) tested the impact of biochar produced from different feedstocks on soilborne pathogen *R. solani* infecting beans and cucumber. Biochar concentration up to 3% by weight resulted in higher plant biomass in healthy plants, but it was beneficial only up to 1% in diseased plants. He termed it as the "Shifted R_{max}-effect," where R_{max} refers to the biochar dose at which there is maximum growth response with maximum disease reduction. Similar experiment by Akhter et al. (2015) showed 3% biochar (green waste as feedstock) that increased shoot and root length of tomato plants grown in disease-free growth medium, while the same concentration was conducive to disease caused by *F. oxysporium lycopersici*. Experiments conducted by Viger et al. (2015) on *Arabidopsis thaliana* indicated that upregulation of genes involved in brassinosteroids and auxin pathway were responsible for the stimulation of growth in the root length, leaf area, and rosette width (50–150%), when plant was exposed to higher quantities of biochar blended with soil. The shifted R_{max} paradigm could thus explain the contradictory results reported on strawberry plant growth by Harel et al. (2012) and at low biochar doses (1% and 3% by weight).

Biochar contains a number of macro- and micromolecules which individually or in combination may show a dose-dependent effect. This is quite similar to hormesis effect exhibited by a wide range of materials. Variation in the biochar composition depends on the feedstock and the temperature at which biochar is formed (Spokas et al. 2012a). This could also be one of the reasons as to why foliar pathogens express low sensitivity to higher biochar concentration than soilborne pathogens.

The frequent u-shaped biochar dose/response curves indicate the extensive use of biochar as replacement of peat needs careful evaluation (Frenkel et al. 2017). Firstly, because of the variable nature of biochar feedstocks that have different concentration of macro- and micromolecules subject to different production conditions. It is thus not advisable to generalize their rates or concentration of application for disease suppression. Hence proper standardization procedures/techniques have to be followed to get reliable and reproducible results. Secondly, further studies need to be carried out on more crop-pathogen biochar systems to validate if biochar can be applied along with other alternative growth media. Thirdly, to increase the efficacy, pre-treatment of the biochar substrates with enhancers (physical/chemical) and then combining it with alternative growing media is another aspect that needs to be studied to standardize the safe limits of biochar application.

10.5 Effect of Biochar on Plant Growth and Productivity

Multiple reports have been enunciated to describe the positive role of biochar on plant's productivity (Baronti et al. 2014). Further, Hass et al. (2012) reported that biochar amendment adds on to crop's yield due to its soil liming effect. There are many underlying mechanisms which promote plant productivity on biochar input, i.e., changes physical conditions of plant, and thermal dynamics is altered and fosters rapid germination (Genesio et al. 2015). Muhammad et al. (2017) reported that biochar addition to soil increases carbon sequestration in soil for longer duration. Reason behind its beneficial role for plant is enhancement in water holding capacity of soil resulting in thinner and extensively branched roots in the soil. It also lessened leaching of nutrients, namely, nitrogen (N) and phosphorus (P) (Bruun et al. 2014). Apart from this, it reduces bioavailability of phytotoxic elements in soils to crop (Al-Wabel et al. 2017). Moreover, Steiner et al. (2007) suggested that single-use of biochar in field can have profitable impact on yield for subsequent growing seasons. Biochar application does not need to be applied in all seasons as it is done for fertilizers. Biochar is endorsed as potential soil amendment; Table 10.5 illustrates response of biochar to crop's productivity.

10.6 Role of Biochar in the Reduction of Greenhouse Gas Emissions from the Soil

Anthropogenic activities are major cause in the rising concentration of greenhouse gases (GHG) to the atmosphere. In the last century, global surface temperature has increased by 0.87 °C due to increased emission of carbon dioxide (CO_2), nitrous oxide (N_2O), and methane (CH_4) ultimately leading to climate change. It has been proposed that biochar application is a promising option in reducing soil greenhouse gas emissions and mitigating climate change. Porous structure of biochar prevents the drainage of compost from the soil thereby reducing the environmental pollution (Cao et al. 2018). Carbon sequestration property of biochar bound most of the carbon in the soil. Plants fix this carbon by photosynthesis and check the emission of CO_2 in the atmosphere. N_2O emission is lowered, and CH_4 consumption is enhanced from fertile temperate prairie soils treated with shrub willow biochar (Hangs et al. 2016). Nitrification, NO_3 ammonification, and denitrification processes generate nitrous

Table 10.5 Effect of biochar from different sources on crop productivity

S. no.	Crop	Biochar source	Effect on crop yield	References
1	Rice and sorghum	Teak and rose-wood biochar	Improves plant growth and yield two to three times	Steiner et al. (2007)
2	Rice	*Tectona grandis* (teak) wood	Grain yield increased	Partey et al. (2016)
3	Pepper and tomato	Citrus wood biochar	Increased yield and canopy dry weight	Graber et al. (2010)
4	Apple	Green waste of acacia	Increase in tree trunk girth	Eyles et al. (2015)
5	Maize	Mango wood	Increases plant biomass and grain yield	Rajkovich et al. (2012)
6	Rice	Green waste compost	Increase in grain yield	Bakar et al. (2015)
7	Wheat	Pine needle	Increases grain yield significantly by 24.2%	Bhattacharjya et al. (2016)
8	Maize	Poultry manure biochar	Increased the biomass of maize	Abbasi and Anwar (2015)
9	Maize	Paper mill biochar	Significantly increased corn yields than the control treatment	Van Zwieten et al. (2015a)
10	Rice	Bamboo	Promoted K contents of rice grains	Lu et al. (2017)
11	Rice	Rice straw	Improved the total nutrient uptake by grain	Lu et al. (2017)
12	Bean	Poultry manure	Increased the growth of plants in calcareous soil	Taskin et al. (2019)
13	Groundnut	Maize cobs	Better crop yield	Martinsen et al. (2014)
14	Maize	Bark of *Acacia mangium*	Yield increased by 50%	Yamato et al. (2006)
15	Maize	Acacia	Increased maize yields by 20%	Arif et al. (2016)
16	Chinese cabbage	Kunai grass	BC with urea significantly increased the yield	Baiga and Rajashekhar Rao (2017)
17	Rice	Wheat straw	Grain yield increased by 18.3%	Bian et al. (2014a, b)
18	Radish	Green waste	Biomass increased by 91%	Chan et al. (2007)
19	Lettuce	Poultry manure	Increased the lettuce growth significantly in alkaline soil	Gunes et al. (2014)
20	Rice	Pig manure compost	Grain size increased by 13.49%	Qian et al. (2014)
21	Rice	Peanut husk	Grain yield increased by 28.1%	Qian et al. (2014)
22	Lettuce	Rice husk char	Increase in biomass	Carter et al. (2013)
23	Rice-wheat	Rice straw	Grain yield increased by 14.8%	Zhao et al. (2014)
24	Radish	Green waste	Increase in biomass by 130%	Chan et al. (2007)

(continued)

Table 10.5 (continued)

S. no.	Crop	Biochar source	Effect on crop yield	References
25	*Phaseolus vulgaris*	*Eucalyptus deglupta*	Seed yield increased by 46%	Rondon et al. (2007)
26	Choy sum and amaranth	Maize straw	Increases crop yield by 48%	Jia et al. (2012)
27	Rice husk	Rice	Grain yield increased by 35%	Haefele et al. (2011)
28	Cotton	Corn straw	21% yield increased over control	Tian et al. (2018)
29	Rice	Rice straw	7% yield increased over control	Si et al. (2018)
30	Maize	Hardwood	48% yield increased over control	Rogovska et al. (2014)
31	Rapeseed	Wheat straw	36% yield increased over control	Liu et al. (2014)
32	Maize	Willow wood	29% yield increased over control	Agegnehu et al. (2016)
33	Grape	Orchard pruning	20% yield increased over control	Genesio et al. (2015)

oxide in the soil through microbial activity (Gillam et al. 2008). N-cycling processes are regulated due to variation in the physicochemical properties of biochar augmented soil (Spokas et al. 2012b). It has been reported that biochar-amended soils having greater porosity and high surface area adsorb more organic C, and inorganic N reduces the bioavailability of carbon to denitrifiers which lessens the production of N_2O to 50–90% (Ameloot et al. 2016).

It has been studied that surface sorption capacity of biochar decreases the methane release from the soil (Jeffrey et al. 2016; Yaghoubi et al. 2014). Improvement of soil aeration due to developed porosity of biochar may increase the adsorption and oxidation of CH_4 (Liu et al. 2011; Yoo and Kang 2012). It was stated that the action of methanogenic archaea is stopped whereas improved that of methanotrophic bacteria in the habitat augmented with biochar leads to the suppression of CH_4 emission and enhancing its oxidation (Van Zwieten et al. 2009; Karhu et al. 2011; Qin et al. 2016).

10.7 Conclusion

In this chapter, we discussed the impact of biochar on soil quality and fertility improvement, its potential against plant diseases, reduction of greenhouse gas emission, heavy metal remediation, and crop growth and productivity in detail. Interactions among biochar and microbes play a crucial role in crop production and environment. Biochar is a recalcitrant form of carbon and a better option for mitigating GHG emission through long-term carbon storage and generation of renewable energy. Rate of biochar application and its characteristics depends on the feedstock type used, pyrolysis temperature, and vaporization conditions. However, there are some research gaps related to use as soil amendment, severe climatic effect on biochar, and strategies for the generation and use of biochar. But it is clear that biochar application could be a supportable approach for sustainable crop and bioenergy production and satisfactory research in this area will be one of the most valuable steps people can take and focus for future studies.

References

Abbasi MK, Anwar AA (2015) Ameliorating effects of biochar derived from poultry manure and white clover residues on soil nutrient status and plant growth promotion greenhouse experiments. PloS One 10(6):e0131592

Abid M, Danish S, Zafar-ul-Hye M, Shaaban M, Iqbal MM, Rehim A, Qayyum MF, Naqqash MN (2017) Biochar increased photosynthetic and accessory pigments in tomato (Solanum lycopersicum L.) plants by reducing cadmium concentration under various irrigation waters. Environ Sci Pollut Res 24(27):22111–22118

Abujabhah IS, Doyle R, Bound SA, Bowman JP (2016) The effect of biochar loading rates on soil fertility, soil biomass, potential nitrification, and soil community metabolic profiles in three different soils. J Soil Sediment 16(9):2211–2222

Agegnehu G, Bass AM, Nelson PN, Muirhead B, Wright G, Bird MI (2015) Biochar and biochar-compost as soil amendments: effects on peanut yield, soil properties and greenhouse gas emissions in tropical North Queensland, Australia. Agr Ecosyst Environ 213:72–85

Agegnehu G, Bass AM, Nelson PN, Bird MI (2016) Benefits of biochar, compost and biochar–compost for soil quality, maize yield and greenhouse gas emissions in a tropical agricultural soil. Sci Total Environ 543:295–306

Ahmad M, Ok YS, Kim BY, Ahn JH, Lee YH, Zhang M, Moon DH, Al-Wabel MI, Lee SS (2016) Impact of soybean stover-and pine needle-derived biochars on Pb and As mobility, microbial community, and carbon stability in a contaminated agricultural soil. J Environ Manage 166: 131–139

Ahmad Z, Mosa A, Zhan L, Gao B (2021) Biochar modulates mineral nitrogen dynamics in soil and terrestrial ecosystems: a critical review. Chemosphere 278:130378

Akhter A, Hage-Ahmed K, Soja G, Steinkellner S (2015) Compost and biochar alter mycorrhization, tomato root exudation, and development of Fusarium oxysporum f.sp lycopersici. Front Plant Sci 6:529

Ali A, Guo D, Jeyasundar PGSA, Li Y, Xiao R, Du J, Li R, Zhang Z (2019) Application of wood biochar in polluted soils stabilized the toxic metals and enhanced wheat (Triticum aestivum) growth and soil enzymatic activity. Ecotoxicol Environ Saf 184:109635

Al-Wabel MI, Sallam AEAS, Usman AR, Ahmad M, El-Naggar AH, El-Saeid MH, Al-Faraj A, El-Enazi K, Al-Romian FA (2017) Trace metal levels, sources, and ecological risk assessment in a densely agricultural area from Saudi Arabia. Environ Monit Assess 189(6):252

Ameloot N, Maenhout P, De Neve S, Sleutel S (2016) Biochar-induced N_2O emission reductions after field incorporation in a loam soil. Geoderma 267:10–16

Arif M, Ali K, Jan MT, Shah Z, Jones DL, Quilliam RS (2016) Integration of biochar with animal manure and nitrogen for improving maize yields and soil properties in calcareous semi-arid agroecosystems. Field Crop Res 195:28–35

Arunkumar BR, Thippeshappa GN (2020) Effect of levels of biochar and FYM on physico-chemical properties, nutrients release and carbon dioxide evolution from the soil. IJCS 8(1): 2389–2395

Azeem M, Hayat R, Hussain Q, Ahmed M, Pan G, Tahir MI, Imran M, Irfan M (2019) Biochar improves soil quality and N2-fixation and reduces net ecosystem CO2 exchange in a dryland legume-cereal cropping system. Soil Tillage Res 186:172–182

Azeem M, Hassan TU, Tahir MI, Ali A, Jeyasundar PGSA, Hussain Q, Bashir S, Mehmood S, Zhang Z (2021) Tea leaves biochar as a carrier of Bacillus cereus improves the soil function and crop productivity. Appl Soil Ecol 157:103732

Baharum NA, Nasir HM, Ishak MY, Isa NM, Hassan MA, Aris AZ (2020) Highly efficient removal of diazinon pesticide from aqueous solutions by using coconut shell-modified biochar. Arab J Chem 13(7):6106–6121

Baiga R, Rajashekhar Rao BK (2017) Effects of biochar, urea and their co-application on nitrogen mineralization in soil and growth of Chinese cabbage. Soil Use Manage 33(1):54–61

Bakar RA, Razak ZA, Ahmad SH, Seh-Bardan BJ, Tsong LC, Meng CP (2015) Influence of oil palm empty fruit bunch biochar on floodwater pH and yield components of rice cultivated on acid sulphate soil under rice intensification practices. Plant Prod Sci 18(4):491–500

Bamminger C, Zaiser N, Zinsser P, Lamers M, Kammann C, Marhan S (2014) Effects of biochar, earthworms, and litter addition on soil microbial activity and abundance in a temperate agricultural soil. Biol Fertil Soils 50(8):1189–1200

Baronti S, Vaccari FP, Miglietta F, Calzolari C, Lugato E, Orlandini S, Pini R, Zulian C, Genesio L (2014) Impact of biochar application on plant water relations in Vitis vinifera (L.). Eur J Agron 53:38–44

Bashir S, Rehman M, Yousaf M, Salam A, Gulshan AB, Iqbal J, Aziz I, Azeem M, Rukh S, Asghar RMA (2019) Comparative efficiency of wheat straw and sugarcane bagasse biochar reduces the cadmium bioavailability to spinach and enhances the microbial activity in contaminated soil. Int J Phytoremediation 21(11):1098–1103

Beusch C, Cierjacks A, Böhm J, Mertens J, Bischoff WA, de Araújo Filho JC, Kaupenjohann M (2019) Biochar vs. clay: Comparison of their effects on nutrient retention of a tropical Arenosol. Geoderma 337:524–535

Bhattacharjya S, Chandra R, Pareek N, Raverkar KP (2016) Biochar and crop residue application to soil: effect on soil biochemical properties, nutrient availability and yield of rice (Oryza sativa L.) and wheat (Triticum aestivum L.). Arch Agron Soil Sci 62(8):1095–1108

Bian R, Joseph S, Cui L, Pan G, Li L, Liu X, Zhang A, Rutlidge H, Wong S, Chia C, Marjo C (2014a) A three-year experiment confirms continuous immobilization of cadmium and lead in contaminated paddy field with biochar amendment. J Hazard Mater 272:121–128

Bian R, Zhang A, Li L, Pan G, Zheng J, Zhang X, Zheng J, Joseph S, Chang A (2014b) Effect of municipal biowaste biochar on greenhouse gas emissions and metal bioaccumulation in a slightly acidic clay rice paddy. BioResources 9(1):685–703

Biederman LA, Harpole WS (2013) Biochar and its effects on plant productivity and nutrient cycling: a meta-analysis. GCB Bioenergy 5:202–214

Blanco-Canqui H (2021) Does biochar improve all soil ecosystem services? GCB Bioenergy 13(2): 291–304

Bonanomi G, Ippolito F, Scala F (2015) A "black" future for plant pathology? Biochar as a new soil amendment for controlling plant diseases. J Plant Pathol 97(2):223–234

Borchard N, Schirrmann M, Cayuela ML, Kammann C, Wrage-Mönnig N, Estavillo JM, Fuertes-Mendizábal T, Sigua G, Spokas K, Ippolito JA, Novak J (2019) Biochar, soil and land-use interactions that reduce nitrate leaching and N2O emissions: a meta-analysis. Sci Total Environ 651:2354–2364

Bruun EW, Petersen CT, Hansen E, Holm JK, Hauggaard-Nielsen H (2014) Biochar amendment to coarse sandy subsoil improves root growth and increases water retention. Soil Use Manage 30(1):109–118

Cabrera A, Cox L, Spokas KURT, Hermosín MC, Cornejo J, Koskinen WC (2014) Influence of biochar amendments on the sorption–desorption of aminocyclopyrachlor, bentazone and pyraclostrobin pesticides to an agricultural soil. Sci Total Environ 470:438–443

Cao Y, Gao Y, Qi Y, Li J (2018) Biochar-enhanced composts reduce the potential leaching of nutrients and heavy metals and suppress plant-parasitic nematodes in excessively fertilized cucumber soils. Environ Sci Pollut Res 25(8):7589–7599

Carter S, Shackley S, Sohi S, Suy TB, Haefele S (2013) The impact of biochar application on soil properties and plant growth of pot grown lettuce (Lactuca sativa) and cabbage (Brassica chinensis). Agronomy 3(2):404–418

Castrillo G, Teixeira PJPL, Paredes SH, Law TF, de Lorenzo L, Feltcher ME, Finkel OM, Breakfield NW, Mieczkowski P, Jones CD, Paz-Ares J (2017) Root microbiota drive direct integration of phosphate stress and immunity. Nature 543(7646):513–518

Cayuela ML, Van Zwieten L, Singh BP, Jeffery S, Roig A, Sánchez-Monedero MA (2014) Biochar's role in mitigating soil nitrous oxide emissions: A review and meta-analysis. Agr Ecosyst Environ 191:5–16

Cayuela ML, Jeffery S, van Zwieten L (2015) The molar H: Corg ratio of biochar is a key factor in mitigating N2O emissions from soil. Agr Ecosyst Environ 202:135–138

Cely P, Tarquis AM, Paz-Ferreiro J, Méndez A, Gascó G (2014) Factors driving the carbon mineralization priming effect in a sandy loam soil amended with different types of biochar. Solid Earth 5(1):585–594

Chan KY, Van Zwieten L, Meszaros I, Downie A, Joseph S (2007) Agronomic values of greenwaste biochar as a soil amendment. Soil Res 45(8):629–634

Chen Z, Zhang J, Liu M, Wu Y, Yuan Z (2018) Immobilization of metals in contaminated soil from E-waste recycling site by dairy-manure-derived biochar. Environ Technol 39(21):2801–2809

Chintala R, Schumacher TE, McDonald LM, Clay DE, Malo DD, Papiernik SK, Clay SA, Julson JL (2014) Phosphorus sorption and availability from biochars and soil/biochar mixtures. CLEAN Soil Air Water 42(5):626–634

Choudhary TK, Khan KS, Hussain Q, Ashfaq M (2021) Nutrient availability to maize crop (Zea mays L.) in biochar amended alkaline subtropical soil. J Soil Sci Plant Nutr 21:1293–1306

Clough TJ, Condron LM, Kammann C, Müller C (2013) A review of biochar and soil nitrogen dynamics. Agronomy 3(2):275–293

Cui H, Ma K, Fan Y, Peng X, Mao J, Zhou D, Zhang Z, Zhou J (2016) Stability and heavy metal distribution of soil aggregates affected by application of apatite, lime, and charcoal. Environ Sci Pollut Res 23(11):10808–10817

Dai Z, Barberan A, Li Y, Brookes PC, Xu J (2017a) Bacterial community composition associated with pyrogenic organic matter (biochar) varies with pyrolysis temperature and colonization environment. mSphere 2(2):e00085

Dai Z, Zhang X, Tang C, Muhammad N, Wu J, Brookes PC, Xu J (2017b) Potential role of biochars in decreasing soil acidification-a critical review. Sci Total Environ 581:601–611

Delwiche KB, Lehmann J, Walter MT (2014) Atrazine leaching from biochar-amended soils. Chemosphere 95:346–352

Egene CE, Van Poucke R, Ok YS, Meers E, Tack FMG (2018) Impact of organic amendments (biochar, compost and peat) on Cd and Zn mobility and solubility in contaminated soil of the Campine region after three years. Sci Total Environ 626:195–202

Elad Y, David DR, Harel YM, Borenshtein M, Kalifa HB, Silber A, Graber ER (2010) Induction of systemic resistance in plants by biochar, a soil-applied carbon sequestering agent. Phytopathology 100(9):913–921

Elad Y, Cytryn E, Harel YM, Lew B, Graber ER (2011) The biochar effect: plant resistance to biotic stresses. Phytopathol Mediterr 50(3):335–349

Eyles A, Bound SA, Oliver G, Corkrey R, Hardie M, Green S, Close DC (2015) Impact of biochar amendment on the growth, physiology and fruit of a young commercial apple orchard. Trees 29(6):1817–1826

Feng Z, Zhu L (2017) Impact of biochar on soil N2O emissions under different biochar-carbon/fertilizer-nitrogen ratios at a constant moisture condition on a silt loam soil. Sci Total Environ 584:776–782

Forján R, Rodríguez-Vila A, Cerqueira B, Covelo EF (2018) Comparison of compost with biochar versus technosol with biochar in the reduction of metal pore water concentrations in a mine soil. J Geochem Explor 192:103–111

Frenkel O, Jaiswal AK, Elad Y, Lew B, Kammann C, Graber ER (2017) The effect of biochar on plant diseases: what should we learn while designing biochar substrates? J Environ Eng Landsc Manag 25(2):105–113

Fungo B, Lehmann J, Kalbitz K, Thionĝo M, Okeyo I, Tenywa M, Neufeldt H (2017) Aggregate size distribution in a biochar-amended tropical Ultisol under conventional hand-hoe tillage. Soil Tillage Res 165:190–197

Futa B, Oleszczuk P, Andruszczak S, Kwiecińska-Poppe E, Kraska P (2020) Effect of natural aging of biochar on soil enzymatic activity and physicochemical properties in long-term field experiment. Agronomy 10(3):449

García-Jaramillo M, Cox L, Cornejo J, Hermosín MC (2014) Effect of soil organic amendments on the behavior of bentazone and tricyclazole. Sci Total Environ 466:906–913

Gascó G, Álvarez ML, Paz-Ferreiro J, Méndez A (2019) Combining phytoextraction by Brassica napus and biochar amendment for the remediation of a mining soil in Riotinto (Spain). Chemosphere 231:562–570

Genesio L, Miglietta F, Baronti S, Vaccari FP (2015) Biochar increases vineyard productivity without affecting grape quality: Results from a four years field experiment in Tuscany. Agr Ecosyst Environ 201:20–25

Getahun A, Muleta D, Assefa F, Kiros S, Hungria M (2020) Biochar and other organic amendments improve the physicochemical properties of soil in highly degraded habitat. Eur J Eng Technol Res 5(3):331–338

Gillam KM, Zebarth BJ, Burton DL (2008) Nitrous oxide emissions from denitrification and the partitioning of gaseous losses as affected by nitrate and carbon addition and soil aeration. Can J Soil Sci 88(2):133–143

Graber ER, Harel YM, Kolton M, Cytryn E, Silber A, David DR, Tsechansky L, Borenshtein M, Elad Y (2010) Biochar impact on development and productivity of pepper and tomato grown in fertigated soilless media. Plant and Soil 337(1):481–496

Graber ER, Frenkel O, Jaiswal AK, Elad Y (2014a) How may biochar influence severity of diseases caused by soilborne pathogens? Carbon Manag 5(2):169–183

Graber ER, Tsechansky L, Lew B, Cohen E (2014b) Reducing capacity of water extracts of biochars and their solubilization of soil Mn and Fe. Eur J Soil Sci 65(1):162–172

Gul S, Whalen JK (2016) Biochemical cycling of nitrogen and phosphorus in biochar-amended soils. Soil Biol Biochem 103:1–15

Gul S, Whalen JK, Thomas BW, Sachdeva V, Deng H (2015) Physico-chemical properties and microbial responses in biochar-amended soils: mechanisms and future directions. Agr Ecosyst Environ 206:46–59

Gunes A, Inal A, Taskin MB, Sahin O, Kaya EC, Atakol ARDA (2014) Effect of phosphorus-enriched biochar and poultry manure on growth and mineral composition of lettuce (Lactuca sativa L. cv.) grown in alkaline soil. Soil Use Manage 30(2):182–188

Guo J, Chen B (2014) Insights on the molecular mechanism for the recalcitrance of biochars: interactive effects of carbon and silicon components. Environ Sci Technol 48(16):9103–9112

Gwenzi W, Chaukura N, Mukome FN, Machado S, Nyamasoka B (2015) Biochar production and applications in sub-Saharan Africa: Opportunities, constraints, risks and uncertainties. J Environ Manage 150:250–261

Haefele SM, Konboon Y, Wongboon W, Amarante S, Maarifat AA, Pfeiffer EM, Knoblauch CJFCR (2011) Effects and fate of biochar from rice residues in rice-based systems. Field Crop Res 121(3):430–440

Hagemann N, Harter J, Kaldamukova R, Guzman-Bustamante I, Ruser R, Graeff S, Kappler A, Behrens S (2017) Does soil aging affect the N2O mitigation potential of biochar? A combined microcosm and field study. GCB Bioenergy 9(5):953–964

Hale L, Luth M, Crowley D (2015) Biochar characteristics relate to its utility as an alternative soil inoculum carrier to peat and vermiculite. Soil Biol Biochem 81:228–235

Han L, Ro KS, Wang Y, Sun K, Sun H, Libra JA, Xing B (2018) Oxidation resistance of biochars as a function of feedstock and pyrolysis condition. Sci Total Environ 616:335–344

Hangs RD, Ahmed HP, Schoenau JJ (2016) Influence of willow biochar amendment on soil nitrogen availability and greenhouse gas production in two fertilized temperate prairie soils. Bioenergy Res 9(1):157–171

Hardy B, Sleutel S, Dufey JE, Cornelis JT (2019) The long-term effect of biochar on soil microbial abundance, activity and community structure is overwritten by land management. Front Environ Sci 7:110

Harel YM, Elad Y, Rav-David D, Borenstein M, Shulchani R, Lew B, Graber ER (2012) Biochar mediates systemic response of strawberry to foliar fungal pathogens. Plant and Soil 357(1): 245–257

Harter J, Weigold P, El-Hadidi M, Huson DH, Kappler A, Behrens S (2016) Soil biochar amendment shapes the composition of N_2O-reducing microbial communities. Sci Total Environ 562:379–390

Harter J, El-Hadidi M, Huson DH, Kappler A, Behrens S (2017) Soil biochar amendment affects the diversity of nosZ transcripts: Implications for N2O formation. Sci Rep 7(1):1–14

Hass A, Gonzalez JM, Lima IM, Godwin HW, Halvorson JJ, Boyer DG (2012) Chicken manure biochar as liming and nutrient source for acid Appalachian soil. J Environ Qual 41(4): 1096–1106

Hawthorne I, Johnson MS, Jassal RS, Black TA, Grant NJ, Smukler SM (2017) Application of biochar and nitrogen influences fluxes of CO2, CH4 and N2O in a forest soil. J Environ Manage 192:203–214

He L, Liu Y, Zhao J, Bi Y, Zhao X, Wang S, Xing G (2016) Comparison of straw-biochar-mediated changes in nitrification and ammonia oxidizers in agricultural oxisols and cambosols. Biol Fertil Soils 52(2):137–149

He L, Zhong H, Liu G, Dai Z, Brookes PC, Xu J (2019) Remediation of heavy metal contaminated soils by biochar: mechanisms, potential risks and applications in China. Environ Pollut 252: 846–855

Hemowng S, Sangrit C, Phunthupan P, Butnan S, Vityakon P (2021) Rice-derived biochars enhance the yield of spring onion (allium cepa l. var. aggregatum), while reducing pesticide contamination in soil and plant. Appl Ecol Environ Res 19(1):349–358

Inyang MI, Gao B, Yao Y, Xue Y, Zimmerman A, Mosa A, Pullammanappallil P, Ok YS, Cao X (2016) A review of biochar as a low-cost adsorbent for aqueous heavy metal removal. Crit Rev Environ Sci Technol 46(4):406–433

Jain MS, Jambhulkar R, Kalamdhad AS (2019) Biochar amendment for batch composting of nitrogen rich organic waste: effect on degradation kinetics, composting physics and nutritional properties. Bioresour Technol 253:204–213

Jain S, Khare P, Mishra D, Shanker K, Singh P, Singh RP, Das P, Yadav R, Saikia BK, Baruah BP (2020) Biochar aided aromatic grass [Cymbopogon martini (Roxb.) Wats.] vegetation: a sustainable method for stabilization of highly acidic mine waste. J Hazard Mater 390:121799

Jaiswal AK, Frenkel O, Elad Y, Lew B, Graber ER (2015) Non-monotonic influence of biochar dose on bean seedling growth and susceptibility to Rhizoctonia solani: the "Shifted R max-Effect". Plant and soil 395(1):125–140

Jeffrey S, Verheijen FGA, Kammann C, Abalos D (2016) Biochar effects on methane emissions from soils: a meta-analysis. Soil Biol Biochem 101:251–258

Jia J, Li B, Chen Z, Xie Z, Xiong Z (2012) Effects of biochar application on vegetable production and emissions of N2O and CH4. Soil Sci Plant Nutr 58(4):503–509

Jin J, Li Y, Zhang J, Wu S, Cao Y, Liang P, Zhang J, Wong MH, Wang M, Shan S, Christie P (2016) Influence of pyrolysis temperature on properties and environmental safety of heavy metals in biochars derived from municipal sewage sludge. J Hazard Mater 320:417–426

Jin Z, Zhang X, Chen X, Du Z, Ping L, Han Z, Tao P (2021) Dynamics of soil organic carbon mineralization and enzyme activities after two months and six years of biochar addition. Biomass Convers Biorefin 13:1153–1162

Jing F, Sohi SP, Liu Y, Chen J (2018) Insight into mechanism of aged biochar for adsorption of PAEs: reciprocal effects of ageing and coexisting Cd2+. Environ Pollut 242:1098–1107

Joseph S, Kammann CI, Shepherd JG, Conte P, Schmidt HP, Hagemann N, Rich AM, Marjo CE, Allen J, Munroe P, Mitchell DR (2018) Microstructural and associated chemical changes during the composting of a high temperature biochar: mechanisms for nitrate, phosphate and other nutrient retention and release. Sci Total Environ 618:1210–1223

Kameyama K, Miyamoto T, Shiono T, Shinogi Y (2012) Influence of sugarcane bagasse-derived biochar application on nitrate leaching in calcaric dark red soil. J Environ Qual 41(4): 1131–1137

Kammann CI, Schmidt HP, Messerschmidt N, Linsel S, Steffens D, Müller C, Koyro HW, Conte P, Joseph S (2015) Plant growth improvement mediated by nitrate capture in co-composted biochar. Sci Rep 5(1):1–13

Karami N, Clemente R, Moreno-Jiménez E, Lepp NW, Beesley L (2011) Efficiency of green waste compost and biochar soil amendments for reducing lead and copper mobility and uptake to ryegrass. J Hazard Mater 191(1-3):41–48

Karer J, Zehetner F, Dunst G, Fessl J, Wagner M, Puschenreiter M, Stapkēviča M, Friesl-Hanl W, Soja G (2018) Immobilisation of metals in a contaminated soil with biochar-compost mixtures and inorganic additives: 2-year greenhouse and field experiments. Environ Sci Pollut Res 25(3): 2506–2516

Karhu K, Mattila T, Bergström I, Regina K (2011) Biochar addition to agricultural soil increased CH4 uptake and water holding capacity–results from a short-term pilot field study. Agr Ecosyst Environ 140(1-2):309–313

Karhu K, Auffret MD, Dungait JA, Hopkins DW, Prosser JI, Singh BK, Subke JA, Wookey PA, Ågren GI, Sebastia MT, Gouriveau F (2014) Temperature sensitivity of soil respiration rates enhanced by microbial community response. Nature 513(7516):81–84

Keiluweit M, Nico PS, Johnson MG, Kleber M (2010) Dynamic molecular structure of plant biomass-derived black carbon (biochar). Environ Sci Technol 44(4):1247–1253

Khalid S, Shahid M, Murtaza B, Bibi I, Naeem MA, Niazi NK (2020) A critical review of different factors governing the fate of pesticides in soil under biochar application. Sci Total Environ 711: 134645

Khorram MS, Lin D, Zhang Q, Zheng Y, Fang H, Yu Y (2017) Effects of aging process on adsorption–desorption and bioavailability of fomesafen in an agricultural soil amended with rice hull biochar. J Environ Sci 56:180–191

Krause HM, Hüppi R, Leifeld J, El-Hadidi M, Harter J, Kappler A, Hartmann M, Behrens S, Mäder P, Gattinger A (2018) Biochar affects community composition of nitrous oxide reducers in a field experiment. Soil Biol Biochem 119:143–151

Kuranchie-Mensah H, Atiemo SM, Palm LMND, Blankson-Arthur S, Tutu AO, Fosu P (2012) Determination of organochlorine pesticide residue in sediment and water from the Densu river basin, Ghana. Chemosphere 86(3):286–292

Kuzyakov Y, Subbotina I, Chen H, Bogomolova I, Xu X (2009) Black carbon decomposition and incorporation into soil microbial biomass estimated by 14C labeling. Soil Biol Biochem 41(2): 210–219

Laghari M, Naidu R, Xiao B, Hu Z, Mirjat MS, Hu M, Kandhro MN, Chen Z, Guo D, Jogi Q, Abudi ZN (2016) Recent developments in biochar as an effective tool for agricultural soil management: a review. J Sci Food Agric 96(15):4840–4849

Li H, Ye X, Geng Z, Zhou H, Guo X, Zhang Y, Zhao H, Wang G (2016) The influence of biochar type on long-term stabilization for Cd and Cu in contaminated paddy soils. J Hazard Mater 304: 40–48

Li H, Li Z, Khaliq MA, Xie T, Chen Y, Wang G (2019) Chlorine weaken the immobilization of Cd in soil-rice systems by biochar. Chemosphere 235:1172–1179

Lian F, Xing B (2017) Black carbon (biochar) in water/soil environments: molecular structure, sorption, stability, and potential risk. Environ Sci Technol 51(23):13517–13532
Liang B, Lehmann J, Sohi SP, Thies JE, O'Neill B, Trujillo L, Gaunt J, Solomon D, Grossman J, Neves EG, Luizão FJ (2010) Black carbon affects the cycling of non-black carbon in soil. Org Geochem 41(2):206–213
Liao N, Li Q, Zhang W, Zhou G, Ma L, Min W, Ye J, Hou Z (2016) Effects of biochar on soil microbial community composition and activity in drip-irrigated desert soil. Eur J Soil Biol 72: 27–34
Lin Y, Ding W, Liu D, He T, Yoo G, Yuan J, Chen Z, Fan J (2017) Wheat straw-derived biochar amendment stimulated N2O emissions from rice paddy soils by regulating the amoA genes of ammonia-oxidizing bacteria. Soil Biol Biochem 113:89–98
Liu Y, Yang M, Wu Y, Wang H, Chen Y, Wu W (2011) Reducing CH 4 and CO 2 emissions from waterlogged paddy soil with biochar. J Soil Sediment 11(6):930–939
Liu Z, Chen X, Jing Y, Li Q, Zhang J, Huang Q (2014) Effects of biochar amendment on rapeseed and sweet potato yields and water stable aggregate in upland red soil. Catena 123:45–51
Liu Y, Lu H, Yang S, Wang Y (2016) Impacts of biochar addition on rice yield and soil properties in a cold waterlogged paddy for two crop seasons. Field Crop Res 191:161–167
Liu Z, He T, Cao T, Yang T, Meng J, Chen W (2017) Effects of biochar application on nitrogen leaching, ammonia volatilization and nitrogen use efficiency in two distinct soils. J Soil Sci Plant Nutr 17(2):515–528
Liu Q, Zhang Y, Liu B, Amonette JE, Lin Z, Liu G, Ambus P, Xie Z (2018) How does biochar influence soil N cycle? A meta-analysis. Plant and Soil 426(1):211–225
Liu D, Feng Z, Zhu H, Yu L, Yang K, Yu S, Zhang Y, Guo W (2020) Effects of corn straw biochar application on soybean growth and alkaline soil properties. BioResources 15(1):1463–1481
Lone AH, Najar GR, Ganie MA, Sofi JA, Ali T (2015) Biochar for sustainable soil health: a review of prospects and concerns. Pedosphere 25(5):639–653
Lu K, Yang X, Gielen G, Bolan N, Ok YS, Niazi NK, Xu S, Yuan G, Chen X, Zhang X, Liu D (2017) Effect of bamboo and rice straw biochars on the mobility and redistribution of heavy metals (Cd, Cu, Pb and Zn) in contaminated soil. J Environ Manage 186:285–292
Luo Y, Durenkamp M, De Nobili M, Lin Q, Brookes PC (2011) Short term soil priming effects and the mineralisation of biochar following its incorporation to soils of different pH. Soil Biol Biochem 43(11):2304–2314
Luo L, Xu C, Chen Z, Zhang S (2015) Properties of biomass-derived biochars: combined effects of operating conditions and biomass types. Bioresour Technol 192:83–89
Luo S, Wang S, Tian L, Li S, Li X, Shen Y, Tian C (2017) Long-term biochar application influences soil microbial community and its potential roles in semiarid farmland. Appl Soil Ecol 117:10–15
Malik Z, Yutong Z, ShengGao L, Abassi GH, Ali S, Kamran M, Jamil M, Al-Wabel MI, Rizwan M (2018) Effect of biochar and quicklime on growth of wheat and physicochemical properties of ultisols. Arab J Geosci 11(17):1–12
Malinowski M, Wolny-Koładka K, Vaverková MD (2019) Effect of biochar addition on the OFMSW composting process under real conditions. Waste Manag 84:364–372
Mandal A, Singh N, Purakayastha TJ (2017) Characterization of pesticide sorption behaviour of slow pyrolysis biochars as low cost adsorbent for atrazine and imidacloprid removal. Sci Total Environ 577:376–385
Manolikaki I, Diamadopoulos E (2017) Ryegrass yield and nutrient status after biochar application in two Mediterranean soils. Arch Agron Soil Sci 63(8):1093–1107
Martin SM, Kookana RS, Van Zwieten L, Krull E (2012) Marked changes in herbicide sorption–desorption upon ageing of biochars in soil. J Hazard Mater 231:70–78
Martinsen V, Mulder J, Shitumbanuma V, Sparrevik M, Børresen T, Cornelissen G (2014) Farmer-led maize biochar trials: effect on crop yield and soil nutrients under conservation farming. J Plant Nutr Soil Sci 177(5):681–695

Mia S, Dijkstra FA, Singh B (2018) Enhanced biological nitrogen fixation and competitive advantage of legumes in mixed pastures diminish with biochar aging. Plant and Soil 424(1): 639–651

Meller HY, Elad Y, Rav-David D, Borenstein M, Shulchani R, Lew B, Graber ER (2012) Biochar mediates systemic response of strawberry to foliar fungal pathogens. Plant and Soil 357:245-257

Mian IA, Ahmad B, Khan S, Khan B, Dawar K, Tariq M, Mussarat M, Muhammad MW, Ali S, Bibi H, Muhammad F (2021) Improving wheat productivity and soil quality through integrated phosphorous management with residual effect of biochar. J Saudi Chem Soc 25(1):101175

Mohamed I, Ali M, Ahmed N, Chen F (2019) Cadmium immobilization and alleviation of its toxicity for soybean grown in a clay loam contaminated soil using sugarcane bagasse-derived biochar. Environ Sci Pollut Res 26(21):21849–21857

Muhammad N, Aziz R, Brookes PC, Xu J (2017) Impact of wheat straw biochar on yield of rice and some properties of Psammaquent and Plinthudult. J Soil Sci Plant Nutr 17(3):808–823

Muter O, Berzins A, Strikauska S, Pugajeva I, Bartkevics V, Dobele G, Truu J, Truu M, Steiner C (2014) The effects of woodchip-and straw-derived biochars on the persistence of the herbicide 4-chloro-2-methylphenoxyacetic acid (MCPA) in soils. Ecotoxicol Environ Saf 109:93–100

Netherway P, Reichman SM, Laidlaw M, Scheckel K, Pingitore N, Gascó G, Méndez A, Surapaneni A, Paz-Ferreiro J (2019) Phosphorus-rich biochars can transform lead in an urban contaminated soil. J Environ Qual 48(4):1091–1099

Novak JM, Ippolito JA, Ducey TF, Watts DW, Spokas KA, Trippe KM, Sigua GC, Johnson MG (2018) Remediation of an acidic mine spoil: Miscanthus biochar and lime amendment affects metal availability, plant growth, and soil enzyme activity. Chemosphere 205:709–718

Oladele S, Adeyemo A, Awodun M, Ajayi A, Fasina A (2019) Effects of biochar and nitrogen fertilizer on soil physicochemical properties, nitrogen use efficiency and upland rice (Oryza sativa) yield grown on an Alfisol in Southwestern Nigeria. Int J Recycl Org Waste Agric 8(3): 295–308

Oo AZ, Sudo S, Akiyama H, Win KT, Shibata A, Yamamoto A, Sano T, Hirono Y (2018) Effect of dolomite and biochar addition on N2O and CO2 emissions from acidic tea field soil. PloS One 13(2):e0192235

Pan D, Liu C, Yu H, Li F (2019) A paddy field study of arsenic and cadmium pollution control by using iron-modified biochar and silica sol together. Environ Sci Pollut Res 26(24):24979–24987

Pandey V, Patel A, Patra DD (2016) Biochar ameliorates crop productivity, soil fertility, essential oil yield and aroma profiling in basil (Ocimum basilicum L.). Ecol Eng 90:361–366

Partey ST, Saito K, Preziosi RF, Robson GD (2016) Biochar use in a legume–rice rotation system: effects on soil fertility and crop performance. Arch Agron Soil Sci 62(2):199–215

Penido ES, Martins GC, Mendes TBM, Melo LCA, do Rosário Guimarães I, Guilherme LRG (2019) Combining biochar and sewage sludge for immobilization of heavy metals in mining soils. Ecotoxicol Environ Saf 172:326–333

Pituello C, Dal Ferro N, Francioso O, Simonetti G, Berti A, Piccoli I, Pisi A, Morari F (2018) Effects of biochar on the dynamics of aggregate stability in clay and sandy loam soils. Eur J Soil Sci 69(5):827–842

Ponnam V, Katari NK, Mandapati RN, Nannapaneni S, Tondepu S, Jonnalagadda SB (2020) Efficacy of biochar in removal of organic pesticide, Bentazone from watershed systems. J Environ Sci Health B 55(4):396–405

Puga AP, Abreu CA, Melo LCA, Beesley L (2015) Biochar application to a contaminated soil reduces the availability and plant uptake of zinc, lead and cadmium. J Environ Manage 159:86–93

Purakayastha TJ, Bera T, Bhaduri D, Sarkar B, Mandal S, Wade P, Kumari S, Biswas S, Menon M, Pathak H, Tsang DC (2019) A review on biochar modulated soil condition improvements and nutrient dynamics concerning crop yields: pathways to climate change mitigation and global food security. Chemosphere 227:345–365

Qambrani NA, Rahman MM, Won S, Shim S, Ra C (2017) Biochar properties and eco-friendly applications for climate change mitigation, waste management, and wastewater treatment: a review. Renew Sustain Energy Rev 79:255–273

Qian X, Shen G, Wang Z, Guo C, Liu Y, Lei Z, Zhang Z (2014) Co-composting of livestock manure with rice straw: characterization and establishment of maturity evaluation system. Waste Manag 34(2):530–535

Qin X, Wang H, Liu C, Li J, Wan Y, Gao Q, Fan F, Liao Y (2016) Long-term effect of biochar application on yield-scaled greenhouse gas emissions in a rice paddy cropping system: a four-year case study in south China. Sci Total Environ 569:1390–1401

Rajkovich S, Enders A, Hanley K, Hyland C, Zimmerman AR, Lehmann J (2012) Corn growth and nitrogen nutrition after additions of biochars with varying properties to a temperate soil. Biol Fertil Soils 48(3):271–284

Rechberger MV, Kloss S, Rennhofer H, Tintner J, Watzinger A, Soja G, Lichtenegger H, Zehetner F (2017) Changes in biochar physical and chemical properties: accelerated biochar aging in an acidic soil. Carbon 115:209–219

Ren H, Huang B, Fernández-García V, Miesel J, Yan L, Lv C (2020) Biochar and rhizobacteria amendments improve several soil properties and bacterial diversity. Microorganisms 8(4):502

Rogovska N, Laird DA, Rathke SJ, Karlen DL (2014) Biochar impact on Midwestern Mollisols and maize nutrient availability. Geoderma 230:340–347

Rondon MA, Lehmann J, Ramírez J, Hurtado M (2007) Biological nitrogen fixation by common beans (Phaseolus vulgaris L.) increases with bio-char additions. Biol Fertil Soils 43(6):699–708

Salam A, Bashir S, Khan I, Hu H (2019) Two years impacts of rapeseed residue and rice straw biochar on Pb and Cu immobilization and revegetation of naturally co-contaminated soil. Appl Geochem 105:97–104

Sánchez-García M, Alburquerque JA, Sánchez-Monedero MA, Roig A, Cayuela ML (2015) Biochar accelerates organic matter degradation and enhances N mineralisation during composting of poultry manure without a relevant impact on gas emissions. Bioresour Technol 192:272–279

Sänger A, Reibe K, Mumme J, Kaupenjohann M, Ellmer F, Roß CL, Meyer-Aurich A (2017) Biochar application to sandy soil: effects of different biochars and N fertilization on crop yields in a 3-year field experiment. Arch Agron Soil Sci 63(2):213–229

Sarfraz R, Li S, Yang W, Zhou B, Xing S (2019) Assessment of physicochemical and nutritional characteristics of waste mushroom substrate biochar under various pyrolysis temperatures and times. Sustainability 11(1):277

Scheifele M, Hobi A, Buegger F, Gattinger A, Schulin R, Boller T, Mäder P (2017) Impact of pyrochar and hydrochar on soybean (Glycine max L.) root nodulation and biological nitrogen fixation. J Plant Nutr Soil Sci 180(2):199–211

Shaaban M, Wu Y, Khalid MS, Peng QA, Xu X, Wu L, Younas A, Bashir S, Mo Y, Lin S, Zafar-ul-Hye M (2018) Reduction in soil N2O emissions by pH manipulation and enhanced nosZ gene transcription under different water regimes. Environ Pollut 235:625–631

Si L, Xie Y, Ma Q, Wu L (2018) The short-term effects of rice straw biochar, nitrogen and phosphorus fertilizer on rice yield and soil properties in a cold waterlogged paddy field. Sustainability 10(2):537

Singh BP, Cowie AL, Smernik RJ (2012) Biochar carbon stability in a clayey soil as a function of feedstock and pyrolysis temperature. Environ Sci Technol 46(21):11770–11778

Singh C, Tiwari S, Gupta VK, Singh JS (2018) The effect of rice husk biochar on soil nutrient status, microbial biomass and paddy productivity of nutrient poor agriculture soils. Catena 171: 485–493

Sopeña F, Semple K, Sohi S, Bending G (2012) Assessing the chemical and biological accessibility of the herbicide isoproturon in soil amended with biochar. Chemosphere 88(1):77–83

Spokas KA, Cantrell KB, Novak JM, Archer DW, Ippolito JA, Collins HP, Boateng AA, Lima IM, Lamb MC, McAloon AJ, Lentz RD (2012a) Biochar: a synthesis of its agronomic impact beyond carbon sequestration. J Environ Qual 41(4):973–989

Spokas KA, Novak JM, Venterea RT (2012b) Biochar's role as an alternative N-fertilizer: ammonia capture. Plant and Soil 350(1):35–42

Steiner C, Teixeira WG, Lehmann J, Nehls T, de Macêdo JLV, Blum WE, Zech W (2007) Long term effects of manure, charcoal and mineral fertilization on crop production and fertility on a highly weathered Central Amazonian upland soil. Plant and Soil 291(1):275–290

Sui Y, Gao J, Liu C, Zhang W, Lan Y, Li S, Meng J, Xu Z, Tang L (2016) Interactive effects of straw-derived biochar and N fertilization on soil C storage and rice productivity in rice paddies of Northeast China. Sci Total Environ 544:203–210

Sui F, Zuo J, Chen D, Li L, Pan G, Crowley DE (2018) Biochar effects on uptake of cadmium and lead by wheat in relation to annual precipitation: a 3-year field study. Environ Sci Pollut Res 25(4):3368–3377

Sui F, Wang J, Zuo J, Joseph S, Munroe P, Drosos M, Li L, Pan G (2020) Effect of amendment of biochar supplemented with Si on Cd mobility and rice uptake over three rice growing seasons in an acidic Cd-tainted paddy from central South China. Sci Total Environ 709:136101

Sui M, Li Y, Jiang Y, Wang L, Zhang W, Sathishkumar K, Zakaria H (2021) Sediment-based biochar facilitates highly efficient nitrate removal: physicochemical properties, biological responses and potential mechanism. Chem Eng J 405:126645

Taskin MB, Kadioglu YK, Sahin O, Inal A, Gunes A (2019) Effect of acid modified biochar on the growth and essential and non-essential element content of bean, chickpea, soybean, and maize grown in calcareous soil. Commun Soil Sci Plant Anal 50(13):1604–1613

Tatarková V, Hiller E, Vaculík M (2013) Impact of wheat straw biochar addition to soil on the sorption, leaching, dissipation of the herbicide (4-chloro-2-methylphenoxy) acetic acid and the growth of sunflower (Helianthus annuus L.). Ecotoxicol Environ Saf 92:215–221

Tian X, Li C, Zhang M, Li T, Lu Y, Liu L (2018) Controlled release urea improved crop yields and mitigated nitrate leaching under cotton-garlic intercropping system in a 4-year field trial. Soil Tillage Res 175:158–167

Van Zwieten L, Kimber S, Morris S, Chan KY, Downie A, Rust J, Joseph S, Cowie A (2010) Effects of biochar from slow pyrolysis of papermill waste on agronomic performance and soil fertility. Plant and Soil 327(1):235–246

Van Zwieten L, Singh B, Joseph S, Kimber S, Cowie A, Chan KY, (2009) Biochar and emissions of non–CO2 greenhouse gases from soil. In: Lehmann J, Joseph S (Eds.) Biochar for Environmental Management: Science and Technology. Earthscan, London & Sterling, VA, UK, pp 227–250

Van Zwieten L, Singh BP, Kimber SWL, Murphy DV, Macdonald LM, Rust J, Morris S (2014) An incubation study investigating the mechanisms that impact N_2O flux from soil following biochar application. Agr Ecosyst Environ 191:53–62

Van Zwieten L, Kammann C, Cayuela ML, Singh BP, Joseph S, Kimber S, Donne S, Clough T, Spokas KA (2015a) Biochar effects on nitrous oxide and methane emissions from soil. In: Biochar for environmental management. Routledge, London, pp 521–552

Van Zwieten L, Rose T, Herridge D, Kimber S, Rust J, Cowie A, Morris S (2015b) Enhanced biological N_2 fixation and yield of faba bean (Vicia faba L.) in an acid soil following biochar addition: dissection of causal mechanisms. Plant and Soil 395(1):7–20

Viger M, Hancock RD, Miglietta F, Taylor G (2015) More plant growth but less plant defence? First global gene expression data for plants grown in soil amended with biochar. GCB Bioenergy 7(4):658–672

Wang Y, Liu Y, Liu R, Zhang A, Yang S, Liu H, Zhou Y, Yang Z (2017) Biochar amendment reduces paddy soil nitrogen leaching but increases net global warming potential in Ningxia irrigation, China. Sci Rep 7(1):1–10

Wang H, Yan S, Ren T, Yuan Y, Kuang G, Wang B, Yun F, Feng H, Ji X, Yuan X, Liu G (2020) Novel environmental factors affecting microbial responses and physicochemical properties by sequentially applied biochar in black soil. Environ Sci Pollut Res 27(30):37432–37443

Wu X, Sun Y, Deng L, Meng Q, Jiang X, Bello A, Sheng S, Han Y, Zhu H, Xu X (2020) Insight to key diazotrophic community during composting of dairy manure with biochar and its role in nitrogen transformation. Waste Manag 105:190–197

Xiao X, Chen B, Zhu L (2014) Transformation, morphology, and dissolution of silicon and carbon in rice straw-derived biochars under different pyrolytic temperatures. Environ Sci Technol 48(6):3411–3419

Xiu L, Zhang W, Wu D, Sun Y, Zhang H, Gu W, Wang Y, Meng J, Chen W (2021) Biochar can improve biological nitrogen fixation by altering the root growth strategy of soybean in Albic soil. Sci Total Environ 773:144564

Xu WH, Zhang YK, Wang GB, Ruan HH (2015) Response of carbon metabolism by soil microbes to different fertilization regimes in a poplar plantation in coastal area of northern Jiangsu, China. Chin J Ecol 34(7):1791

Xu P, Sun CX, Ye XZ, Xiao WD, Zhang Q, Wang Q (2016) The effect of biochar and crop straws on heavy metal bioavailability and plant accumulation in a Cd and Pb polluted soil. Ecotoxicol Environ Saf 132:94–100

Yaghoubi P, Yargicoglu EN, Reddy KR (2014) Effects of biochar-amendment to landfill cover soil on microbial methane oxidation: initial results. In: Geo-Congress 2014: geo-characterization and modeling for sustainability, pp 1849–1858

Yamato M, Okimori Y, Wibowo IF, Anshori S, Ogawa M (2006) Effects of the application of charred bark of Acacia mangium on the yield of maize, cowpea and peanut, and soil chemical properties in South Sumatra, Indonesia. Soil Sci Plant Nutr 52(4):489–495

Yao Y, Gao B, Zhang M, Inyang M, Zimmerman AR (2012) Effect of biochar amendment on sorption and leaching of nitrate, ammonium, and phosphate in a sandy soil. Chemosphere 89(11):1467–1471

Yao T, Zhang W, Gulaqa A, Cui Y, Zhou Y, Weng W, Wang X, Liu Q, Jin F (2021) Effects of peanut shell biochar on soil nutrients, soil enzyme activity, and rice yield in heavily saline-sodic paddy field. J Soil Sci Plant Nutr 21:1–10

Yin D, Wang X, Peng B, Tan C, Ma LQ (2017) Effect of biochar and Fe-biochar on Cd and As mobility and transfer in soil-rice system. Chemosphere 186:928–937

Yoo G, Kang H (2012) Effects of biochar addition on greenhouse gas emissions and microbial responses in a short-term laboratory experiment. J Environ Qual 41(4):1193–1202

Yoo G, Lee YO, Won TJ, Hyun JG, Ding W (2018) Variable effects of biochar application to soils on nitrification-mediated N2O emissions. Sci Total Environ 626:603–611

Yu H, Zou W, Chen J, Chen H, Yu Z, Huang J, Tang H, Wei X, Gao B (2019) Biochar amendment improves crop production in problem soils: a review. J Environ Manage 232:8–21

Yuan H, Lu T, Wang Y, Chen Y, Lei T (2016) Sewage sludge biochar: nutrient composition and its effect on the leaching of soil nutrients. Geoderma 267:17–23

Zahra MB, Aftab ZEH, Akhter A, Haider MS (2021) Cumulative effect of biochar and compost on nutritional profile of soil and maize productivity. J Plant Nutr:1–13

Zhan F, Zeng W, Yuan X, Li B, Li T, Zu Y, Jiang M, Li Y (2019) Field experiment on the effects of sepiolite and biochar on the remediation of Cd-and Pb-polluted farmlands around a Pb–Zn mine in Yunnan Province, China. Environ Sci Pollut Res 26(8):7743–7751

Zhang Y, Peng S, Song Q, Zhang W, Guo J, Wang G, Ruan H (2016) Effects of different fertilizers regimes on the functional diversity of soil microbes under poplar plantation. J Nanjing For Univ (Nat Sci Ed) 40(5):1–8

Zhang Q, Saleem M, Wang C (2019) Effects of biochar on the earthworm (Eisenia foetida) in soil contaminated with and/or without pesticide mesotrione. Sci Total Environ 671:52–58

Zhao L, Cao X, Mašek O, Zimmerman A (2013) Heterogeneity of biochar properties as a function of feedstock sources and production temperatures. J Hazard Mater 256:1–9

Zhao J, Ni T, Li Y, Xiong W, Ran W, Shen B, Shen Q, Zhang R (2014) Responses of bacterial communities in arable soils in a rice-wheat cropping system to different fertilizer regimes and sampling times. PloS One 9(1):e85301

Zhao J, Tao Q, Li B, Luo J, Zhang H, Lu C, Li Q, Xu Q, Huang R, Li H, Li B (2021) Low-pyrolysis-temperature biochar promoted free-living N2-fixation in calcareous purple soil by affecting diazotrophic composition. Geoderma 388:114969

Zheng M, Chen H, Li D, Zhu X, Zhang W, Fu S, Mo J (2016) Biological nitrogen fixation and its response to nitrogen input in two mature tropical plantations with and without legume trees. Biol Fertil Soils 52(5):665–674

Zheng H, Wang X, Chen L, Wang Z, Xia Y, Zhang Y, Wang H, Luo X, Xing B (2018) Enhanced growth of halophyte plants in biochar-amended coastal soil: roles of nutrient availability and rhizosphere microbial modulation. Plant Cell Environ 41(3):517–532

Zheng X, Song W, Guan E, Wang Y, Hu X, Liang H, Dong J (2020) Response in physicochemical properties of tobacco-growing soils and N/P/K accumulation in tobacco plant to tobacco straw biochar. J Soil Sci Plant Nutr 20(2):293–305

Zhou H, Zhang D, Wang P, Liu X, Cheng K, Li L, Zheng J, Zhang X, Zheng J, Crowley D, van Zwieten L (2017) Changes in microbial biomass and the metabolic quotient with biochar addition to agricultural soils: a meta-analysis. Agr Ecosyst Environ 239:80–89

Zielińska A, Oleszczuk P (2015) The conversion of sewage sludge into biochar reduces polycyclic aromatic hydrocarbon content and ecotoxicity but increases trace metal content. Biomass Bioenergy 75:235–244

Zimmerman AR (2010) Abiotic and microbial oxidation of laboratory-produced black carbon (biochar). Environ Sci Technol 44(4):1295–1301

Zimmerman AR, Gao B, Ahn MY (2011) Positive and negative carbon mineralization priming effects among a variety of biochar-amended soils. Soil Biol Biochem 43(6):1169–1179

Chapter 11
Production of Alternative Fuel from Lignocellulosic Kitchen Waste Through Pyrolysis

Neelanjan Bhattacharjee, Akanksha Majumder, and Asit Baran Biswas

Abstract Lignocellulosic organic kitchen waste (LOKW) feedstock can be transformed into fuels through combustion, pyrolysis, and gasification. The product yield rests on the lignocellulosic composition and operating parameters of the thermochemical process. This chapter highlights the role of LOKW composition in achieving better yield, quality, low acid, high density, and calorific value bio-oil from various pyrolysis processes. The effect of different pyrolysis processes and reactors in processing feedstocks has been considered more in this chapter. The H_2 from syngas can be utilized as fuel in the transportation sector. This chapter also highlights the different feed pretreatment and product upgradation techniques, available in the literature, to obtain increased quantities of bio-oil and to make it suitable for commercial use. The presence of oxygenated compounds in the bio-oil needs further refining stages such as deoxygenation and solvent fractionation to refine the quality of the bio-oil. LOKW can be seen as an energy-efficient biomass which can solve present defiance by narrowing down CO_2 emissions from the atmosphere.

Keywords Bio-oil · Biomass · Lignocellulosic · Kitchen waste · Pyrolysis · Feedstocks

11.1 Introduction

In a few decades, the exhaustion of nonrenewable energy assets has risen by many folds due to insatiable energy demand from the rising population and industrial development. Factors like global warming, pollution, health, and economic inflation have increased due to the prolonged usage of nonrenewable energy assets. The exigency for alternative energy can be met through solar, wind, hydropower, geothermal, and oceanic sources. However, geographical limitations, higher upfront

N. Bhattacharjee · A. Majumder · A. B. Biswas (✉)
Department of Chemical Engineering, University of Calcutta, Kolkata, India
e-mail: abbschemengg@caluniv.ac.in

D. B. Pal, A. K. Tiwari (eds.), *Sustainable Valorization of Agriculture & Food Waste Biomass*, Clean Energy Production Technologies,
https://doi.org/10.1007/978-981-99-0526-3_11

costs, and inconsistent availability limit the utilization of these resources as alternative forms of energy. European Union (EU) has pledged to reduce greenhouse gas emissions to 55% by the end of 2030 and net zero by 2050 (Sieradzka et al. 2022). Worldwide scientists and technologists are now focusing on sustainable renewable energy sources obtained through waste. Waste is converted mostly into value-added fuels through *biochemical and thermochemical* processes. This chapter is intended for the conversion of LOKW to a value-added product/fuel through pyrolysis. The thermal instability of feedstock is converted into char, bio-oil, and syngas through slow, intermediate, fast, and flash pyrolysis. Alternative forms of energy (liquid fuel) can be derived through bio-oil only. So, this chapter is mainly concerned with the product bio-oil production gradation of bio-oil produced to make it suitable for commercial use. A comparative study of the "value and extent" of bio-oil obtained over the various pyrolysis processes has also been provided here.

11.2 Biomass Availability and Composition

The definition of biomass varies because of the heterogeneity of materials, the applications, and the origin. Except for plastics derived from fossil materials and petrochemicals, all organic waste generated from plants and animals is considered waste/residue (McKendry 2002).

In 2005, the UNFCCC (United Nations Framework Convention on Climate Change) defined it as: "A non-fossilized and bio-degradable organic material originating from plants, animals and micro-organisms" (Garba 2020). The list included products, by-products, and remains from agriculture, forestry, and linked industries in addition to the non-fossilized and biodegradable organic fractions of industrial and municipal wastes.

Biomass is recognized as the third largest renewable energy resource consisting mostly from forestry and agricultural biomass (Sieradzka et al. 2022). It is predicted to be the principal energy resource with a substantial total energy load of approximately 10–15% worldwide (Garba 2020). Since the CO_2 released through the combustion and utilization of biomass is of biogenic origin, it does not cause a surge in atmospheric CO_2. Plants use this CO_2, for growth and various metabolic pathways (Tkemaladze and Makhashvili 2016; Tursi 2019). EU members have sanctioned to bind least of 32% share of the biomass energy consumption (Sieradzka et al. 2022).

Waste materials from plants (land, forest, and aquatic), crops, mills, sewage, municipal solid waste (human, marine, and animal), and industries are taken as the major source of biomass feedstock (Tursi 2019). Figure 11.1 reflects almost all the available sources of biomass. Few studies have estimated that the total biomass reserve available on aquatic and land accounts for 4 billion and 1.8 trillion tons. The total biomass present on our planet has the possibility of producing 3.3×10^{16} (MJ) which is 80% higher than the world's annual energy exhaustion (Anon 2018).

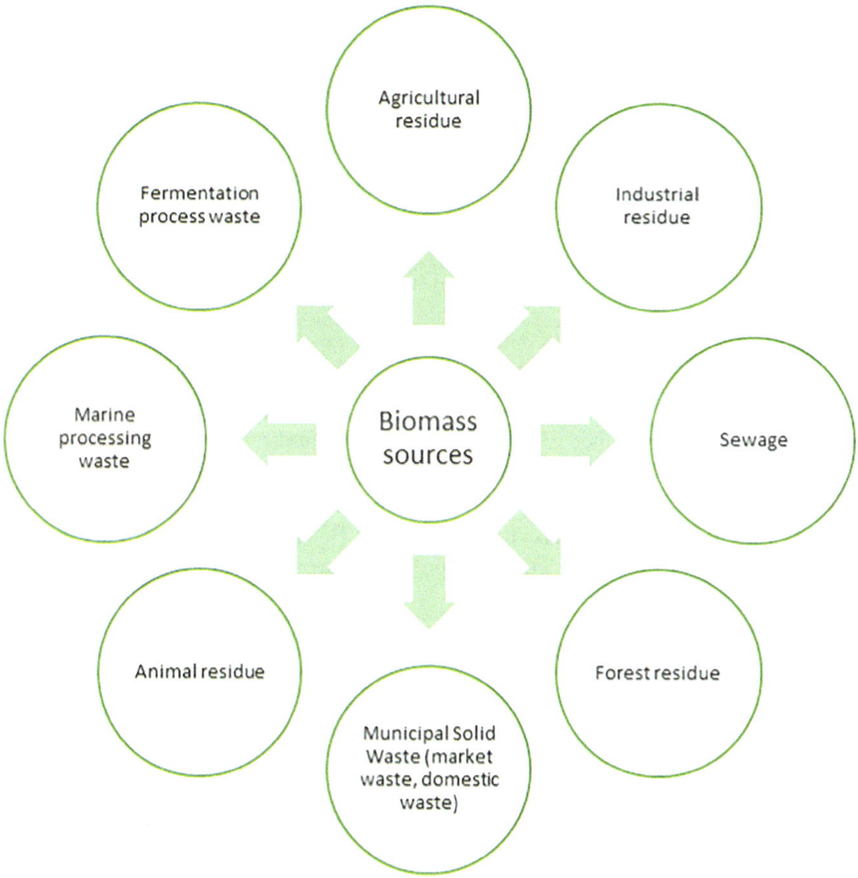

Fig. 11.1 Different sources of biomass

The chemical composition of biomass is largely varied. Let's look into each component that is generally present in biomass.

11.2.1 Cellulose

Cellulose, a linear polymeric polysaccharide, is made of a maximum of 10,000 β-D-glucose units. These glucose units are attached by glycosidic bonds between C_1 of one glucose unit and C_4 of the adjacent glucose unit. It is a homopolymer whose composition varies between 9 and 80 % of the plant cell wall. Both the National Renewable Energy Laboratory (NREL) and Technical Association of the Pulp and Paper Industry (TAPPI) procedures are employed to determine cellulosic content.

11.2.2 Hemicellulose

Hemicellulose, a chain heteropolymer of plant cell wall, consists of D-pentose sugars along with rarely slight quantities of L-sugars. The various molecular groups that make up hemicellulose include mannans, xylans, *Galatians*, and arabinogalactans. Lignocellulosic biomass consists of 19–35% of hemicellulose. Hemicellulose content can be determined with "high-performance liquid chromatography".

11.2.3 Lignin

Lignin is a vastly branched, 3D, heterogeneous phenolic polymer providing resistance and structural support to a plant. It is the largest noncarbohydrate component consists of 15–40% of lignocellulosic structure. The most popular methods used for determining lignin content are Klason and Ac-Br methods (Moreira-Vilar et al. 2014). The lignocellulosic composition can also be determined by methods described elsewhere (Li et al. 2004).

11.2.4 Starch

Starch is a polysaccharide polymer composed of several glucose units linked by glycosidic bonds and called energy storage house of plant tissues. Amylose (20–25 wt.%) and amylopectin (75–80 wt.%) are two polysaccharide polymers present in the starch. Amylose is soluble in water, while amylopectin is water-insoluble.

11.2.5 Other Organic Components

From the literature, we have gathered the information that apart from cellulose, hemicellulose, lignin, and starch; additional organic components such as lipids, proteins, acetyls, nucleic acids, etc. are also present in small quantities.

11.2.6 Inorganic Components

The lignocellulosic waste contains minute quantities of inorganic matter, the quantity of which is dependent on the variety of raw matter. Metals such as magnesium, calcium, potassium, silicon, iron, aluminium, phosphorus, and sodium are usually discovered in feedstocks.

11.2.7 Fluid Content

Fluid content refers to the water quantity contained in lignocellulosic waste. The observed moisture content can be related to that present in living cells. It typically ranges from less than 15% to more than 90% depending on the type of waste.

11.3 LOKW

The significance of choosing LOKW lies in its easy availability and abundance. Figure 11.2 reflects the different classifications of LOKW. LOKW provides a variation in the phrase of cellulose, hemicellulose, and lignin from vegetable skins, seeds, leaves, vegetarian food remains, and fruits husk, peel, bagasse, and coir. From the literature study, we can state the fact that feed composition greatly affects the quality and yield of products obtained from waste conversion techniques. Lignocellulosic waste is mostly made of cellulose, hemicellulose, and lignin. There is an additional organic and inorganic matter that exists in the lignocellulosic structure. The chemical components present in our selected feed are those mentioned in the composition of biomass residue. Table 11.1 shows a short overview of the composition of biomass residue found in LOKW.

11.4 Waste Conversion Processes

The thermochemical and biochemical processes are generally employed for the waste to energy conversion.

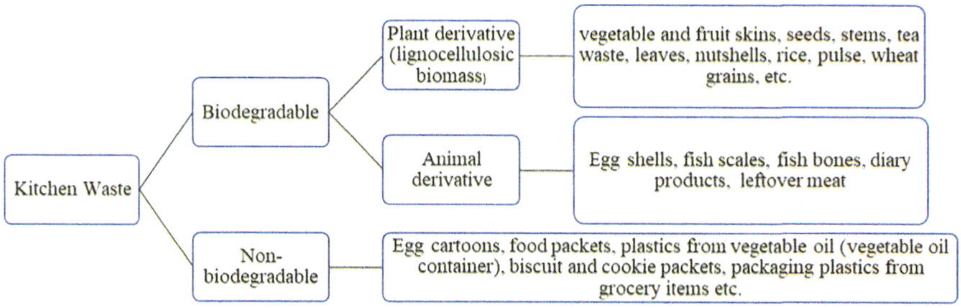

Fig. 11.2 Classification of kitchen waste

Table 11.1 Composition of LOKW (Jahirul et al. 2012)

Composition	Cellulose (%)	Hemicellulose (%)	Lignin (%)
Hard stems	40–45	24–40	28–35
Short stems	45–50	25–35	25–35
Almond shells	29–31.1	28–38	27.7–35
Tea waste	33.2	23.3	43.5
Corncobs	45	35	15
Leaves	15–32	80–85	0
Grasses	25–40	35–50	10–30
Potato peel	8.3	7.41	32.88
Onion skins	41.1	16.2	38.9
Other vegetables' skins, stems, roots, etc.	40–50	30	9–18
Spent coffee grains	13	42	25
Wheat bran	10.5–14.8	35.5–39.2	8.3–12.5
Spent grains	17–25	21–28	12–28
Coconut fibre	35–60	15–28	20–48
Walnut shell	23.3	20.4	53.5
Ground nut shell	37	18.7	28
Sugarcane peel	41.11	26.40	24.31
Banana peels	13.2	14.8	14.2
Coconut coir	44.2	22.1	32.8
Sugarcane peel	41.11	26.40	24.31

11.4.1 Thermochemical Processes

Thermochemical processes are performed in high temperatures for bond breakage and reformation of the organic components present in the waste matter to produce char, bio-oil, and syngas fuels. Thermochemical processes are becoming more popular due to short processing time, low water usage, availability of highly developed industrial infrastructure, and their capability of converting plastic wastes to energy since plastic wastes cannot be digested by microbial activity (Lee et al. 2019). Thermochemical processes like combustion, gasification, pyrolysis, and liquefaction depend on the amount of oxygen required to complete the process. The option of the process is reliant on the composition and amount of the feedstock, available energy, environmental principles, desired products, economic issues, etc. The value and extent of each product depend on the process chosen.

11.4.1.1 Combustion

Combustion simply refers to feedstock burning in the presence of O_2. The products obtained from the combustion of lignocellulosic waste are CO_2, heat, and water. The quantity of heat generated depends on feedstock properties, combustion atmosphere, and temperature (Shekhar Silori and Kumar Mishra 2001). The temperature required for carrying out this process usually ranges from 700 to 1400 °C. One important condition required for combustion is that the moisture content of the feedstock ought to be less than 50%.

11.4.1.2 Gasification

This process is carried out under inert atmosphere in a reactor heated generally between 500 and 1400 °C to generate H_2, CH_4, CO, CO_2, and low hydrocarbon gases (Tursi 2019). The gasification reactor generally operates between 1 and 33 atm pressure (Tursi 2019). The functioning parameters for gasification are temperature, flow rate of feed, type and amount of catalyst, feed properties, design of gasifier, energy and moisture content of feed, etc. The gasification is the utmost preferred for producing H_2 from waste biomass feedstock.

11.4.1.3 Pyrolysis

Pyro means fire and lysis means separating in Greek analogy. Most of the pyrolysis/thermolysis processes are carried out in a closed reactor purge with inert gas and heated between 350 and 700 °C to obtain predefined products such as char, bio-oil, and syngas. The rate of pyrolysis increases with temperature. Briefly, pyrolysis is carried out in three stages: (1) feeding of raw materials, (2) conversion of organic mass, and (3) separation of the products. Depolymerization, vaporization, repolymerization, and cross-linking reactions occur during pyrolysis (Babu and Chaurasia 2003; Gupta and Demirbas 2010).

11.4.1.4 Liquefaction

Liquefaction generates bio-oil at low temperatures (250–374 °C) and high pressure (40–220 bar) (Balat 2008). The product of liquefaction has a high calorific value with low oxygen content, and that makes it a chemically stable fuel (Tursi 2019). Thus, it can be said that liquefaction produces a marketable liquid product, but it requires further refining to be used as a commercial fuel. In liquefaction, big feedstock particles are degraded into smaller ones, with or without using a catalyst, in an organic solvent or aqueous medium.

11.4.2 Biochemical Processes

Biochemical conversion processes are carried out at lower temperatures in addition to lower reaction rates, compared to thermochemical processes (CalRecycle 2022). Biochemical conversion denotes the decomposition of waste into energy with the help of a microbial activity. The processes included under this conversion technology are anaerobic/aerobic digestion, fermentation, different photobiological methods, and acid/enzymatic hydrolysis (Lee et al. 2019; Chandraratne and Daful 2021). These processes can treat a feed with high moisture content and are much slower than thermochemical processes. In the case of anaerobic digestion, due to the absence of atmospheric oxygen, microbes take oxygen from the feed. Anaerobic digestion produces a combination of CO_2 and CH_4 (bio-gas) and solid residue. The anaerobic bacteria can digest not more than 5–10% of the feed, inside an anaerobic digester. The residue mainly consists of the remaining material, which is indigestible. Unlike anaerobic digestion, aerobic digestion, also called composting, is performed in the presence of O_2. Different kinds of microbes that are capable of accessing oxygen from the atmosphere are used in this process. The products of aerobic digestion are CO_2, solid residue (compost), and heat. Fermentation refers to the conversion of starch into sugar using enzymes or acids, and then ethanol or other products are produced from sugar by using yeast. The fermentation needs the lignocellulosic waste to undergo hydrolysis to decompose the cellulose and hemicellulose into easy sugars. The lignin is generally not converted through biochemical conversion and is thus left for thermochemical conversion techniques. The biological pretreatment of feedstock lignin is performed through earthworms before the biochemical process. Photobiological methods refer to the production of H_2 gas by some biomass like microalgae (Lee et al. 2019), naturally in the presence of light.

11.5 Advantages of Thermochemical Conversion Processes

Generally, thermochemical conversions offer several advantages over physicochemical and biochemical conversion techniques. As mentioned in Sect. 11.4.2, highly developed industrial infrastructure is available for carrying out thermochemical processes. Unlike the biochemical process, the thermochemical process can handle a wide range of feedstocks, doesn't require feedstock pretreatment, has higher conversion efficiencies with shorter residence times, and has less conversion time. Because of the above-mentioned reasons, recently, thermochemical processes have gained more attention for the production of biofuels (Chandraratne and Daful 2021; Liang et al. 2021). Only a few thermochemical processes, such as liquefaction and pyrolysis, yield bio-oil as the main product. Both physicochemical and biochemical processes are incapable of producing the desired quality and yield of bio-oil. Since the chapter intends to study the production of bio-oil from waste matter, so that it can be used as an alternative fuel, we have decided to proceed with a very popular thermochemical process, pyrolysis.

11.6 Advantages of Pyrolysis

Unlike pyrolysis, the liquefaction process also produces bio-oil as its by-product. The disadvantage of liquefaction over pyrolysis involves low-quality bio-oil yield (20–55 wt%), high capital investment, the production of high oxygen content and low calorific value bio-oil, and the need for additional catalysts or other reactants for process upgradation. Due to these reasons, pyrolysis is a much-preferred thermochemical process for waste-to-energy conversion.

The benefits of pyrolysis over other thermochemical processes are listed below:

(a) The thermal decomposition in pyrolysis releases lesser oxides of nitrogen and sulphur into the atmosphere compared to combustion. Using a cyclone separator or activated carbon filter may further reduce the emission of harmful substances, pollutants, and flue gas (Lue 2019).
(b) Pyrolysis does not require the feed to be shredded before it is fed into the reactor (Haiqi Machinery 2019).
(c) It gives bio-oil, char, and syngas as products. The quantity of each product can be controlled by varying the type of pyrolysis.
(d) Bio-oil can be utilized as a replacement for commercial fuels, with the help of pretreatment and upgradation.
(e) The bio-oil can be utilized further for the manufacturing of different valuable chemicals.
(f) The residue of pyrolysis is devoid of toxic organic matter and can thus avoid the elution of metal substances (Lue 2019).
(g) All the by-products can be reused.
(h) The bio-oil can be directly stored and easily transported, in contrast to fuel gas, manufactured by the process of gasification (Dhyani and Bhaskar 2018).
(i) Pyrolysis is an inexpensive process with a low maintenance cost and a small footprint (Lue 2019; Grycová et al. 2016).
(j) Pyrolysis requires less time than the other thermochemical processes.

11.7 Pyrolysis Products and Their Applications as Fuels

Waste pyrolysis is performed with the motive of energy recovery. The products obtained from the pyrolysis process often have good fuel characteristics. Portions of the pyrolysis products can be utilized to meet the energy essential for the process itself. Pyrolysis can transform waste matter into an energy resource on a domestic scale along with an industrial scale (Czajczyńska et al. 2017). Char, bio-oil, and syngas are the three main by-products produced after pyrolysis. The bio-oil produced at ambient temperature after condensation is a viscous dark brown colour liquid and can be upgraded into fuels and chemicals (Jahirul et al. 2012).

11.7.1 Char

Char is the solid leftover, left at the end of pyrolysis, and is composed of different ratios of charcoal (highly carbonaceous matter) and ash, which contains various inorganic residues. Generally, about 15–25 wt.% of the overall products obtained from pyrolysis is char. The carbon content of char is typically around 50 wt.% of the overall carbon content, initially found in the feedstock. Char can be utilized as boiler fuel, activated carbon in water treatment (55 pop), manufacturing of carbon nanotubes, reactive feed for further gasification technologies, and organic fertilizer (high retention of water and nutrients in the soil) (Czajczyńska et al. 2017; Goyal et al. 2008; Li et al. 2013).

11.7.2 Syngas

The syngas, obtained from pyrolysis, has a complex composition and the quality and quantity of the gases depend on heating rate, type of feedstock, vapour residence time, feed particle size, etc. Generally, syngas consists of different proportions of various gases such as CH_4, H_2, CO, CO_2, ethane, ethene, propane, propylene, butane, other organic derivatives, and water vapour (Vamvuka 2011).

11.7.3 Bio-oil

The major outcome of pyrolysis is the bio-oil. Most pyrolysis processes are designed in a way to maximize the yield of bio-oil because it has the highest potential for being a substitute for commercial fuel. Bio-oil is produced by the condensation of gaseous vapours formed in a pyrolysis reaction. The major advantages of pyrolysis oil are:

(a) The heating values of bio-oils are comparable to nearly 40–50 % of the heating values of hydrocarbon fuels.
(b) Potential of being used in both small-scale systems and large power stations.
(c) Ease of transportation and storage.
(d) High-energy density in comparison to fuel produced from gasification.

The fuel characteristics of bio-oil are mentioned in the next section.

11.8 Fuel Characteristics of Bio-oil

The characteristics of bio-oil have significant differences from those of petroleum fuels. The differences exist because the bio-oil is highly viscous, having high oxygen along with a high water (moisture) content. More importantly, the heating value of oil is much lower than petroleum fuels. The viscosity of bio-oil can be compared to that of heavy fuel oil (Gupta and Demirbas 2010). High viscosity rest on the feedstock, temperature profile of the process, amounts of light components and water present, degree of thermal degradation, and catalytic cracking (Gupta and Demirbas 2010). Decreasing the moisture content of bio-oil is difficult and cannot be performed by typical techniques, like distillation. Moreover, this high moisture content is also accountable for reducing the energy density and ignition temperature and also causes ignition problems (Gupta and Demirbas 2010). Unlike nonpolar fuels (diesel and gasoline), bio-oil is highly polar and can gladly absorb water equal to 35 wt% (Demirbas 2007). Previous studies and experiments reveal that the heating value of bio-oil is 16–19 MJ/kg, whereas that for conventional petroleum fuels is 40–45 MJ/kg (Gupta and Demirbas 2010).

11.8.1 Chemical Composition

Bio-oils derived through pyrolysis can contain even up to 300 different kinds of organic compounds, reliant on the pyrolysis process and the feedstock structure (Mohan et al. 2006). The bio-oil produced at 450 °C consists largely of organic compounds, such as 1-hydroxy-2-butanone, propanone, acetic acid, methanol, methoxy phenol, cyclic aliphatic compounds, and so on. With the increase in temperature, some of the above-mentioned compounds transform via hydrolysis (Gupta and Demirbas 2010). Phenolic compounds may also be existing in concentrations even up to 50 wt.%, with moderately low quantities of cresols, eugenol, phenol, xylenols, and ample quantities of alkylated (poly-)phenols (Li et al. 2013). The bio-oils also consist of furfural, levoglucosan, guaiacol, formic acid, and their alkylated phenol derivatives.

11.8.2 Viscosity

The viscosity of bio-oil rests on the kind of feedstock, temperature, and time of the pyrolysis process (Diebold 2000). The viscosity of the bio-oil increases during storage owing to different physical and chemical variations as numerous reactions take place and the volatiles are lost because of ageing (Jahirul et al. 2012; Oasmaa and Kuoppala 2003; Aho et al. 2008). Researchers have put efforts to find out ways for reducing and/or stabilizing the viscosity of bio-oil. Studies show that char removal can decrease the tertiary reactions and, thus, slow the rise in viscosity (ageing), and ageing can be further reduced by adding various organic compounds, for example, methanol, acetone, ethanol, ethyl acetate, etc. The degree of success in reducing ageing varies for each organic compound. The high viscosity of the bio-oil would increase the cost of the equipment due to high-pressure drop, as well as pumping costs, and would reduce atomization (Li et al. 2013).

11.8.3 Density

The density of bio-oil (~1.2 kg l^{-1}) is significantly higher than fossil-based crude oil (0.85 kg l^{-1}) due to high moisture content and heavy molecule contamination (Jahirul et al. 2012; Gupta and Demirbas 2010).

11.8.4 Acidity

The reason for the low pH (2, 3) and high total acid content is due to the existence of carboxylic acids in bio-oil. This acidic nature leads to the corrosiveness of bio-oil, and this is more effective at higher temperatures. This results in the corrosion of vessels and pipes and also makes the oil highly unstable (Gupta and Demirbas 2010).

11.8.5 Water Content

The high water content of bio-oil (15–30 wt.%) originates from the moisture content of the feedstock (Gupta and Demirbas 2010). The increased moisture content lowers the heating value, density, stability, and acidity (Li et al. 2013). The reduction in heating value lowers the flame temperature. Alternatively, the existence of water also increases the fluidity of the bio-oil and decreases its viscosity (Gupta and Demirbas 2010).

11.8.6 Oxygen Content

Based on the category and structure of feedstock and the nature of the pyrolysis process, the oxygen content of bio-oil can be said to be approximately 35–40 wt.%, which includes oxygen contained in more than 300 oxygenated compounds (Gupta and Demirbas 2010). The heating value and stability of the oil get lowered due to its high oxygen content. Another notable disadvantage of the high oxygen content is that it causes immiscibility with hydrocarbons.

11.8.7 Solid Content

The larger char particles are separated from bio-oil, with the help of cyclone separators, whereas the fine-sized char particles get passed over and accumulated in the bio-oil. Usually, the solid content present in bio-oil is approximately 1 wt.%. Since the particles are minor in size, they have a high surface area and, thus, can catalyse several reactions at the time of storage (Gupta and Demirbas 2010). Moreover, the presence of char causes the ageing of oil, catalysis, filter blockage, engine injector blockage, erosion, damage to turbines, and increased emissions due to incomplete combustion (Gupta and Demirbas 2010; Li et al. 2013).

11.9 Types of Pyrolysis

There are various kinds of pyrolysis, namely, slow, torrefaction, intermediate, fast, and flash pyrolysis. The value and extent of the products depend on the type of pyrolysis. The distribution of products is determined by a few parameters such as final temperature, heating rate and pressure, and residence time. Let's briefly discuss the different types of pyrolysis.

11.9.1 Slow Pyrolysis

This type of pyrolysis is carried out $<$ or $\geq$ 400 °C. This is a slow procedure with a high residence time of 5–30 min and a slow heating rate ranging from 0.1 to 0.8 °C s^{-1} (Chandraratne and Daful 2021); the feed can be held at a fixed temperature for a small period (Gupta and Demirbas 2010). The main products are bio-oil, char, and syngas. Biochar yield dominates over the other two products. In slow pyrolysis, char production is favoured at the expense of bio-oil production. This type of pyrolysis has been practised for decades mainly to yield char/coke (Gupta and Demirbas 2010). The overall process can be exothermic due to the existence of secondary reactions (Chandraratne and Daful 2021). The typical particle size, for slow pyrolysis, ranges from 5 to 50 mm (Chandraratne and Daful 2021).

11.9.2 Torrefaction

Torrefaction is an incomplete form of pyrolysis in which dried feed is heated at 200 and 400 °C in presence of the N_2, He, Ar, etc. (Asia Development Bank 2020),. The residence time of this process is around 1–2 h (Salimbeni 2019). On heating biomass at the prescribed temperature levels, the moisture content reduces and low calorific or superfluous volatile components contained in the biomass are removed. Torrefaction reduces the biomass weight to around 20–30%, whereas the energy loss is only 10–15% and produces the so-called torrified bio-waste (Asia Development Bank 2020; Salimbeni 2019). Torrified bio-waste is a coal-like substance, and it has better fuel characteristics than the original biomass used as the feed. A high-grade biofuel can be formed by the torrefaction of feedstock and can be considered as a replacement for coal in electricity and also as feed for gasification methods. Torrefaction is used as a thermal pretreatment process that alters both the physical and chemical composition of the original feed (Asia Development Bank 2020). After torrefaction, biomass becomes more brittle, and this makes it easier to grind, and the size reduction process becomes less energy-intensive. This process is also used for primary moisture removal from the feed, thus reducing the cost of storage and transportation. Thus, from quite a few studies, we have gathered that torrefaction increases the grindability and hydrophobicity and lessens the biodegradability, in comparison to crude biomass (Chandraratne and Daful 2021).

11.9.3 Intermediate Pyrolysis

This type of pyrolysis is performed at moderate temperatures around 500 °C. It is generally conducted in fixed-bed reactors with a residence time of 10–30 s (Chandraratne and Daful 2021). The operating conditions offer a wide range of process variations. This technique works on feedstocks with a moisture content of up to 40% (Chandraratne and Daful 2021). Intermediate pyrolysis has better product distribution, compared to other pyrolysis techniques (Kazawadi et al. 2021). It produces high-quality dry biochar but in lesser quantity as compared to slow pyrolysis along with two-phase separable bio-oil with a tall calorific value (Kazawadi et al. 2021). Generally, the product distribution of this process is 15–25% of biochar, 20–30% of syngas, and 40–60% of bio-oil (Chandraratne and Daful 2021).

11.9.4 Fast Pyrolysis

This kind of pyrolysis is usually performed out in temperatures between 577 and 977 °C and has a residence time of less than 2 s and a high heating rate of 200 °C/s (Lee et al. 2019). The typical product distribution for fast pyrolysis is 20% of char, 20% of syngas, and 60% of bio-oil (Tursi 2019). The main product obtained from the fast pyrolysis of feedstock is raw bio-oil, which is a dark brown, viscous liquid with a foul smell (Kan et al. 2016; Patel and Kumar 2016; Saber et al. 2016). But this crude bio-oil, produced from fast pyrolysis, is not capable of being directly used as a commercial transportation fuel (Kan et al. 2016; Patel and Kumar 2016; Perkins et al. 2018). The major drawback behind this is that the crude bio-oil, generated from this process, has a lower calorific value and higher oxygen content than the commonly used fuels such as gasoline, heavy oil, etc. (Chen et al. 2014). Since fast pyrolysis is a constant procedure, having a shorter residence time, it can avoid secondary reactions of cracking and repolymerization. These are the reasons why the oil yield is the highest for this process, compared to the previous ones. The moisture content of the feed for fast pyrolysis is preferred to be low (<10%) so that there is minimum water in the bio-oil produced (Chandraratne and Daful 2021). Fast pyrolysis is gaining incredible acceptance for certain advantages over the other types of pyrolysis (Zaman et al. 2017; Venderbosch and Prins 2010; Chiaramonti et al. 2007):

1. Fast pyrolysis has the capability of confirming the primary breakdown of the simple oligosaccharide and lignin fractions from the feedstock with consecutive improvement.
2. Scale-up is economically plausible.
3. Fast pyrolysis can use second-generation bio-oil feedstocks such as municipal solid wastes, industrial wastes, and forest residue.
4. It produces liquid fuels which are easy to store and transport.

11.9.5 Flash Pyrolysis

Flash pyrolysis is a rapid form of pyrolysis with a short residence time (<0.5 s) and high heating rate (>1000 °C/s) than a fast pyrolysis process (Chandraratne and Daful 2021; Kurian et al. 2022). This procedure is performed at temperatures varying between 800 and 1000 °C (Chandraratne and Daful 2021). This procedure intends to escalate the yield of bio-oil. Typically, the product delivery is 75–80 wt% of bio-oil and 12–13 wt% of char. Bio-oil production by flash pyrolysis is being investigated industrially. The disadvantages of this procedure are its inferior thermal stability and because of the catalytic effect of char, which produces more viscous bio-oil and sometimes the bio-oil contains solid residue (Zaman et al. 2017; Cornelissen et al. 2008).

11.10 A Comparative Study of Bio-oil Generated by the Different Types of Pyrolysis

Table 11.2 shows the ability of various pyrolysis processes in producing separate yields of bio-oil, char, and syngas.

Table 11.2 The extent and value of bio-oil depend upon the type of pyrolysis

Type of pyrolysis	Quantity of bio-oil	Quality of bio-oil
Slow	20–30 wt% (Tursi 2019)	Mainly consists of aliphatic hydrocarbons (<70%), a heating value similar to that of diesel, and moderate water content (Czajczyńska et al. 2017)
Intermediate	40–60 wt% (Jahirul et al. 2012)	Two-phase separable bio-oil with a high calorific value (Jahirul et al. 2012)
Fast	60–75 wt% (Li et al. 2013)	Mainly consists of 63.5% aliphatic hydrocarbons (along with 44.1% alkenes) and 23.5% of aromatic compounds, with minimum water content (Czajczyńska et al. 2017); satisfactory heating value but lower than gasoline, heavy oil, etc., oxygen content higher than these conventional fuels, easy and economical transportability and storability (Jahirul et al. 2012)
Flash	75–80 wt% (Jahirul et al. 2012)	Low water content, poor thermal stability, corrosive, presence of solids in oil (Jahirul et al. 2012)

11.11 Types of Pyrolysis Reactors

In pyrolysis, the reactor type and design are of cardinal importance. A considerable amount of research has been done in this domain. Reactors have been outlined in a mode to mollify certain situations considering specifications such as product residence time, temperature, pressure, etc. (Zaman et al. 2017). Studies reveal some reactors have produced an elevated bio-oil yield of 70–80 wt.% from feedstocks (Gupta and Demirbas 2010). Reactors can be identified as fixed-bed reactors, fluidized bed reactors, rotary kilns, ablative fast pyrolysis, vacuum pyrolysis, auger type, plasma type, microwave, and solar reactors (Czajczyńska et al. 2017; Mohan et al. 2006; Zaman et al. 2017).

11.11.1 Fixed-Bed Reactor

This type of reactor gives priority to bio-oil production. These reactors consist of two basic parts: the gas cooling section and the cleaning setup (cyclone, wet scrubbers, and dry filters). Fixed-bed reactors can be manufactured using either firebricks, steel, or concrete (Zaman et al. 2017). Heat usually diffuses from the outer wall of the reactor towards the stack particle bed height. The expanding volatiles are channelled out of the bed by a constant velocity inert medium. This also reduces the residence time of the volatiles in the reaction zone. The temperature control, temperature gradients, loss of catalyst, development of ineffective region due to blockage of the gas stream, the occurrence of side reactions, and improper heat transfer in the reactor region.

11.11.2 Fluidized Bed Reactor

This kind of reactor generates high bio-oil (≈75 wt%) due to a high rate of heat transfer, better control of vapour holding time, high surface area, and the relative velocity between the two phases (Jahirul et al. 2012; Lv et al. 2004). Several kinds of fluidized bed reactors used in pyrolysis/gasification processes are listed below.

11.11.3 Bubbling Fluidized Bed Reactor

The modest fluidized bed reactor in which the bed is kept suspended by controlling the gas velocity is known as a bubbling fluidized bed reactor (Gupta and Demirbas 2010). The operation and fabrication of this type of reactor are very straightforward (Zaman et al. 2017). The high loading of solid matter in the bed is responsible for uniform temperature control, good storage capacity, and efficient transfer of heat (Zaman et al. 2017). The residence time of the feed and the products is dependent on the flow rate of the fluidizing gas. The char has the highest residence time but is further separated with the help of a cyclone separator, and the vapour is liquefied into bio-oil. The bubbling fluidized bed reactor yields a high-quality bio-oil in a large amount (Gupta and Demirbas 2010).

11.11.4 Circulating Fluidized Bed Reactor

When the flow of the fluidizing gas is higher, bed (sand), vapour, and char particles are passed out of the reactor, by the fluidizing gas, to the cyclone separator. This is why a circulating fluidized bed reactor was designed. In this type of reactor, sand particles are circulated back to the reactor. The basic aspect of this type of reactor is alike to that of a bubbling fluidized bed reactor. The only difference is that in a circulating fluidized bed reactor, the residence time of the char and vapour is shorter which affects high gas velocity. The char particles and vapour have more or less the same amount of residence time, which guides to higher coal content in the produced bio-oil. The heat transfer rate is less when related to that in a bubbling fluidized reactor, and the solid content is diluted in a circulating fluidized bed reactor (Gupta and Demirbas 2010).

11.11.5 Ablative Reactor

In this kind of reactor, the heating is mainly performed in the absence of any fluidizing gas and along a molten layer on the heated reactor surface. Feedstock melting is carried out by pressing the feed mechanically against the hot wall of a reactor (Zaman et al. 2017), and this provides an increased heating rate, which in turn encourages quick melting and sublimation of the feed (Gupta and Demirbas 2010). Faster heat transfer requires good particle-particle surface contact, and thus it is important for the feed particles in fast pyrolysis to be small in size or to have a large surface area. But this mechanism of extruding the feed against the heated reactor wall solves the problem of size reduction for large particles, and, thus, this type of reactor allows particles up to a size of 20 mm. The feedstocks do not need extreme grinding (Jones et al. 2009). Due to the mechanical nature of the procedure and rapid heating, the reaction setup is more complex and intense.

11.11.6 Vacuum Pyrolysis Reactor

A vacuum pyrolysis reactor is a slow pyrolysis reactor and has an immensely low rate of heat transfer, thus resulting in a low yield of bio-oil (35–50 wt%). The feed is transported into a high-temperature vacuum chamber with the help of a conveyor belt (Roy and Yang 1997). While doing this, the feed is stirred periodically by mechanical agitation. A burner along with an induction heater and molten salts as heat carriers is used to heat the feed (Roy and Yang 1997). The fabrication of this kind of reactor is very complicated, and the requirement of high expenditure and maintenance costs makes it economically unsuitable. The key advantage of this kind of reactor is that it can deal with big-sized feed particles (even up to 2–5 cm) compared to fluidized bed reactors. Because of their function in a vacuum, these reactors need special solids feedstock and execute devices to keep up effective seals all the time (Jahirul et al. 2012).

11.11.7 Rotating Cone Reactor

The rotating cone reactor needs intense mechanical mixing of feedstock along with hot sand and does not have the requirement of inert gas. The feed along with the hot sand is injected from the cone bottom and is transported to the cone lip during spinning, with the help of centrifugal force, and as the feed and sand mixture reaches the tip, the generated vapour is liquified by the condenser (Pattiya et al. 2008). The pyrolysis of the feedstock particles starts instantly after plunging into the reactor. The char mixed with sand is heated up again and reinjected to mix with fresh feed at the cone bottom. Even though they might have a complex design, these types of reactors are desirable because of the escalated yield of bio-oil. Sand is added with the feed particles to prevent the cone wall from getting fouled and also to surge the wall-to-particle heat transfer. Rotating cone reactors operate at atmospheric pressure and also have a high feed capacity, but these reactors need very small feed particle sizes (usually 50–100 μm), and one more disadvantage is that the process scale-up is problematic (Gupta and Demirbas 2010).

11.11.8 Auger Reactor

In this type of reactor, augers (screw-type conveying) are utilized to pass the feedstock across a heated cylindrical tube, which is devoid of oxygen. While proceeding across the tube, the temperature of the feedstock particles gets increased to the chosen pyrolysis temperature, usually ranging from 400 to 800 °C, and this causes the feedstock to gasify (Jahirul et al. 2012). With the generation of char, the condensable gases are liquefied to obtain bio-oil, while the noncondensable vapour is accumulated as the syngas. In this type of reactor, the vapour residence time can be altered by varying the heated zone of the tube across which the vapour moves before moving into the condenser train, and this in turn can control the secondary pyrolysis reactions (Jahirul et al. 2012). Auger reactor doesn't require a carrier gas medium since it operates uninterruptedly. The energy required to operate this type of reactor is generally lesser than the other types of reactor.

11.11.9 Plasma Reactors

The plasma reactors are made up of a quartz cylindrical tube which is enclosed by two Cu electrodes. Feed is injected in the middle of the tube with the help of a shifting-speed screw feeder, which is situated on the uppermost region of the tube. The two Cu electrodes are joined with power to generate thermal energy for gas that passes across the tube. An inert gas, included in the reactor, removes oxygen and also helps as an active gas to produce plasma. The pyrolysis vapour products are pumped out of the reactor using a changeable-speed vacuum pump (Liang et al. 2015). Plasma pyrolysis of organic waste between 5000 and 10,000 K elevated the yield of H_2 to 14.9 vol.%. With the steam injection, the yield of H_2 elevated to 27.06 vol.% (Huang and Tang 2007). Besides, plasma pyrolysis is effective in generating char and syngas. Together these product streams are easy to handle and are very useful.

11.11.10 Microwave Reactors

In the electromagnetic spectrum, microwaves generally lie among infrared and radiofrequencies, wavelengths ranging between 1 mm to m. However, their equivalent frequencies range between 300 GHz and MHz, respectively. Most domestic microwave ovens operate at a frequency of 2.45 GHz (Li et al. 2002). This kind of reactor is one of the current advances in pyrolysis (Zaman et al. 2017). Lam and Chase studied microwave-assisted pyrolysis of waste (Lam and Chase 2012). The scientists concluded that a dearth of the conception of microwave systems and the technical knowledge required for fabricating viable equipment for this type of pyrolysis hinders the industrial growth of microwave-assisted applications. In microwave reactors, energy transfer occurs due to the interaction between the atoms and the molecules using a microwave (Zaman et al. 2017). The entire process of drying and pyrolysis of the feedstock is performed in a microwave oven chamber. Inert gas is chosen as the carrier gas and is also used to create an anaerobic atmosphere (Jahirul et al. 2012). The advantages of microwave reactors include effective heat transfer, the capacity to manage the heating process effectively, the ability to reduce the formation of unwanted by-products, and the potential of being used effectively on an industrial basis. Additionally, it is likely to process an extensive range of biomass and industrial waste matter and obtain escalated yields of bio-oil and syngas (Lam et al. 2010).

11.11.11 Solar Reactors

The application of solar reactors in pyrolysis processes offers an appropriate way of classifying solar energy as chemical energy. Solar reactors are made of quartz tubes with opaque outside walls, which are generally exposed to solar radiation of high concentration. To increase the concentration of solar radiation, a parabolic solar concentrator is added to the reactor. The highly concentrated solar radiation has the capability of producing elevated temperatures (>700 °C) inside the reactor (Kalinci et al. 2009; Noushad et al. 2014). This kind of reactor has a few edges over conventional slow reactors. In the case of slow pyrolysis, a portion of the feedstock is utilized to generate heat for the process, and thus the quantity of feedstock available is reduced and also causes pollution (Czajczyńska et al. 2017). Thus, the use of solar energy for pyrolysis does not reduce the quantity of feedstock accessible and at the same time solves the pollution problem. Additionally, solar reactors have the capability of more rapid start-up and shut-down stages as related to slow reactors (Confederation of European Paper Industries 2015). Moreover, using a renewable energy resource for providing energy makes pyrolysis eco-friendlier, and energy efficiency also rises.

11.12 Pretreatment of the Feed

In this study, our objective is to escalate the quantity of bio-oil produced by pyrolysis and also upgrade the characteristics of the bio-oil to make it suitable for commercial usage. For achieving this purpose, studies have shown that the feedstock needs certain pretreatment. The term "pretreatment" refers to the treatment of feed before use. Pretreatment methods alter the structure of the feed molecules. Thus, it is vital to study the type of pretreatment.

A prerequisite for bio-oil production from lignocellulose, through most conversion techniques, is to break down the hindrance offered due to the structure of biomass by removing lignin and making cellulose more available by altering its structure. Some of the requirements of an effective and economically fit pretreatment method are (1) the use of low-cost chemicals, (2) reduced chemical consumption, (3) preventing hemicellulose and cellulose from getting denatured, (4) reduced energy consumption, and (5) the low-cost size reduction process. These are the factors that want to be noticed before choosing a pretreatment process. Pretreatment of lignocellulosic waste is not an informal process as it appears to be after the fitting of the power source. At the industrial level, pretreatment is the second most costly process. It contributes to the rate of the final product (≈18–20%) (Aftab et al. 2019).

In the event of lignocellulosic biomass, it should be pretreated to increase its chances of being used as transportation fuels or other commercial fuels (Agbor et al. 2011; Dalena et al. 2018). The pretreatment methods concurrently decrease the crystallinity index of cellulose and increase the surface area by getting rid of lignin and hemicellulose layers. This leads to more effective cellulose hydrolysis (Mosier et al. 2005).

Pretreatment methods, applied for lignocellulosic matter, can be distinguished into the physical, chemical, physicochemical, and biological processes. Table 11.3 highlights the pros and cons of the various pretreatment process.

Table 11.3 includes different pretreatment methods, and lignocellulosic feedstocks need to undergo, reliant on the kind of conversion technique and the quality and quantity of the desired product. As we have mentioned before, this chapter aims to learn the ways of obtaining the maximum amount of bio-oil from lignocellulosic waste through a thermochemical conversion technique called pyrolysis. Let's talk about the pretreatment methods that influence the products obtained through pyrolysis. Physical properties such as particle size, shape, and density have a significant impact on the foremost properties for a continuous feeding arrangement like the flowability of particles and material handling (Tannous et al. 2013), and chemical properties like proximate analyses affect the dispersal and value of pyrolysis products. Homogeneous feedstock with uniform physical and chemical properties plays an important part in the pyrolysis process and the value of its products. Table 11.4 shows the significance of different pretreatments required to increase the bio-oil yield with low oxygen content and high heating value.

11.13 Bio-oil Upgradation

The molecules present in bio-oil have molecular weights, ranging between 30 and 1000 g/mol, while the boiling points are between 19 and 386 °C (Branca et al. 2006). The polarity of the compounds present in bio-oil varies broadly, making bio-oil a heterogeneous mixture (Gupta and Demirbas 2010). Due to the poor qualities and

Table 11.3 Classification of pretreatment methods

Pretreatment process	Different types of that particular method	Situations	Pros	Cons	References
Physical pretreatment methods					
Milling	Ball Two-mill Hammer Celluloid Vibro energy	Takes place at normal temperature and pressure	Size reduction, cellulose crystallinity decreased, the available surface area increased, degree of polymerization decreased	Energy extensive, no delignification	Agbor et al. (2011), Paulová et al. (2013), Chandra and Bura (2007)
Irradiation	Gamma-ray Electron-beam Microwave		Size reduction of feed		Kassim et al. (2016)
Others	Grinding, hydrothermal, high-pressure steaming, expansion, and extrusion				
Chemical and physicochemical pretreatment process					
Acid pretreatment	Dilute acid treatment (H_2SO_4, HCl, H_3PO_4, HNO_3)	Temperature 140–215 °C, concentration <4%, pressure 0.5 MPa	High reaction rates, lignin disruption, demineralization (when washed with dilute acid and hot water) augmented availability of cellulose enhanced digestibility, reasonable temperatures	The quantity of lignin removed is less, sugar decomposition, hemicellulose dissolved, acid recycling needed, pH adjustment	Agbor et al. (2011), Paulová et al. (2013), Liu et al. (2011a), Modenbach and Nokes (2012)
	Concentrated acid hydrolysis (H_2SO_4, H_3PO_4)	Concentration 70–77%, temperature 40–100 °C	Destroys the crystalline structure of cellulose, amorphous cellulose achieved, low-temperature process	Strong acids corrode equipment damage (corrosion), hemicellulose diluted, sugar breakdown, acid recycling needed, pH alterations, environmental concerns	Liu et al. (2011a)

Alkali pretreatment	NaOH, KOH, Ca $(OH)_2$	Temperature ranges between 25°C and 130 °C	Cellulose crystallinity decreased, demineralization (when washed with dilute acid and hot water), polymerization decreased, and significant lignin removal, fewer inhibitors	Hemicellulose dissolved, pH adjustment	Agbor et al. (2011), Modenbach and Nokes (2012)
	Ammonia pretreatment	Temperature ranges between 25 °C and 60 °C, the reaction takes several days	High lignin removal, the volatility of ammonia is high, low-cost ammonia	No change in cellulose crystallinity, environmental concerns	Chandra and Bura (2007), Liu et al. (2011b)
Explosion	Steam	High-pressure saturated steam was inserted into the reactor occupied with feedstock followed by a sudden pressure reduction	Delignification and degradation of hemicellulose	Formation of inhibitors (g)	Tursi (2019), Rezaei et al. (2018)
	CO_2 explosion	High-pressure CO_2 is injected into a batch reactor occupied with feedstock followed by pressure reduction	An explosive decompression occurs due to the pressure reduction and this improves hydrolysis	Inhibitor formation	Tursi (2019), Rezaei et al. (2018)
Oxidative pretreatment	Ozonolysis	Normal pressure and ambient temperature	No inhibitor formation, takes place at room temperature, lignin degradation, increasing the digestibility of biomass, reduced environmental pollution	Hemicellulose dissolved in a dangerous process due to the highly corrosive and toxic characteristics of ozone, high cost	Paulová et al. (2013), Travaini et al. (2016)
	Wet oxidation	Addition of oxidizing agents such as O_2, H_2O, H_2O_2	Low inhibitor concentration	High cost of chemicals and equipment, high-temperature and pressure	Paulová et al. (2013)

(continued)

Table 11.3 (continued)

Pretreatment process	Different types of that particular method	Situations	Pros	Cons	References
Biological pretreatment method					
Biological hydrolysis	Fungi and actinomycetes	Takes place at normal temperature and pressure	Lignin and hemicellulose degradation, cheap, low energy consumption	Large quantity of microbes mandatory for effective pretreatment need for large space, controlled and careful growth conditions, and long residence time (almost 10–14 days)	Agbor et al. (2011), Chandra and Bura (2007), Liu et al. (2011b), Rezaei et al. (2018)

Table 11.4 Significance of various pretreatments

Pretreatment	Significance
Moisture reduction	The higher moisture content of the feedstock is responsible for reducing the heating value of the fuel produced and also for shifting the ignition temperature of the fuel to a higher point. Moisture content should be reduced to 10% or less (Bridgwater and Peacocke 2000). Pyrolysis of dry feed produces bio-oil with a higher heating value, lower ignition point, lower amount of water and a higher rate of combustion
Size reduction	The minor particles have an added available surface area in the elevated-temperature setting, and this increases the rate of mass as well as heat transfer among the particles and surroundings. Research showed that bigger particles (diameter ranging from 1 to 5 mm) have an interruption in both mass and heat transfer and thus the rate of pyrolysis is reduced (Rezaei et al. 2018)
Demineralization	The presence of minerals facilitates the catalytic decomposition of levoglucosan into hydroxyl acetaldehyde compounds (Agblevor and Besler 1996; Scott et al. 2001). Levoglucosan is a desired part of volatiles, while hydroxyl acetaldehyde compounds are an unwanted part of volatiles
Torrefaction	It is a physicochemical method which alters both physical and chemical properties of the feed by removing moisture and hemicelluloses and partially removing cellulose and lignin (Garba 2020)

disadvantages of crude bio-oil mentioned in this chapter, currently, it is mostly used in boilers as low-grade fuels with the purpose of heating or generating power, instead of being directly used in the form of high-grade transportation fuels (Wang et al. 2020).

To use bio-oil as high-grade fuels, modification and upgrading of crude bio-oil are important. Bio-oil should be parted and concentrated into compounds of comparable polarities. The oxygen content should be lowered for increasing the heating value of the bio-oil. To achieve the above-mentioned requirements, solvent fractionation and deoxygenation are two techniques, used for upgrading bio-oil.

11.13.1 Solvent Fractionation

Bio-oil can be parted into an organic fraction and an aqueous fraction easily by adding water. The water-soluble fraction, which is rich in carbohydrate-derived compounds, gets collected as the top phase, and a viscous oligomeric fraction, rich in lignin, gets settled as the lowermost phase (Gupta and Demirbas 2010). The lignin-rich fraction consists of around 25–30% of the bio-oil. The oxygenated compounds distribute themselves among the two phases.

11.13.2 Deoxygenation

The drawbacks of bio-oil such as high viscosity, low heating value, corrosiveness, and storage instability, can be upgraded by lowering the oxygen content in the bio-oil. Thus, the upgradation of bio-oil is required before final use. Catalytic cracking of pyrolysis vapours, hydro-processing, etc. is a few recent upgradation techniques (Zhang et al. 2007).

In the case of hydro-processing, bio-oil is hydrogenated with the help of a catalyst (e.g. Co-Mo, Ni-Mo-Al), and oxygen is detached as H_2O and CO_2. In the case of catalytic cracking, the oxygenated compounds in bio-oil are decomposed to form hydrocarbons along with oxygen removal in the form of H_2O and CO/CO_2. A few examples of catalysts used are ZnO, Al-MCM-41, Cu/Al-MCM-41, etc. Additionally, this method improves the yield of acetic acid and furans and removes levoglucosan and big phenolic compounds. Even though catalytic cracking is a low-priced route for performing deoxygenation, the disadvantage of this process is high coke formation. The quest for an effective catalyst with low coke formation and high conversion is still going on.

11.14 Current Status of Pyrolysis Technology

Presently, there is great hope in the abundantly available lignocellulosic kitchen waste as a source of feed for producing bio-fuels through the pyrolysis techniques process though some teething challenges need to be overcome (Hill 2007). There is a dire prerequisite for improved technical knowledge and consideration of the pyrolysis process, so that waste matter can be effectively used for large-scale commercial production of biofuels. Table 11.5 shows various compact and pilot-scale pyrolysis plants operated by various research organizations over the world (Jahirul et al. 2012). These setups are dependent on fuel price, availability of feed and cost, local expenses, workers, etc. Most of these plants are used for research purposes, while a few of them are being operated on an industrial scale. Recently, researchers are giving more attention to increasing the overall yield of oil from pyrolysis techniques instead of being concerned with the value and upgradation of bio-oil.

11.15 Future Challenges and Limitations

Despite rapid development over the past few years and clear knowledge of the pyrolysis process, numerous aspects such as the commercialization of pyrolysis products need to overcome several technical and economic barriers. Some of the potential challenges are listed below:

Table 11.5 Various pyrolysis plants

Name of the organization/location	Reactor used	Main product
Bio-alternative, USA	Fixed-bed	Char
THEE	Bubbling fluidized bed	Syngas
DynaMotive, Canada	Bubbling fluidized bed	Bio-oil
BEST Energy, Australia	Bubbling fluidized bed	Bio-oil
Wellman, UK	Bubbling fluidized bed	Bio-oil
Union Fenosa, Spain	Bubbling fluidized bed	Bio-oil
Zhejiang University, China	Bubbling fluidized bed	Bio-oil
RTI, Canada	Bubbling fluidized bed	Bio-oil
Waterloo university	Bubbling fluidized bed	Bio-oil
Zhejiang University, China	Bubbling fluidized bed	Bio-oil
Red Arrow, WI; Ensyn	Circulating fluidized bed	Chemicals
Ensyn Engineering	Circulating fluidized bed	Bio-oil
VTT, Finland, Ensyn	Circulating fluidized bed	Bio-oil
BTG, Netherlands	Rotating cone	Bio-oil
University of Twente	Rotating cone	Bio-oil
Privacy, Canada	Vacuum	Bio-oil
Laval University	Vacuum	Bio-oil
Solar Energy Research Ins.	Vortex	Bio-oil
Fortum, Finland	Other types	Bio-oil
University Zaragoza	Other types	Syngas
Georgia Tech. Research Ins.	Other types	Bio-oil

- The initial feed pretreatment process lacks a methodical and even approach.
- The quality of pyrolysis products is another challenge. The bio-oil produced is not appropriate for head-on application in automobile engines because of the high oxygen and moisture content. Thus, a suitable cost-effective upgradation method has to be performed after the main pyrolysis process.
- The connection between the feed matter and the operation of the pyrolysis plant is still not studied well.
- The pyrolysis reactor should be fabricated in such a path that it operates at the lowest possible temperature while producing an optimum quantity of bio-oil.
- Another concern is the environmental and safety issues while handling, transporting, and usage of products.
- The reliability of pyrolysis needs to be warranted, and this can be achieved by upgrading material and energy balances for sustainable and gainful applications (Yogalakshmi et al. 2022).
- The technology must be economically optimized to harvest high-quality bio-oil to a bulk extent (Yogalakshmi et al. 2022).
- Researchers are focusing on the usage of catalysts in the pyrolysis process to advance the value and extent of bio-oil. But the adoption of this methodology on an industrial scale will demand related advances in catalyst separation, recovery, and regeneration approach.

11.16 Conclusion

The purpose of this chapter is to present a view of achieving value-added fuel from LOKW through pyrolysis. The thermochemical processes took lesser time than biochemical processes in converting feedstock into value-added fuels. Among the different thermochemical processes, pyrolysis proved to be effective in generating desired products for fuel or chemical application. In pyrolysis tar/bio-oil is simultaneously release with light gas species from the thermal decomposition of functional groups. Depolymerization and thermal decomposition of functional groups leads to stabilized H_2 generation. Various pyrolysis reactors showed their efficacy in volatizing the feedstock for the generation of products. The pretreatment methods highlighted the necessity in upgrading the bio-oil by reducing its oxygen and acid content.

References

Aftab MN, Iqbal I, Riaz F, Karadag A, Tabatabaei M (2019) Different pretreatment methods of lignocellulosic biomass for use in biofuel production. In: Abomohra AE-F (ed) Biomass for bioenergy. IntechOpen, London. https://doi.org/10.5772/intechopen.84995

Agblevor FA, Besler S (1996) Inorganic compounds in biomass feedstocks. 1. Effect on the quality of fast pyrolysis oils. Energy Fuel 10:293–298

Agbor VB, Cicek N, Sparling R, Berlin A, Levin DB (2011) Biomass pretreatment: fundamentals toward application. Biotechnol Adv 29:675–685

Aho A et al (2008) Catalytic pyrolysis of woody biomass in a fluidized bed reactor: influence of the zeolite structure. Fuel 87:2493–2501

Anon (2018) WBA global bioenergy statistics. Summary report. www.worldbioenergy.org; www.worldbioenergy.org

Asia Development Bank (2020) Waste to energy in the age of the circular economy. https://www.adb.org/publications/waste-to-energy-age-circular-economy-handbook

Babu BV, Chaurasia AS (2003) Modeling for pyrolysis of solid particle: kinetics and heat transfer effects. Energy Convers Manag 44:2251–2275

Balat M (2008) Mechanisms of thermochemical biomass conversion processes. Part 1: reactions of pyrolysis. Energy sources, part a recover. Util Environ Eff 30:620–635

Branca C, Di Blasi C, Elefante R (2006) Devolatilization of conventional pyrolysis oils generated from biomass and cellulose. Energy Fuel 20:2253–2261

Bridgwater AV, Peacocke GVC (2000) Fast pyrolysis processes for biomass. Renew Sust Energ Rev 4:1–73

CalRecycle (2022) Biochemical pathway and processes for conversion of organic materials to energy. https://calrecycle.ca.gov/organics/conversion/pathways/biochem

Chandra RP, Bura R (2007) Substrate pretreatment: the key to effective enzymatic hydrolysis of lignocellulosics? In: Olsson L (ed) Biofuels. Springer, Berlin, pp 67–93. https://doi.org/10.1007/10_2007_064

Chandraratne MR, Daful AG (2021) Recent advances in thermochemical conversion of biomass. In: Bartoli M, Giorcelli M (eds) Recent perspectives in pyrolysis research. IntechOpen, London. https://doi.org/10.5772/intechopen.100060

Chen D, Zhou J, Zhang Q, Zhu X (2014) Evaluation methods and research progresses in bio-oil storage stability. Renew Sust Energ Rev 40:69–79

Chiaramonti D, Oasmaa A, Solantausta Y (2007) Power generation using fast pyrolysis liquids from biomass. Renew Sust Energ Rev 11:1056–1086

Confederation of European Paper Industries (2015) ERPC monitoring report. www.cepi.org/erpc-monitering-report-2015/

Cornelissen T et al (2008) Flash co-pyrolysis of biomass with polyhydroxybutyrate: part 1. Influence on bio-oil yield, water content, heating value and the production of chemicals. Fuel 87: 2523–2532. https://doi.org/10.1016/j.fuel.2008.02.024

Czajczyńska D et al (2017) Potential of pyrolysis processes in the waste management sector. Therm Sci Eng Prog 3:171–197

Dalena F et al (2018) Advances in methanol production and utilization, with particular emphasis toward hydrogen generation via membrane reactor technology. Membranes 8:98

Demirbas A (2007) Combustion of biomass. Energy sources, part A recover. Util Environ Eff 29: 549–561

Dhyani V, Bhaskar T (2018) A comprehensive review on the pyrolysis of lignocellulosic biomass. Renew Energy 129:695–716

Diebold JP (2000) A review of the chemical and physical mechanisms of the storage stability of fast pyrolysis bio-oils. http://www.doe.gov/bridge

Garba A (2020) Biomass conversion technologies for bioenergy generation: an introduction. In: Basso TP, Basso TO, Basso LC (eds) Biotechnological applications of biomass. IntechOpen, London. https://doi.org/10.5772/intechopen.93669

Goyal HB, Seal D, Saxena RC (2008) Bio-fuels from thermochemical conversion of renewable resources: a review. Renew Sust Energ Rev 12:504–517

Grycová B, Koutník I, Pryszcz A (2016) Pyrolysis process for the treatment of food waste. Bioresour Technol 218:1203–1207

Gupta RB, Demirbas A (2010) Bio-oil from biomass pyrolysis. In: Gasoline, diesel, and ethanol biofuels from grasses and plants. Cambridge University Press, Cambridge, pp 140–157. https://doi.org/10.1017/CBO9780511779152.010

Haiqi Machinery (2019) Henan Haiqi Environmental Protecton Technology Co., Ltd

Hill J (2007) Environmental costs and benefits of transportation biofuel production from food- and lignocellulose-based energy crops. A review. Agron Sustain Dev 27:1–12

Huang H, Tang L (2007) Treatment of organic waste using thermal plasma pyrolysis technology. Energy Convers Manag 48:1331–1337

Jahirul MI, Rasul MG, Chowdhury AA, Ashwath N (2012) Biofuels production through biomass pyrolysis-a technological review. Energies 5:4952–5001

Jones SB et al (2009) Production of gasoline and diesel from biomass via fast pyrolysis, hydrotreating and hydrocracking: a design case

Kalinci Y, Hepbasli A, Dincer I (2009) Biomass-based hydrogen production: a review and analysis. Int J Hydrog Energy 34:8799–8817

Kan T, Strezov V, Evans TJ (2016) Lignocellulosic biomass pyrolysis: a review of product properties and effects of pyrolysis parameters. Renew Sust Energ Rev 57:1126–1140

Kassim MA et al (2016) Irradiation pretreatment of tropical biomass and biofiber for biofuel production. In: Monteiro WA (ed) Radiation effects in materials. IntechOpen, London. https://doi.org/10.5772/62728

Kazawadi D, Ntalikwa J, Kombe G (2021) A review of intermediate pyrolysis as a technology of biomass conversion for coproduction of biooil and adsorption biochar. J Renew Energy 2021: 5533780

Kurian V, Gill M, Dhakal B, Kumar A (2022) Chapter 21 - recent trends in the pyrolysis and gasification of lignocellulosic biomass. In: Gurunathan B, Sahadevan R (eds) Biofuels and bioenergy. Elsevier, Amsterdam, pp 511–552. https://doi.org/10.1016/B978-0-323-90040-9.00028-X

Lam SS, Chase HA (2012) A review on waste to energy processes using microwave pyrolysis. Energies 5:4209–4232

Lam SS, Russell AD, Chase HA (2010) Microwave pyrolysis, a novel process for recycling waste automotive engine oil. Energy 35:2985–2991
Lee SY et al (2019) Waste to bioenergy: a review on the recent conversion technologies. BMC Energy 1:4
Li S-Q, Yan J-H, Li R-D, Chi Y, Cen K-F (2002) Axial transport and residence time of MSW in rotary kilns: part I. Experimental. Powder Technol 126:217–227
Li S, Xu S, Liu S, Yang C, Lu Q (2004) Fast pyrolysis of biomass in free-fall reactor for hydrogen-rich gas. Fuel Process Technol 85:1201–1211
Li L, Rowbotham JS, Christopher Greenwell H, Dyer PW (2013) Chapter 8 - an introduction to pyrolysis and catalytic pyrolysis: versatile techniques for biomass conversion. In: Suib SL (ed) New and future developments in catalysis. Elsevier, Amsterdam, pp 173–208. https://doi.org/10.1016/B978-0-444-53878-9.00009-6
Liang S, Han Y, Wei L, McDonald AG (2015) Production and characterization of bio-oil and bio-char from pyrolysis of potato peel wastes. Biomass Convers Biorefinery 5:237–246
Liang X et al (2021) Experimental investigation on supercritical water gasification of organic-rich shale with low maturity for syngas production. Energy Fuel 35:7657–7665
Liu C, Huang J, Huang X, Li H, Zhang Z (2011a) Theoretical studies on formation mechanisms of CO and CO_2 in cellulose pyrolysis. Comput Theor Chem 964:207–212
Liu WJ, Zeng FX, Jiang H, Yu HQ (2011b) Total recovery of nitrogen and phosphorus from three wetland plants by fast pyrolysis technology. Bioresour Technol 102:3471–3479
Lue S (2019) Principle and advantages of pyrolysis
Lv PM et al (2004) An experimental study on biomass air–steam gasification in a fluidized bed. Bioresour Technol 95:95–101
McKendry P (2002) Energy production from biomass (part 1): overview of biomass. Bioresour Technol 83:37–46
Modenbach AA, Nokes SE (2012) The use of high-solids loadings in biomass pretreatment-a review. Biotechnol Bioeng 109:1430–1442. https://doi.org/10.1002/bit.24464
Mohan D, Pittman CU, Steele PH (2006) Pyrolysis of wood/biomass for bio-oil: a critical review. Energy Fuel 20:848–889
Moreira-Vilar FC et al (2014) The acetyl bromide method is faster, simpler and presents best recovery of lignin in different herbaceous tissues than Klason and thioglycolic acid methods. PLoS One 9:1–7
Mosier N et al (2005) Features of promising technologies for pretreatment of lignocellulosic biomass. Bioresour Technol 96:673–686
Noushad M, Ab Rahman I, Che Zulkifli NS, Husein A, Mohamad D (2014) Low surface area nanosilica from an agricultural biomass for fabrication of dental nanocomposites. Ceram Int 40: 4163–4171. https://doi.org/10.1016/j.ceramint.2013.08.073
Oasmaa A, Kuoppala E (2003) Fast pyrolysis of forestry residue. 3. Storage stability of liquid fuel. Energy Fuel 17:1075–1084
Patel M, Kumar A (2016) Production of renewable diesel through the hydroprocessing of lignocellulosic biomass-derived bio-oil: a review. Renew Sust Energ Rev 58:1293–1307
Pattiya A, Titiloye JO, Bridgwater AV (2008) Fast pyrolysis of cassava rhizome in the presence of catalysts. J Anal Appl Pyrolysis 81:72–79
Paulová L, Patáková P, Rychtera M, Melzoch K (2013) Production of 2nd generation of liquid biofuels. In: Fang Z (ed) Liquid, gaseous and solid biofuels. IntechOpen, London. https://doi.org/10.5772/53492
Perkins G, Bhaskar T, Konarova M (2018) Process development status of fast pyrolysis technologies for the manufacture of renewable transport fuels from biomass. Renew Sust Energ Rev 90: 292–315
Rezaei H, Yazdanpanah F, Lim JC, Lau A, Sokhansanj S (2018) Woody feedstock pretreatments to enhance pyrolysis bio-oil quality and produce transportation fuel. In: Abomohra AE-F (ed) Biomass for bioenergy. IntechOpen, London. https://doi.org/10.5772/intechopen.81818

Roy C, Yang J (1997) Development of a novel vacuum pyrolysis reactor with improved heat transfer potential. In: Bridgwater AV, Boocock DGB (eds) Developments in thermochemical biomass conversion: volume 1/volume 2. Springer, Cham, pp 351–367. https://doi.org/10.1007/978-94-009-1559-6_28

Saber M, Nakhshiniev B, Yoshikawa K (2016) A review of production and upgrading of algal bio-oil. Renew Sust Energ Rev 58:918–930

Salimbeni A (2019) 6 - Organic waste streams upgrading for gasification process optimization. In: Materazzi M, Foscolo PU (eds) Substitute natural gas from waste. Academic, New York, pp 75–103. https://doi.org/10.1016/B978-0-12-815554-7.00006-4

Scott DS, Paterson L, Piskorz J, Radlein D (2001) Pretreatment of poplar wood for fast pyrolysis: rate of cation removal. J Anal Appl Pyrolysis 57:169–176

Shekhar Silori C, Kumar Mishra B (2001) Assessment of livestock grazing pressure in and around the elephant corridors in Mudumalai Wildlife Sanctuary, south India. Biodivers Conserv 10: 2181–2195. https://doi.org/10.1023/A:1013285910650

Sieradzka M, Kirczuk C, Kalemba-rec I, Mlonka-mędrala A, Magdziarz A (2022) Pyrolysis of biomass wastes into carbon materials. Energies 15:1–12

Tannous K, Lam PS, Sokhansanj S, Grace JR (2013) Physical properties for flow characterization of ground biomass from Douglas fir wood. Part Sci Technol 31:291–300

Tkemaladze GS, Makhashvili KA (2016) Climate changes and photosynthesis. Ann Agrar Sci 14: 119–126

Travaini R, Martín-Juárez J, Lorenzo-Hernando A, Bolado-Rodríguez S (2016) Ozonolysis: an advantageous pretreatment for lignocellulosic biomass revisited. Bioresour Technol 199:2–12

Tursi A (2019) A review on biomass: importance, chemistry, classification, and conversion. Biofuel Res J 6:962–979. https://doi.org/10.18331/BRJ2019.6.2.3

Vamvuka D (2011) Bio-oil, solid and gaseous biofuels from biomass pyrolysis processes-an overview. Int J Energy Res 35:835–862. https://doi.org/10.1002/er.1804

Venderbosch RH, Prins W (2010) Fast pyrolysis technology development. Biofuels Bioprod Biorefin 4:178–208

Wang G et al (2020) A review of recent advances in biomass pyrolysis. Energy Fuel 34:15557–15578

Yogalakshmi KN et al (2022) Lignocellulosic biomass-based pyrolysis: a comprehensive review. Chemosphere 286:131824

Zaman CZ et al (2017) Pyrolysis: a sustainable way to generate energy from waste. In: Samer M (ed) Pyrolysis. IntechOpen, London. https://doi.org/10.5772/intechopen.69036

Zhang Q, Chang J, Wang T, Xu Y (2007) Review of biomass pyrolysis oil properties and upgrading research. Energy Convers Manag 48:87–92

Chapter 12
Generation of Bioenergy from Industrial Waste Materials

Rashmi Dhurandhar, Pankaj Parmar, Chandrakant Thakur, Bimal Das, and Nilambar Bariha

Abstract The technology for recovering energy from garbage can be very helpful in reducing the issues. Population growth and the need to improve living conditions have led to an increase in energy demand. As a result, there is a greater need for exploring and depleting energy resources mainly in fossil fuel. The environment is greatly impacted by the use of fossil fuels. Within the process of combustion, greenhouse gases are released that may contribute to global warming. This leads to excessive climatic change and unchecked contamination of the water, air, and soil, which constitute a serious threat to human existence on earth. Based on the idea that environmentally friendly energy sources and systems are necessary for energy sustainability, waste to energy is a relatively recent, efficient, and green method for managing the waste materials. The biodegradable portion of industrial wastes is a sustainable biomass resource, trying to maximize its management through energy valorization, reducing the amount of waste that must be managed (and its associated costs), minimizing the health risks, having a minimal negative environmental impact, and reducing the high reliance of industries on primary sources and fossil fuels. The capacity to create a diversity of bioenergy from garbage and biomass residues is promising. However, compared to petro-fuels and additional renewable energy sources, its cost is high. The two primary methods of transformation for the

R. Dhurandhar (✉) · B. Das
Department of Chemical Engineering, National Institute of Technology Durgapur, Durgapur, West Bengal, India

P. Parmar
Department of Chemical Engineering, Indian Institute of Technology Kharagpur, Kharagpur, West Bengal, India

C. Thakur
Department of Chemical Engineering, National Institute of Technology Raipur, Raipur, Chhattisgarh, India

N. Bariha (✉)
Assam Energy Institute, Sivasagar, Centre of Rajiv Gandhi Institute of Petroleum Technology, Sivasagar, Assam, India
e-mail: nbariha@rgipt.ac.in

D. B. Pal, A. K. Tiwari (eds.), *Sustainable Valorization of Agriculture & Food Waste Biomass*, Clean Energy Production Technologies,
https://doi.org/10.1007/978-981-99-0526-3_12

evolving innovations, namely, thermochemical (i.e., gasification, liquefaction, and pyrolysis) and biological (i.e., fermentation, anaerobic digestion, transesterification, and photobiological H_2 manufacture), are assessed.

Keywords Generation · Bio-energy · Industrial waste · Biological · Pyrolysis · Thermochemical

12.1 Introduction

In general, the amount of solid trash created increases with economic success and the proportion of people living in cities. A critical concern is reducing the volume and bulk of the solid waste, particularly given the scarcity of ultimate disposal locations in many parts of the world. A switch from traditional energy systems to meet the rising demand for energy and address environmental concerns, renewable resources are required. A technologically advanced, commercially viable waste treatment system is crucial for the sustainability of human civilization. There are several ways to turn solid waste into energy sources, ranging from elementary systems for getting rid of trash to more complicated ones that can handle massive amounts of industrial waste. Waste material can be converted to energy using three basic processes: thermochemical, biochemical, and physicochemical (Kim et al. 2022). Waste energy (WTE) is a conception that objects to reduction of the quantity of MSW while converting to additional energy sources like steam, bio-oil, synthetic gas, and heat (Kumar and Samadder 2022). To provide energy for sustainable development, this study investigates how to use garbage when science and technology are adequately utilized. It is advised that a campaign be launched to educate the public about the advantages of using garbage as a source of energy (Muhammad et al. 2013).

Valorizing trash is the best management method in light of legal obligations and circular economy principles. While conventional biomass sources like crops, waste from agriculture and forestry, and wood, food waste, and municipal garbage are affordable, sustainable, and renewable, they are in a food supply war and produce waste throughout the energy production process. Every year, significant amounts of waste are generated by the food business all around the world. The 1.3 billion tonnes of food waste formed yearly occupy approximately 28% of the total farming area, which is equal to 1.4 billion hectares of working cultivable area (Sharma et al. 2021). This article covers studies that suggest using industrial by-products and waste as sustainable, renewable resources, and limitless biomass sources to create power, cold and heat, bioliquids, and fuels. The benefits and drawbacks of different resources kinds are discussed, along with their constraints and the obstacles they must overcome (Gil 2022).

A supply of usable biomass, primarily from the chocolate industry, can be utilized to make new products including food, animal feed, cosmetics, chemicals, and even biofuels. Examples of industrial wastes from the cocoa industry that contain materials useful to many different industries include cocoa pod husk, mucilage, and bean shells. The concentration of lignocellulose in these by-products, however, makes a pretreatment necessary in order to utilize them successfully. As a result, various biofuels can be produced, with the best biomass output being produced using the conversion process. Recent studies have demonstrated that the industrial wastes generated during the manufacturing of cocoa can be converted into solid, liquid, or gaseous biofuels; however, the most common recorded application is as a source for direct burning, which is used to supply the same production plants. Therefore, the goal of this work is to analyze the technological idea used for the conversion and the usage of the by-products produced from cocoa to produce biofuels. The use of catalysts in manufacturing to increase compound conversion reactions and to improve the processing parameters, using statistical models, is also indicated by future trends (Mendoza-Meneses et al. 2021).

Renewable energy fuels that are both environmentally beneficial and economically viable are produced using the cereals, microalgae, and biological wastes—all contain biomass that can be used to create bioenergy. Nowadays, biogases and/or biofuels are the primary bioenergy sources, offering an as opposed to currently used (fossil fuel) energy that can satisfy the rising present and future energy needs of people throughout the world while minimizing the release of greenhouse gases into the atmosphere. Utilizing an efficient and adequate microbial system, sustainable raw substrates, and biological methods (anaerobic digestion or fermentation), bioenergy is typically created from these activities (biological wastes or crop biomass residue). Utilizing microbial catabolic biochemical reactions to generate electric energy allows the use of microbial fuel cells for complicated renewable biomasses and organic wastes. Utilizing microbial fuel cells to combine wastewater handling methods with electric energy synthesis has been helpful for sustainable bioenergy synthesis. Using genetic material modification or metabolic pathway engineering techniques, operative and appropriate microorganism species systems are now produced. Additionally, the improvement of synthetic microbial cells with planned internal gene malfunctions and alternative sources of microalgae species or biotechnological systems (for global cellular data), as well as systems biology and artificial life, can successfully improve biofuel production. The use of various sustainable resources production of bioethanol, biodiesel, biohydrogen, biobutanol, and biogas will be deliberated by Srivastava (2019).

Over the time, the lacks of energy and inefficient production methods have caused environmental destruction, so we must turn to more environmentally friendly practices. Keep in mind that we haven't been able to produce enough energy to supply it in many places, even after so much money has been spent and natural resources have been destroyed. The average person finds it difficult to get energy. We are looking at this to see how we might encourage local fuel production to meet the people's basic requirements in places with no energy access. Many rural communities in India still lack access to power to meet their basic needs. Still, they do have access to household and agricultural wastes that can be used to generate electricity. Conventional fuels are currently the main source of energy globally. They are having finite natural resources and also nonrenewable, Fortunately, cutting-edge technologies are now available that treat MSW while also producing electricity (Lee et al. 2022). To limit the use of conventional fuels and the resulting pollution, we wish to promote bioenergy as one of the sustainable energies that will be used in the future. Producers of waste can add value by turning their trash into biomass energy, a more lucrative use for their waste. The organic ingredients required to make biomass are endless because society will always produce waste and manure. We also want to reduce the amount of waste in landfills and the illicit burning of garbage, which contribute significantly to air pollution (Chavan et al. 2022).

The use of biomass for various reasons, such as foods, feeds, energy fuels, heating, cooling, and, most significantly, feedstock for biorefineries, has been the subject of growing discussion throughout the past century. Due to their renewable nature, biorefinery products were intended to replace chemicals and fossil fuels. Biomass is a biodegradable material made from forestry, aquaculture, and agricultural wastes and residues. A range of raw resources, including wood and by-products from the processing of wood, manure, different parts of organic waste products, and crops, can be used to make biomass. They have the benefits of simple storage, transportation, adaptable load use, and various applications as renewable energy sources. This study will give a general overview of existing thermochemical and biochemical biomass conversion techniques. Biofuel production also received attention due to its potential as a significant market for large-scale green sustainable biomass products (Garba 2020).

Sustainable biofuel production is the most efficient strategy to reduce greenhouse gas emissions caused by fossil fuels while maintaining food security and land usage. Waste biorefineries can be integrated into the waste management system to address the difficulties of waste disposal in the future and produce bioenergy. On the other hand, biomass waste is viewed as a low-quality feedstock for biorefineries with a wide variety of compositions and seasonal unpredictability. More research is required to increase the profitability of waste biorefineries since these constraints limit the conversion technologies that can be used to add value to biomass waste. Bioprocesses need to be combined to process a variation of biomass resources and produce a diversified variety of products to keep waste biorefineries commercially and environmentally sustainable. It is crucial to classify biomass wastes to match the available bioprocesses to reduce extraneous processes that can raise the production costs of the biorefinery and achieve optimal integration. This paper analyses the viability of the most popular waste-to-energy conversion techniques and the development of integrated biorefineries based on biomass classification. Through the integration of feedstocks, processes, platforms, and coexistence of wastes and by-products, the integration of waste biorefineries is examined in this paper. About the various types of biomass wastes, this review aims to create a framework for discovering and incorporating waste-to-energy technologies (Ochieng et al. 2022).

Since they are a desirable replacement for fossil fuels, which have historically been challenging to get, biofuels are becoming more and more significant. They can also assist the European Union nations to achieve their energy and environmental goals. Second-generation biofuels made from municipal solid waste's organic portion are becoming increasingly popular. Such an approach encourages the circular economy as an illustration of an efficient strategy for trash management. A review of various biofuels is presented in the study, focusing on second-generation biofuels made from lignocellulose biomass, and a special mention of one of the technologies for converting garbage into fuel is the BioRen process. It enables the economy to preserve the value of materials. The entire method is described, and material flow is presented. BioRen project's framework calls for the simultaneous production of bioethanol and isobutanol from the organic fraction through a two-stage enzymatic process and glycerol tertiary butyl ether. The hydrothermal carbonization method is used to treat the post-fermentation sludge, boosting the effectiveness of the entire waste management techniques and completing the circle. The bio-coal produced during the carbonization process can be burned to produce higher calorific value gasoline. The inorganic particles in the residual portion are physically, chemically, and catalytically converted into their clean, dry, and inert parts. The findings demonstrate that 78% of the heat produced during the BioRen project's operation is recovered and used to generate energy. Furthermore, 53% of garbage is recovered and reused, thanks to the profitable, effective, ecologically beneficial, and highly secure method. An example of such cutting-edge technology that is highly prepared has achieved results and has experience that may inspire European nations and institutions to put such a successful solution—which helps the growth of bioenergy into practice (Kowalski et al. 2022).

An expanding population's increased need for food and other essential resources has increased industrial and agricultural activity. Due to ineffective or inconsistently used disposal, utilization, and management techniques, agricultural waste is becoming an increasingly severe issue. Most biomass wastes, particularly in developing nations, are either burned outdoors or left in the field to decay, which seriously affects the environment. Similarly, alternative sustainable energy sources and raw material supplies are necessary due to the expanding urbanization of the world and the rising demand for construction products. Since there hasn't been much activity focusing on a "low-carbon" approach for their valorization, biomass wastes are an underutilized resource (for energy and material generation). The current research makes an effort to solve this by analyzing the potential of biomass wastes as feedstock for the production of high-volume construction materials on a worldwide scale. This study may interest individuals working to reduce carbon emissions, even though it is aimed at professionals in the sustainable biomass waste management industry. We provide an overview of the possibilities for CO_2 mitigation in a mineralization process, including the effects of biomass residues on CO_2 capture and utilization (CCU) to create building materials from gaseous and solid wastes. This effort aids in creating reliable waste-derived goods with higher added value and lower embodied carbon. The strategy will reduce carbon emissions and less stress on natural resources (soil, virgin stone) (Tripathi et al. 2019).

The number of biomass wastes is rising quickly, which causes a lot of problems with disposal and governance. As a result, the reuse and recycling of biomass wastes for applications with value addition have gained popularity. The study on the use of biomass waste and the valuable materials produced from it over the previous 5 years is reviewed in this publication. The following three aspects are the primary recent research interests: biomass wastes can be recycled, natural polymers can be extracted from them, and carbon-based materials can be made into functional composites, new adsorbents, catalyst carriers, electrode materials, and electrode materials. As low-cost raw materials, various biomass wastes have been gathered from municipal solid wastes, animal wastes, industrial wastes, and agricultural and forestry wastes. However, further research is needed to assess the quality and safety of the products made from these wastes and to create highly feasible and efficient methods for converting them into fuel for industrial-scale production (Zhou and Wang 2020).

According to scientific studies, bioenergy may be produced from trash and biomass residues, but it is more expensive than petro-fuels and other renewable energy sources. To lower production costs, ongoing efforts are being made to improve conversion technology. In particular, their technological principles, alternatives, and chances for implementation are covered in this review, which concentrates on conversion methods for turning biomass leftovers and trash into biofuels. Anaerobic digestion, alcoholic fermentation, and photobiological hydrogen production are three examples of biochemical conversion procedures. Gasification, liquefaction, and pyrolysis are the three thermochemical conversion pathways. Also covered is transesterification, which is the quickest and most cost-effective way to make vast amounts of biodiesel. Finally, the methods for direct biomass waste and waste residue conversion to bioelectricity are explored. These methods include burning and microbial fuel cells (Lee et al. 2019).

The contribution of biomass energy to the global energy system is substantial and increasing. Energy from biomass can significantly lower carbon emissions, especially in hard-to-decarbonize industries like manufacturing, heavy transportation, and aircraft. However, land-intensive bioenergy frequently results in significant carbon dioxide from changes in land use and from the processes of production, harvesting, and transportation. The land is a resource that is fundamentally scarce and must be used in enormous quantities for land-intensive bioenergy to scale. Due to the limited amount of available land, the fundamentally poor energy returns per unit of land area, and the rapid technological development of competing technologies, land-intensive bioenergy makes the most sense as a transitory component of the global energy mix. It will play a significant before diminishing, most likely after the mid-century, during the following few decades. It will need a unique combination of regulations and incentives to effectively manage a trajectory for land-intensive bioenergy that encourages proper use in the short term while minimizing lock-in in the long term (Reid et al. 2020).

Biomass energy is a renewable energy source made from plant and animal matter. Bioethanol, bio-methanol, and biodiesel are three different types of biomass energy. Bioethanol is one of the top 12 sources of alternative energy that replace fossil fuels. This study aims to synthesize bioethanol from used office paper. Waste papers were used in five lab trials to create bioethanol. The trash paper was chopped into bits after being dried in the oven. Then it underwent the corresponding processes of fermentation, distillation, and hydrolysis with dilute acid. A significant amount of ethanol was found at a liquid-to-solid ratio of 20 to 16 and at the time of 2 h. An economical and cost analysis of producing ethanol from waste papers was done. According to the analysis' findings, a paper-to-ethanol facility is economically feasible, with a rate of return (RR) of 38.61% and a payback period of 2.2 years (Awol and Abate 2021).

12.2 Food Waste Sources, Worldwide Position, and Its Energy Values

The use of energy from various sources that are sustainable and ecologically benign has become more critical as a result of the crisis involving fossil fuels (Li et al. 2010a). Bioenergy from food leftover is a cutting-edge strategy for maintaining a sustainable environment. The majority of the food supply, including vegetables, fruits, meat, dairy, and bread, contributes over 1.6 billion tonnes of the world's total food waste (Braguglia et al. 2017). Because of the growing population, it is predicted that there will be an increase in food waste in Asian countries during the next 25 years.

Because of cheap cost of fossil fuels, high manufacturing expenses, and unsuitable scalability, bioenergy production has yet to catch up to that of the first half of the twentieth century. Food waste can come from any stage, including manufacturing, handling, retail, distribution, and consumption. Food waste is reportedly released in large quantities during developing nations' production, processing, and distribution phases. However, this also happens in wealthy countries' retail and consuming stages. According to the report of European communities, the EU27 produced 37.7 million tonnes of domestic food leftover in 2006 and predisposed of it inefficiently.

According to the FAO (2011), 11 kg of food leftover was created yearly in African and Asian nations, compared to about 115 kg annually in Europe, Oceania, and North America. Therefore, compared to developing countries, industrialized countries have been found to have a higher percentage of food waste. The actual economic, social, as well as environmental consequences of food leftover in the global economy are estimated by the FAO (2014) to be in the range of USD 2.6 trillion annually. The nations of Denmark, Finland, Iceland, Norway, and Sweden produce roughly ten million tonnes of food leftover yearly, excluding cereal straw. Households have five million tonnes of food leftover yearly, which makes up a significant portion of this total.

90% of the waste is collected and used to generate electricity in these nations. Composting, biogas, food/feed, and incineration are the three main uses of food waste. Except for biogas, only 1% of the volume of food waste created is used as fuel for transportation (FAO and FAOSTAT 2014). According to Liebetrau et al. (2017), anaerobic or aerobic biological treatment methods that have recently developed may address the issues of waste management and energy recovery. Methane is produced in Denmark by co-digesting food waste and animal slurry, which aids in slurry digestion, increases the amount of energy produced from all biomass, and lowers emissions of greenhouse gases from farms. The Danish model also demonstrates a sophisticated, straight nutrition-controlling technique for gathering and sorting food leftover from all households before sending it to a central hub for secure handling. The generated biogas is exported when there is excess after being used to meet farm energy needs. This digest furthermore offers a beneficial organic fertilizer that replenishes nutrients needed for crop production as mineral fertilizers.

There needs to be more comprehensive data on the composition of global food waste (Parfitt et al. 2010). An analysis of information on 84 nations' food wastes from 202 independent researchers in 52 different years, from 1933 to 2014, by Xue et al. (2017) confirms the similar result. Most statistics were gathered from developed nations in North America and Europe. According to Banks et al. (2018), there is a lot of confusion in the secondary database currently accessible regarding global food waste. This proves that thorough research based on primary data is required to improve policy-making on reducing food leftover and mitigating its ecofriendly effects.

Vasco-Correa et al. (2018) studied that in the 1970s energy emergency prompted us to focus on implementing renewable energy, but progress is being slowed by the extraordinary. According to estimates, synthesizing hydrogen from food waste high in carbohydrates is 20 times more efficient than that from waste high in protein or fat (Show et al. 2012). According to Patel et al. (2012), estimation of the hydrogen produces for hexose ranges from 0.9 to 8.35 mol hydrogen per mol. Universally, the energy cost of biogas increased 280000 TJ to 1.3 million TJ between 2000 to 2014 and now yearly growth around 3.2 percent. Vegetable wastes have a specific methane energy output ranging from 31.0 to 42.3% (Mozhiarasi et al. 2020).

Also, it was discovered by the authors Chen et al. (2014) and Pilarski et al. (2016) that the methane output in pretreated maize silage and rice straw was 72.2% and 35.3%, respectively. The methane output of vegetable and fruit wastes was estimated to be 530 mL/g by Viturtia et al. (1989) and 180,732 mL/g of volatile solids (VSS), by Gunaseelan (2004). According to Niphadkar et al. (2017), bioethanol production increased to around 41 billion gallons globally. From 2005 to 2010, biodiesel produced worldwide increased from 3.9 to 18.1 billion liters annually. According to predictions by Nystrom et al. (2019), this will amount to 41.4 billion liters by 2025.

The world's production of 647 KT (kilotons) of biodiesel was supported by a maximum lipid output from bio-oil of 0.74 g/g used cooking oil. This might produce 24.5×10^6 GJ of bioenergy annually worldwide (Kiran et al. 2014).

12.3 Bioenergy Production from Food Waste

Renewable feedstocks, including corn, wheat, and sugarcane, are used to make the first generation of biofuel from starch- and sugar-rich food crops. This might directly compete with soil-based food production (Gasparatos et al. 2011). Timilsina and Shrestha (2011) and Refaat (2012) suggested using lignocellulosic wastes other than food to create a new generation of biofuel. This prompted the investigation of studies using cheap wastes and by-products as raw materials. Fishery, dairy, animal, and vegetable wastes produced in kitchens, eateries, cafeterias, and other food-processing facilities are some of the different food waste sources, and they are excellent candidates for the waste-to-wealth strategy. According to Atabani et al. (2019) and Pham et al. (2015), the cost of the used substrate is more expensive than the cost of manufacture, and bioenergy from food waste is becoming increasingly popular.

Depending on where they come from, food wastes' distinctive chemical compositions can vary greatly, but they are typically high in one of three components: proteins, lipids, or carbohydrates. Technologies of many kinds have been developed to use garbage to produce bioenergy. Energy supply generated by various convergence technologies uses a variety of energy sources. The two different technologies are biochemical transesterification and thermochemical transesterification (including biological processes that include anaerobic digestion, fermentation, and photobiological processes, as well as pyrolysis, gasification, combustion, and liquefaction). As a result, utilizing the technologies mentioned above, various biofuels from food waste are created based on their chemical makeup.

In addition to value-added by-products made from food waste, the broad range of approaches for bioenergy reclamation from waste contain light-driven biohydrogen manufacture or dark fermentation processes, biogas (methane) manufacturing using anaerobic digestion, bioethanol manufacturing by fermentation of carbohydrates, and production of biodiesel using transesterification of fats (Sindhu et al. 2019).

The excellent biodegradability of food waste also makes it possible to produce bioenergy on an industrial scale. Kondusamy and Kalamdhad (2014) suggested the use of anaerobic treatment to produce sustainable energy from food waste. Among all solid food waste residues as an origin-isolated organic segment, anaerobic digestion is a developed invention that may be practically applied to various biodegradable substrates. Anaerobic assimilation has been evaluated for its ability to produce methane through the biological transformation of carbon dioxide and hydrogen from multiple sources (Burkhardt et al. 2015). In this way, anaerobic digestion addresses a straightforward technique that may be used as the final conversion technique in a biorefinery sequence for the entire substrates that are not other transformable to valuable goods. Plans for anaerobic absorption are considered the best option for producing bioenergy (Guo et al. 2010).

Banks et al. (2018) suggested that different factors are included in examining various food wastes' chemical and biological composition. Proteins, carbs, fiber, lipids, trace nutrients, micronutrients, and possibly hazardous nutrients make up the biochemical composition. The numerical examination of numerous food leftovers (from restaurants and homes) reveals a significant variance between samples by 24% from varied geographic backgrounds (Fisgativa et al. 2016). However, it was discovered by Micolucci et al. (2018) that there were no significant variations between restaurant food waste that has been source-separated or homes on front side and organic parts of municipal solid trash on opposite side.

Although the look of food from various geographical areas varies, such as in Northern Europe and Southeast Asia, potatoes' primary source of commonality is carbs. Similarly, protein predominates in many sources, including eggs, lentils, and animals in Central Europe and India. Depending on the source, the biochemical makeup of food waste differs by geographic area. According to Pleissner et al. (2014), food waste from commercial restaurants made up 33%, 15%, and 10% of the carbs, lipids, and proteins, respectively.

According to Lin et al. (2013), food waste from homes and fruit and vegetable markets primarily comprises 40% carbs, with modest amounts of 8.5% and 6.9%, respectively, of fats and proteins. Food wastes constitute the primary origin of protein, carbohydrates, and fat in a range of 3.9–21.9%, 35.5–69%, and 10–40%, correspondingly given by Karmee and Lin (2014a, b). This demonstrates the potential for using organic food wastes in biofuel manufacturing while minimizing reliance on conventional fuels. Utilizing food leftover to generate electricity for transportation is made possible by many methods. Technologies for producing anaerobic digestion, fermentation and hydrolysis, hydrogenation, esterification, and thermal gasification are all processes used to produce bioethanol, biodiesel, and biogas.

These cutting-edge methods for digesting food waste can be utilized to produce biogas. Additionally, anaerobic digestion and large-scale hydrogenation esterification have different rates of commercialization. Some cutting-edge techniques work well for specific combinations of food waste, but others, like the anaerobic biogas approach, allow for a fair amount of feedstock adaptability. The newly discovered method by Li et al. (2010b) of producing biohydrogen combines the water splitting using photo electrochemistry, fermentation, and solar thermal energy of sugar-rich feedstocks.

This is a good alternative for producing hydrogen because more carbohydrate resources are available, and they are also less expensive. The two-stage bioconversion method in synthesizing H_2 results in the creation of biogas. The nutrient-rich digestate end product created during this procedure can be used as fertilizer or soil enhancer. By using yeast fermentation and hydrolysis, biofuel ethanol is produced. Davis (2007), Oberoi et al. (2011), and Huang et al. (2015) suggested that to generate high ethanol output from food wastes by glucose valorization, many strategies fed-batch fermentation, continuous fermentation, and solid-state fermentation are all parts of this process. According to Yaakob et al. (2013), fat-enriched food waste, waste cooking oil, and oil are desirable feedstocks which are used for the straight transesterification method that produces biodiesel utilizing chemical enzymes or catalysts.

12.3.1 Biohydrogen

In order to ease the strain of finite resources and encourage the adoption of ecologically friendly technology, biohydrogen serves as a beneficial energy source. In this article, the foundations and restrictions of the various processes including biophotolysis of water, dark anaerobic fermentation and photofermentation, and microbial water-gas shift reaction of carbon monoxide (CO) are discussed. Light is converted into H_2 during photosynthesis by green algae or cyanobacteria using hydrogenase or nitrogenase, either directly or indirectly photolyzing water.

The main obstacles to making photosynthesis a commercially viable technology are the enzymes' susceptibility to oxygen inactivation and the low light saturation level of the process. High hydrogen yields are produced from the substrate during photofermentation, but the process is hindered by low light conversion efficiency and nitrogenase's sensitivity to oxygen. Dark anaerobic fermentation appears to be a more practical method since it permits the use of wastes, which boosts the economic advantages due to its high production rate and ease of usage. But because of the partial oxidation, intermediate chemicals are produced. The mass transfer of CO from a gas phase to a liquid bacterial suspension continues to be the key obstacle despite the fact that the thermodynamics of the microbial water-gas shift reaction is favorable to evolve hydrogen.

12.3.2 Biogas

A widely used approach for managing food waste is the anaerobic digestion of food waste to produce biogas, mainly methane. The simultaneous generation of CH_4 and H_2 is an enticing two-stage procedure because the waste might be utilized as a nutrient-enriched compost or soil conditioner. The components of biogas include methane, hydrogen sulfide carbon dioxide, siloxanes, and moisture (Sindhu et al. 2019). The biological breakdown of organic waste occurs during anaerobic digestion, which produces biogas. According to Lee et al. (2014) and Van et al. (2020), the following reactions occur in order during the conversion of biomass to biogas: hydrolysis, acid genesis, methanogenesis, and acetogenesis. Additionally, for producing biogas, it also results in nutrient-enriched compost. Due to high carbohydrate content of food waste, biogas production from it represents a practical and organic biodegradable material.

The use of anaerobic digestion has increased recently all around the world. The use of renewable resources to produce electricity was encouraged by numerous national programs (Clarke and Alibardi 2010). The pretreatment of substrates considerably changes how much biogas is produced from both pretreated and untreated food waste which is studied by Deepanraj et al. (2017). Their investigation used the microwave, autoclave, and several pretreatment techniques for food waste. Out of these, ultra-sonication produced the most biogas, 10.12%. The anaerobic digestion method was also used to pretreat poultry wastes to produce biogas.

According to Viturtia et al. (1989), by-products from fruit and vegetable anaerobic digestion produce 530 mL/g of methane. Correspondingly, it is observed that food trash with 70% VSs made 440 mL/g of methane (Lee et al. 1999). Methane generation (180,732 mL/g) was investigated by Gunaseelan (2004) utilizing various waste from fruits and vegetables. Wu et al. (2016) demonstrated biogas manufacturing from food waste consuming several digesters, including mesophilic, temperature-phased anaerobic with recycling digesters, and temperature-phased anaerobic.

12.3.3 Bioethanol

Globally, there is a high need for industrial ethanol applications. The various uses of ethanol are used, including as a transportation fuel to substitute gasoline, as fuel for thermal combustion power production, and as feedstock for the chemical industry. Therefore, domestic and other food scraps serve as a substitute raw material source while creating bioethanol. Traditionally, crops high in polysaccharides, like sugarcane, potatoes, and rice, are used to make bioethanol (Kiran et al. 2014). Using *Saccharomyces cerevisiae* for fermentation, polysaccharide cellulose is treated with cellulose enzymes to create glucose and ethanol. Recent years have seen the investigation of plentiful food wastes as alternative substrates for ethanol synthesis.

Because of its copious source, reusability, and affordability, the manufacture of bioethanol from food waste is a promising practical, ecologically friendly strategy (Kiran et al. 2014). It has been investigated how to produce bioethanol via fermentation on various potential substrates.

Food waste fermentation is a crucial bioenergy-generating process that depends on several variables, including the reactor type, temperature, time, and pH conditions (Ren et al. 2011). To produce bioethanol, anaerobic bacteria and yeast are used for hydrolysis and fermentation (Dhiman et al. 2017). According to Chintagunta et al. (2016), co-culturing of *Aspergillus niger* and *S. cerevisiae* produced bioethanol from potato peeling waste with a yield of about 6.18%. For bioethanol generation, waste bananas or other rotten fruit wastes have been widely explored by Arumugam and Manikandan (2011) and Bello et al. (2014). Simultaneous fermentation procedure is used to produce 7.12% bioethanol utilizing simultaneous saccharification and fermentation from the leftover pineapple leaf waste.

12.3.4 Biodiesel

Chen et al. (2009) and Mahmood and Hussain (2010) find that the automotive fuels can be created through transesterification to produce biodiesel. To make biodiesel, Refaat (2012) proposed use of fatty acids from natural lipids. Yaakob et al. (2013) used food waste and cooking oil waste to make biodiesel, which is carbon-neutral, nontoxic, and eco-friendly. Its applicability is hampered by its high viscosity and limited volatility. It consists of transesterified mono-alkyl esters with long chains of fatty acids and low density. Additionally, combining will enhance the characteristics of the fuel, and B20 is a well-known example of a popular blend. Food scraps that are readily accessible and inedible can be used to produce biodiesel at a lower cost.

The primary components of food waste include proteins, carbohydrates, and lipids which are among the macromolecules present; however, the amount varies based on the type of waste and its components. The amount of lipid available in biodiesel manufacturing can be used to evaluate the trash (Nizami et al. 2017). One potential feedstock for biodiesel is kitchen garbage, and it is suggested by Wang et al. (2017) and Kavitha et al. (2017). Cooking oil waste is used by Panadare and Rathod (2016), Gupta et al. (2015), Jain et al. (2011), and Karmee and Lin (2014a, b) in mechanical, chemical, physical, and biological pretreatment procedures to produce biodiesel. Additionally, studies have shown that mixed food waste's low productivity makes it challenging to produce biodiesel from it. Transesterification and extraction phase is also required to obtain straight biodiesel from the food leftovers. The residual H_2O in the food leftover poses a serious barrier to transesterification. However, according to the waste management strategy, the consumption of animal wastes or nonconsumable vegetable oils has recently been a major worry; hence it is considered as a more useful and effective component in the manufacturing of biodiesel by Gasparatos et al. (2011) and Pirozzi et al. (2012).

12.4 Various Methods for Bioenergy Production

The biological conversion and thermochemical conversion are the two different technologies which is mainly used for energy generation from food waste is suggested by Lee et al. (2019). The comprehensive entire method of energy manufacturing is defined in the subsequent subdivision.

12.4.1 Biological Conversion

In the biological conversion process, specialized bacteria or yeast is used for conversion of waste or biomass into valuable energy.

12.4.2 Anaerobic Digestion

In anaerobic digestion method, the food waste is converted into biogas in presence of microorganisms, and the biogas mainly contains methane, carbon dioxide, and traceable amount of hydrogen sulfide. Moist biomass having up to 90% moisture content can accommodate using anaerobic digestion (Brennan and Owende 2010). The anaerobic digestion process has three most important phases, i.e., methanogenesis, acidogenesis and acetogenesis, and hydrolysis (Cantrell et al.

2008). In the hydrolysis method, the intricate biomolecules of the biomass are converted into simple molecules (Anukam et al. 2019). In the consecutive stage suggested by Karuppiah and Azariah (2019) acidogenesis an acidogenic bacterium, fermentation of substrate carried out which gives a volatile product, specifically fatty acids, beside Hydrogen and carbon dioxide gas mixture. According to Venkiteshwaran et al. (2015), the subsequent process of acetogenesis comprises the transformation of volatile alcohols and fatty acids into hydrogen and acetate by proton-reducing microorganisms. Finally, methanogens absorb the gas and produce biogas.

12.4.3 Fermentation

Ethanol is produced from sugar-enriched substrates which are carried out as designated in the succeeding stages suggested by Gavahian et al. (2019). It may be separated into three different stages: fermentable sugar manufacturing, sugar solution fermentation for ethanol manufacturing, and the ethanol recuperating from the fermented solution. In this method, meals' waste material is first of all grounded and hydrolyzed via way of means of the usage of acidic, thermal, or enzymatic strategies for the conversion of starch into easy fermentable sugars. At the same time, bioethanol manufacturing is executed the usage of fed-batch, batch, and continuous techniques. In the ultimate step, ethanol is distilled and recuperated from the fermented broth.

12.4.4 Dark Fermentation and Photobiological

Some biomass together with microalgae has the natural capacity to produce hydrogen. Gas is withinside the existence of light. Transformation of water molecules into H^+ and oxygen has carried out in the photosynthesis. Under the anaerobic state H+ is reduced into hydrogen gas using hydrogenase enzymes. The oxygen released during photosynthesis speedily prevents the hydrogenase enzymes and disturbs the discharge of hydrogen gas. This means that anaerobic state is vital for the culturing microalgae for the hydrogen gas manufacturing (Cantrell et al. 2008).

Dark fermentation method is used to ferment rich amount of sugar feedstock into biohydrogen and fatty acid. The strategies for biohydrogen manufacturing encompass dark fermentation and photofermentation. Initially the former is accomplished under light (aerobic), while the latter proceeds in anaerobic conditions. To provoke biohydrogen manufacturing, the wastes are to start with transformed into fermentable sugars via way of means of appropriate pretreatment technique. In dark fermentation, microorganism transforms glucose into pyruvic acid via way of means of the glycolytic pathway. Later, they are transformed into carbon di oxide and hydrogen with the help of enzymatic method, that is, hydrogenase and pyruvate ferredoxin oxidoreductase (Sarangi and Nanda 2020).

According to Hallenbeck (2011), photofermentation of organic substrates is carried out with the help of photosynthetic microorganism. Further, the cellular decrease of nitrogenase action in the form of proton gradient is carried out using the produced chemical energy. Thus decreased nitrogenase can usage the ATP comprised of the photophosphorylation method, that is, deduction of metabolically derived protons into H_2.

12.4.5 Transesterification

The biofuel production from the prospective biomass such as cellulosic biomass is more complicated because the performance and properties of the mined oil requires to be modified so that it grabs the hydrocarbon-based fuel properties. The undertaking lies within the transformation of the fat and oil acquired from those biomasses into appropriate biofuels to be able to be successfully used alternatively to the traditional fuel. Biofuels acquired from biomass along with lignocellulosic substances frequently have the problem of low vitality, excessive viscosity, and polyunsaturated qualities. These problems may be solving through numerous pretreatment strategies in which the most possible approach is transesterification. Transesterification is the response in which oils and fat are transformed to form glycerol and esters withinside the presence of catalytic agent. The physical traits of the fatty acid methyl ester (FAME) formed are then equated with the industrial petroleum fuel, and the spinoff glycerol additionally has commerce value. Enzyme, acid, and base group catalysts are used for biodiesel production. The improvement of supercritical transesterification is an easy methodology where there is a no need of any reagent (Deshpande et al. 2017).

12.4.6 Thermochemical Transformation

In the thermochemical transformation, heat is used for the disintegration of organic components in the biomass. High temperature is required for the reformation and initiation for breaking the bond and conversion of organic matter into syngas, solid (biochar), and highly oxygenated liquid (bio-oil) (Overend and Chum 1993). There are mainly three types of the conversion method, that is, gasification, pyrolysis, and liquefaction (Bhaskar et al. 2011).

12.4.6.1 Pyrolysis

In the pyrolysis process, thermal decomposition of biomass is carried out in the absence of oxygen gas. The operating temperature for the pyrolysis is between 350 and 550 °C, and it is an endothermic process (Dhyani and Bhaskar 2018). It is the very important process as it includes both the combustion and gasification rocess. The products obtained from the pyrolysis are carbon dioxide, carbon monoxide methane, hydrogen, biochar, and bio-oil. This process is very fast as it occurs within very first 2 s (Oochit et al. 2017).

12.4.6.2 Gasification

Gasification process of biomass involves of burning of biomass at appropriate temperature in a very restricted quantity of air. It generally produces ignitable gases like methane carbon monoxide, carbon dioxide, hydrogen, and nitrogen, along with other contaminants. The cleaning of the produced gas is done, and after that it can be used in turbines, boilers, and engines for the generation of heat and power. The very basic idea of gasification process comprises of methods like combustion, devolatilization, and reduction. The above-described method of gasification provides different forms of energy than conventional combustion method of biomass (Neubauer 2013). Gasification occurs between 800 and 1800 °C to yield combustible gas.

12.4.6.3 Liquefaction

High-quality biofuel is obtained from the biomass using liquefaction method. The liquefaction process is one of the major and important methods. There are two types of liquefaction which are direct liquefaction and indirect liquefaction, in which, indirect liquefaction trails Fischer-Tropsch method which uses the syngas obtained from biomass, and it is the raw material to yield amount of liquid fuel which contains of dimethyl ether, methyl alcohol, and ethyl alcohol. In the direct liquefaction method, the foremost used technology for the conversion of bio-oil from the biomass is thermodynamic liquefaction and hydrolysis fermentation (Zhang et al. 2019).

12.5 Conclusion

Managing food waste is one of the biggest issues society is currently facing, and an appropriate, long-term solution is needed. A sustainable solution is one which is economically feasible to turn food waste into bioethanol, biogas, and biodiesel. In this context, household food wastes should be utilized as raw materials for the production of biofuels as well as the extraction of value-added goods since they are no longer fit for human ingestion. Resources including land, water, and energy can be significantly lost as a result of untreated food waste. They might harm the environment by emitting ozone-depleting gases and secondary pollution. Therefore, the prevention of food waste necessitates novel approaches. This sector targets cutting-edge environmentally friendly technologies with resource conservation and waste reduction tactics.

Transesterification, thermochemical, and biochemical methods can be used to transform the biomass waste and residue into the transportation fuels and bioelectricity. The desired end product and feedstocks influence the processing method chosen. Thermochemical technology, which employs thermal heat to create biofuels, may be less sensitive to the composition of biomass waste than biochemical approaches are. The manufacturing of biofuels using biomass waste is still viewed as having more advanced material management, conveyance, and transformation technologies than conventional biofuels based on genetically modified food crops. Nevertheless, ongoing research is being conducted to close the technological gaps and increase the efficiency and profitability of the most cutting-edge technology. Still, there are ongoing research studies committed to addressing the shortcomings of the current techniques and enhancing the effectiveness and economics of the used manufacturing techniques.

Waste to energy is a collection of transformative processes that can be utilized to produce energy, reduce waste, improve the environment, and develop healthy secondary markets for materials that have been reused, refurbished, and upcycled. The technology can be used to encourage a more circular economy in both smart cities and prosperous rural areas.

References

Anukam A, Mohammadi A, Naqvi M, Granström K (2019) A review of the chemistry of anaerobic digestion: methods of accelerating and optimizing process efficiency. PRO 7(8):504

Arumugam R, Manikandan M (2011) Fermentation of pretreated hydrolyzates of banana and mango fruit wastes for ethanol production. Asian J Exp Biol Sci 2(2):246–256

Atabani AE, Alaa H, Kumar G, Saratale GD, Aslam M, Khan HA, Said Z, Mahmoud E (2019) Valorization of spent coffee grounds into biofuels and value-added products: pathway towards integrated bio-refinery. Fuel 254:115640

Awol ZS, Abate RT (2021) Production of bio-ethanol from solid waste paper using biological activation. https://doi.org/10.21203/rs.3.rs-1009779/v1

Banks C, Heaven S, Zhang Y, Baier U (2018) Food waste digestion: anaerobic digestion of food waste for a circular economy. IEA Bioenergy, Paris

Bello RH, Linzmeyer P, Franco CMB, Souza O, Sellin N, Medeiros SHW, Marangoni C (2014) Pervaporation of ethanol produced from banana waste. Waste Manag 34(8):1501–1509

Bhaskar T, Bhavya B, Singh R, Naik DV, Kumar A, Goyal HB (2011) Thermochemical conversion of biomass to biofuels. In: Biofuels. Academic Press, New York, pp 51–77

Braguglia CM, Gallipoli A, Gianico A, Pagliaccia P (2017) Anaerobic bioconversion of food waste into energy: a critical review. Bioresour Technol 248(Pt A):37–56

Brennan L, Owende P (2010) Biofuels from microalgae—a review of technologies for production, processing, and extractions of biofuels and co-products. Renew Sustain Energy Rev 14(2): 557–577

Burkhardt M, Koschack T, Busch G (2015) Biocatalytic methanation of hydrogen and carbon dioxide in an anaerobic three-phase system. Bioresour Technol 178:330–333

Cantrell KB, Ducey T, Ro KS, Hunt PG (2008) Livestock waste-to-bioenergy generation opportunities. Bioresour Technol 99(17):7941–7953

Chavan N, Bhor D, Pote R (2022) A short review on biomass production from agricultural wastes. Int J Eng Res Technol 11:75–80

Chen Y, Xiao B, Chang J, Fu Y, Lv P, Wang X (2009) Synthesis of biodiesel from waste cooking oil using immobilized lipase in fixed bed reactor. Energ Conver Manage 50(3):668–673

Chen X, Zhang Y, Gu Y, Liu Z, Shen Z, Chu H, Zhou X (2014) Enhancing methane production from rice straw by extrusion pretreatment. Appl Energy 122:34–41

Chintagunta AD, Jacob S, Banerjee R (2016) Integrated bioethanol and biomanure production from potato waste. Waste Manag 49:320–325

Clarke WP, Alibardi L (2010) Anaerobic digestion for the treatment of solid organic waste: what's hot and what's not. Waste Manag 10(30):1761–1762

Davis RA (2007) Parameter estimation for simultaneous saccharification and fermentation of food waste into ethanol using Matlab Simulink. In: Biotechnology for fuels and chemicals, pp 379–389

Deepanraj B, Sivasubramanian V, Jayaraj S (2017) Effect of substrate pretreatment on biogas production through anaerobic digestion of food waste. Int J Hydrogen Energy 42(42): 26522–26528

Deshpande SR, Sunol AK, Philippidis G (2017) Status and prospects of supercritical alcohol transesterification for biodiesel production. Rev Energy Environ 6(5):252

Dhiman SS, David A, Shrestha N, Johnson GR, Benjamin KM, Gadhamshetty V, Sani RK (2017) Simultaneous hydrolysis and fermentation of unprocessed food waste into ethanol using thermophilic anaerobic bacteria. Bioresour Technol 244:733–740

Dhyani V, Bhaskar T (2018) A comprehensive review on the pyrolysis of lignocellulosic biomass. Renew Energy 129:695–716

FAO (2011) Global food losses and food waste—extent, causes. FAO, Rome, p 29. http://www.fao.org/docrep/014/mb060e/mb060e00.pdf

FAO, FAOSTAT (2014) Eurostat, Swedish Board of Agriculture, FUSIONS EU, LUKE, EPA Denmark, EPA Sweden, Ostfoldforskning, Statistics Sweden, Statistics Finland, Statistics Denmark, Statistics Iceland

FAO of the United Nations (2014) Food wastage footprint—full-cost accounting. http://www.fao.org/3/a-i3991e.pdf

Fisgativa H, Tremier A, Dabert P (2016) Characterizing the variability of food waste quality: a need for efficient valorisation through anaerobic digestion. Waste Manag 50:264–274

Garba A (2020) Biomass conversion technologies for bioenergy generation: an introduction. In: Biotechnological applications of biomass. IntechOpen, London

Gasparatos A, Stromberg P, Takeuchi K (2011) Biofuels, ecosystem services and human wellbeing: putting biofuels in the ecosystem services narrative. Agric Ecosyst Environ 142(3-4):111–128

Gavahian M, Munekata PE, Eş I, Lorenzo JM, Khaneghah AM, Barba FJ (2019) Emerging techniques in bioethanol production: from distillation to waste valorization. Green Chem 21(6):1171–1185

Gil A (2022) Challenges on waste-to-energy for the valorization of industrial wastes: electricity, heat and cold, bioliquids and biofuels. Environ Nanotechnol Monit Manag 17:100615

Gunaseelan VN (2004) Biochemical methane potential of fruits and vegetable solid waste feedstocks. Biomass Bioenergy 26(4):389–399

Guo XM, Trably E, Latrille E, Carrère H, Steyer JP (2010) Hydrogen production from agricultural waste by dark fermentation: a review. Int J Hydrogen Energy 35(19):10660–10673

Gupta AR, Yadav SV, Rathod VK (2015) Enhancement in biodiesel production using waste cooking oil and calcium diglyceroxide as a heterogeneous catalyst in presence of ultrasound. Fuel 158:800–806

Hallenbeck PC (2011) Microbial paths to renewable hydrogen production. Biofuels 2(3):285–302

Huang H, Qureshi N, Chen MH, Liu W, Singh V (2015) Ethanol production from food waste at high solids content with vacuum recovery technology. J Agric Food Chem 63(10):2760–2766

Jain S, Sharma MP, Rajvanshi S (2011) Acid base catalyzed transesterification kinetics of waste cooking oil. Fuel Process Technol 92(1):32–38

Karmee SK, Lin CSK (2014a) Valorisation of food waste to biofuel: current trends and technological challenges. Sustain Chem Process 2(1):1–4

Karmee SK, Lin CSK (2014b) Lipids from food waste as feedstock for biodiesel production: case Hong Kong. Lipid Technol 26(9):206–209

Karuppiah T, Azariah EV (2019) Biomass pretreatment for enhancement of biogas production. In: Anaerobic digestion. IntechOpen, London

Kavitha S, Banu JR, Priya AA, Yeom IT (2017) Liquefaction of food waste and its impacts on anaerobic biodegradability, energy ratio and economic feasibility. Appl Energy 208:228–238

Kim J, Han J, Lee J (2022) A strategy for food waste-to-biofuels: co-production of gasoline alternatives from volatile fatty acids. J Clean Prod 348:131408

Kiran EU, Trzcinski AP, Ng WJ, Liu Y (2014) Bioconversion of food waste to energy: a review. Fuel 134:389–399

Kondusamy D, Kalamdhad AS (2014) Pre-treatment and anaerobic digestion of food waste for high rate methane production—a review. J Environ Chem Eng 2(3):1821–1830

Kowalski Z, Kulczycka J, Makara A, Verhé R, De Clercq G (2022) Assessment of energy recovery from municipal waste management systems using circular economy quality indicators. Energies 15(22):8625

Kumar A, Samadder SR (2022) Assessment of energy recovery potential and analysis of environmental impacts of waste to energy options using life cycle assessment. J Clean Prod 365:132854

Lee JP, Lee JS, Park SC (1999) Two-phase methanization of food wastes in pilot scale. In: Twentieth symposium on biotechnology for fuels and chemicals, pp 585–593

Lee HV, Hamid SBA, Zain SK (2014) Conversion of lignocellulosic biomass to nanocellulose: structure and chemical process. Sci World J 2014:631013

Lee SY, Sankaran R, Chew KW, Tan CH, Krishnamoorthy R, Chu DT, Show PL (2019) Waste to bioenergy: a review on the recent conversion technologies. BMC Energy 1(1):1–22

Lee J, Lin KYA, Jung S, Kwon EE (2022) Hybrid renewable energy systems involving thermochemical conversion process for waste-to-energy strategy. Chem Eng J 452:139218

Li S, Lai C, Cai Y, Yang X, Yang S, Zhu M, Wang J, Wang X (2010a) High efficiency hydrogen production from glucose/xylose by the ldh-deleted Thermoanaerobacterium strain. Bioresour Technol 101(22):8718–8724

Li X, Kim TH, Nghiem NP (2010b) Bioethanol production from corn Stover using aqueous ammonia pretreatment and two-phase simultaneous saccharification and fermentation (TPSSF). Bioresour Technol 101(15):5910–5916

Liebetrau J, Reinelt T, Agostini A, Linke B (2017) Methane emissions from biogas plants: methods for measurement, results and effect on green- house gas balance of electricity produced. In: Murphy JD (ed) IEA bioenergy task, vol 37, p 12

Lin CSK, Pfaltzgraff LA, Herrero-Davila L, Mubofu EB, Abderrahim S, Clark JH, Koutinas AA, Kopsahelis N, Stamatelatou K, Dickson F, Thankappan S (2013) Food waste as a valuable resource for the production of chemicals, materials and fuels. Current situation and global perspective. Energ Environ Sci 6(2):426–464

Mahmood T, Hussain ST (2010) Nanobiotechnology for the production of biofuels from spent tea. Afr J Biotechnol 9(6):858–868

Mendoza-Meneses CJ, Feregrino-Pérez AA, Gutiérrez-Antonio C (2021) Potential use of industrial cocoa waste in biofuel production. J Chem 2021:11

Micolucci F, Gottardo M, Pavan P, Cavinato C, Bolzonella D (2018) Pilot scale comparison of single and double-stage thermophilic anaerobic digestion of food waste. J Clean Prod 171: 1376–1385

Mozhiarasi V, Weichgrebe D, Srinivasan SV (2020) Enhancement of methane production from vegetable, fruit and flower market wastes using extrusion as pretreatment and kinetic modeling. Water Air Soil Pollut 231(3):1–21

Muhammad MU, Abubakar L, Musa A, Kamba AS (2013) Utilization of wastes as an alternative energy source for sustainable development: a review. ChemSearch J 4(1):57–61

Neubauer Y (2013) Biomass combustion science, technology and engineering: 6. In: Biomass gasification. Elsevier, Amsterdam, pp 106–129

Niphadkar S, Bagade P, Ahmed S (2017) Bioethanol production: insight into past, present and future perspectives. Biofuels 9(2):229–238

Nizami AS, Shahzad K, Rehan M, Ouda OKM, Khan MZ, Ismail IMI, Almeelbi T, Basahi JM, Demirbas A (2017) Developing waste biorefinery in Makkah: a way forward to convert urban waste into renewable energy. Appl Energy 186:189–196

Nystrom I, Bokinge P, Franck PA (2019) Production of liquid advanced biofuels-global status. CIT Industriell Energi AB, Gothenburg

Oberoi HS, Vadlani PV, Nanjundaswamy A, Bansal S, Singh S, Kaur S, Babbar N (2011) Enhanced ethanol production from Kinnow mandarin (Citrus reticulata) waste via a statistically optimized simultaneous saccharification and fermentation process. Bioresour Technol 102(2):1593–1601

Ochieng R, Gebremedhin A, Sarker S (2022) Integration of waste to bioenergy conversion systems: a critical review. Energies 15(7):2697

Oochit D, Selvarajoo A, Arumugasamy SK (2017) Pyrolysis of biomass. In: Waste biomass management–a holistic approach. Springer, Cham, pp 215–229

Overend RP, Chum HL (1993) Thermochemical conversion research in the US Department of Energy Biofuels and power programs. In: Advances in thermochemical biomass conversion. Springer, Dordrecht, pp 26–47

Panadare DC, Rathod VK (2016) Microwave assisted enzymatic synthesis of biodiesel with waste cooking oil and dimethyl carbonate. J Mol Catal B: Enzym 133:S518–S524

Parfitt J, Barthel M, Macnaughton S (2010) Food waste within food supply chains: quantification and potential for change to 2050. Philos Trans R Soc B Biol Sci 365(1554):3065–3081

Patel SK, Kumar P, Kalia VC (2012) Enhancing biological hydrogen production through complementary microbial metabolisms. Int J Hydrogen Energy 37(14):10590–10603

Pham TPT, Kaushik R, Parshetti GK, Mahmood R, Balasubramanian R (2015) Food waste-to-energy conversion technologies: current status and future directions. Waste Manag 38:399–408

Pilarski K, Pilarska AA, Witaszek K, Dworecki Z, Żelaziński T, Ekielski A, Makowska A, Michniewicz J (2016) The impact of extrusion on the biogas and biomethane yield of plant substrates. J Ecol Eng 17(4):264–272

Pirozzi D, Yousuf A, Zuccaro G, Aruta RSF, Sannino F (2012) Synthesis of biodiesel from hydrolysates of Arundo donax. Environ Eng Manag J 11:1797–1801

Pleissner D, Kwan TH, Lin CSK (2014) Fungal hydrolysis in submerged fermentation for food waste treatment and fermentation feedstock preparation. Bioresour Technol 158:48–54

Refaat AA (2012) Biofuels from waste materials. Comp Renew Energy 5:217–262

Reid WV, Ali MK, Field CB (2020) The future of bioenergy. Glob Chang Biol 26(1):274–286

Ren N, Guo W, Liu B, Cao G, Ding J (2011) Biological hydrogen production by dark fermentation: challenges and prospects towards scaled-up production. Curr Opin Biotechnol 22(3):365–370

Sarangi PK, Nanda S (2020) Biohydrogen production through dark fermentation. Chem Eng Technol 43(4):601–612

Sharma P, Gaur VK, Sirohi R, Varjani S, Kim SH, Wong JW (2021) Sustainable processing of food waste for production of bio-based products for circular bioeconomy. Bioresour Technol 325: 124684

Show KY, Lee DJ, Tay JH, Lin CY, Chang JS (2012) Biohydrogen production: current perspectives and the way forward. Int J Hydrogen Energy 37(20):15616–15631

Sindhu R, Gnansounou E, Rebello S, Binod P, Varjani S, Thakur IS, Nair RB, Pandey A (2019) Conversion of food and kitchen waste to value-added products. J Environ Manage 241:619–630
Srivastava RK (2019) Bio-energy production by contribution of effective and suitable microbial system. Mater Sci Energy Technol 2(2):308–318
Timilsina GR, Shrestha A (2011) How much hope should we have for biofuels? Energy 36(4): 2055–2069
Tripathi N, Hills CD, Singh RS, Atkinson CJ (2019) Biomass waste utilisation in low-carbon products: harnessing a major potential resource. NPJ Clim Atmos Sci 2(1):1–10
Van DP, Fujiwara T, Tho BL, Toan PPS, Minh GH (2020) A review of anaerobic digestion systems for biodegradable waste: configurations, operating parameters, and current trends. Environ Eng Res 25(1):1–17
Vasco-Correa J, Khanal S, Manandhar A, Shah A (2018) Anaerobic digestion for bioenergy production: global status, environmental and techno-economic implications, and government policies. Bioresour Technol 247:1015–1026
Venkiteshwaran K, Bocher B, Maki J, Zitomer D (2015) Relating anaerobic digestion microbial community and process function: supplementary issue: water microbiology. Microbiol Insights 8:MBI-S33593
Viturtia AM, Mata-Alvarez J, Cecchi F, Fazzini G (1989) Two-phase anaerobic digestion of a mixture of fruit and vegetable wastes. Biol Wastes 29(3):189–199
Wang C, Xie S, Zhong M (2017) Effect of hydrothermal pretreatment on kitchen waste for biodiesel production using alkaline catalyst. Waste Biomass Valor 8(2):369–377
Wu LJ, Kobayashi T, Kuramochi H, Li YY, Xu KQ (2016) Improved biogas production from food waste by co-digestion with de-oiled grease trap waste. Bioresour Technol 201:237–244
Xue L, Liu G, Parfitt J, Liu X, Van Herpen E, Stenmarck Å, O'Connor C, Ostergren K, Cheng S (2017) Missing food, missing data? A critical review of global food losses and food waste data. Environ Sci Technol 51(12):6618–6633
Yaakob Z, Mohammad M, Alherbawi M, Alam Z, Sopian K (2013) Overview of the production of biodiesel from waste cooking oil. Renew Sustain Energy Rev 18:184–193
Zhang S, Yang X, Zhang H, Chu C, Zheng K, Ju M, Liu L (2019) Liquefaction of biomass and upgrading of bio-oil: a review. Molecules 24(12):2250
Zhou C, Wang Y (2020) Recent progress in the conversion of biomass wastes into functional materials for value-added applications. Sci Technol Adv Mater 21(1):787–804

The manufacturer's authorised representative in the EU is Springer Nature Customer Service Centre GmbH, Europaplatz 3, 69115 Heidelberg, Germany. If you have any concerns regarding our products, please contact ProductSafety@springernature.com

Printed and bound by CPI Group (UK) Ltd, Croydon, CR0 4YY
01/10/2026
02225870-0005